T0135616

General Linear Methods
for Integrated Circuit Design

DISSERTATION

zur Erlangung des akademischen Grades
doctor rerum naturalium
(Dr. rer. nat.)
im Fach Mathematik

eingereicht an der
Mathematisch-Naturwissenschaftlichen Fakultät II
Humboldt-Universität zu Berlin

von
Herr Dipl.-Math. Steffen Voigtmann
geboren am 12.07.1976 in Berlin

Präsident der Humboldt-Universität zu Berlin:
Prof. Dr. Christoph Markschies

Dekan der Mathematisch-Naturwissenschaftlichen Fakultät II:
Prof. Dr. Uwe Küchler

Gutachter:

(a) Prof. Dr. John Butcher
(b) Prof. Dr. Roswitha März
(c) Prof. Dr. Caren Tischendorf

eingereicht am: 30. Januar 2006
Tag der mündlichen Prüfung: 26. Juni 2006

Bibliografische Information der Deutschen Nationalbibliothek

Die Deutsche Nationalbibliothek verzeichnet diese Publikation in der Deutschen Nationalbibliografie; detaillierte bibliografische Daten sind im Internet über http://dnb.d-nb.de abrufbar.

©Copyright Logos Verlag Berlin 2006
Alle Rechte vorbehalten.

ISBN 3-8325-1353-1

Logos Verlag Berlin
Comeniushof, Gubener Str. 47,
10243 Berlin
Tel.: +49 030 42 85 10 90
Fax: +49 030 42 85 10 92
INTERNET: http://www.logos-verlag.de

Preface

Today electronic devices play an important part in everybody's life. In particular, there is an ongoing trend towards using mobile devices such as cell phones, laptops or PDAs. Integrated circuits for these kind of applications are mainly produced in CMOS technology (complementary metal-oxide semiconductor). CMOS circuits use almost no power when they are not active and thus, combining negatively and positively charged transistors, they draw power only when switching polarity. Furthermore, advanced CMOS technology is expected to dominate in the future since it allows to manufacture transistors in the nanoscale regime.

Circuit simulation is one of the key technologies enabling a further increase in performance and memory density. A mathematical model is used in order to assess the circuit's behaviour before actually producing it. Thus production starts with an already optimised layout and production costs but also the time-to-market is significantly reduced.

One important analysis type in circuit simulation is the transient analysis of layouts on varying input signals. Based on schematics or netlist descriptions of electrical circuits the corresponding model equations are automatically generated. This network approach preserves the topological structure of the circuit but does not lead to a minimal set of unknowns. Hence the resulting model consists of differential algebraic equations (DAEs). Typically these equations suffer from poor smoothness properties due to the model equations of modern transistors but also due to e.g. piecewise linear input functions. Similarly, time constants of several orders of magnitudes give rise to stiff equations and low order A-stable methods need to be used.

The further miniaturisation of electrical devices drives simulation methods for circuit DAEs to their limits. Due to the reduced signal/noise ratio, stability questions become more and more important for modern circuits. Thus there is a strong need to improve stability properties of existing methods such as the combination of BDF and trapezoidal rule. There are fully implicit Runge-Kutta methods that exhibit much better stability properties. However, these methods are currently not attractive for industrial circuit simulators due to their high computational costs.

General linear methods (GLMs) provide a framework covering, among others, both linear multistep and Runge-Kutta methods. They enable the construction of new methods with improved convergence and stability properties. Up to now little is known about solving DAEs using general linear methods. In particular the application of general linear methods in electrical circuit simulation has not yet been addressed. Hence the object of this thesis is to study general linear methods for integrated circuit design.

The work is organised as follows:

Part I: Using the charge oriented modified nodal analysis the differential algebraic equations describing electrical circuits are derived. Classical methods for solving these equations are briefly addressed and their limitations are investigated. As a means to overcome these shortcomings general linear methods are introduced.

Part II: Linear and nonlinear DAEs of increasing complexity are investigated in detail. Using the concept of the tractability index a decoupling procedure for nonlinear DAEs is derived. This decoupling procedure is the key tool for giving conditions for the existence and uniqueness of solutions but also for studying numerical integration schemes.

Part III: General linear methods are applied to differential algebraic equations. In order to prove convergence for index-2 DAEs it is seminal to investigate GLMs for implicit index-1 equations. Order conditions and further additional requirements on the method's coefficients are derived such that convergence is ensured. Using the decoupling procedure from Part II these results are transferred to the case of index-2 equations.

Part IV: Methods with order p are constructed for $1 \leq p \leq 3$. As different design decisions are possible, the emphasise is on comparing two families of methods: the first one having $p+1$ internal stages while the other one employs just p stages. While the former type of methods allows better stability properties and highly accurate error estimators, the latter family reduces the work per step and is capable of reacting more rapidly to changes of the numerical solution. Implementation issues such as Newton iteration, error estimation and order control are addressed for both families of methods. Extensive numerical tests indicate high potential for general linear methods in integrated circuit design.

Acknowledgement

This work is one result of the close friendship between the numerical analysis group of Prof. Roswitha März and the 'Runge-Kutta Club' headed by Prof. John Butcher.

Roswitha März not only teaches numerical analysis at the Humboldt University in Berlin but she also fills students with enthusiasm about the numerical analysis of differential algebraic equations. I am one of these students and I want to thank her for the motivating, encouraging and supportive atmosphere that I enjoyed at Humboldt University.

After finishing my Master's degree I was fortunate to get the chance to visit Prof. John Butcher at The University of Auckland. This stay in New Zealand was most influential for my future work. I thank John Butcher for letting me become part of the Runge-Kutta Club and teaching me so many things (not only about mathematics and general linear methods). I am honoured that he consented to review this thesis.

Towards the end of my stay in New Zealand a project developed that aimed at combining the two mathematical worlds I lived in so far: the application of general linear methods to differential algebraic equations. My supervisor Prof. Caren Tischendorf (Technical University Berlin) was enthusiastic about this idea from the very beginning. I thank her for realising a project within the Research Center MATHEON. Throughout working on this project I was free to explore my own ideas but Caren offered most valuable help whenever needed. I always trusted her guidance but she never forced me into a certain direction.

While working on this thesis I was fortunate to meet many colleagues and friends influencing my work. I thank Claus Führer (Lund University, Sweden) for many fruitful discussions on DAEs. He not only invited me to Lund but also arranged a visit with Anne Kværnø (NTNU Trondheim, Norway). I thank her for helping me with the convergence proof for general linear methods.

I learned a lot from Helmut Podhaisky (Martin-Luther University Halle-Wittenberg), in particular about the construction of methods and implementation issues. His MATLAB codes formed the basis for developing my own DAE solver. Stepsize prediction and order control were discussed with Gustaf Söderlind (Lund University, Sweden). I thank him for taking interest in my work. René Lamour (Humboldt University Berlin) was always available for discussion and I thank Andreas Bartel (University of Wuppertal) for sending me a copy of his PhD thesis.

I am pleased to acknowledge the financial support of the MATHEON Research Center and the German Research Foundation (Deutsche Forschungsgemeinschaft). I thank my colleagues at the Infineon Technologies AG / Qimonda AG for supporting me in many ways. Special thanks go to Sabine Bergmann and to Sieglinde Jänicke from the Humboldt University for extraordinary support when submitting the thesis.

After all, writing and finishing this work would not have been possible without the loving support of my wife, Sabine. I am lucky to have such a wonderful woman at my side.

> *Kei mai koe ki au*
> *He aha te mea nui te no?*
> *Makue ki atu -*
> *He tangata, he tangata, he tangata.*
>
> Maori proverb

Steffen Voigtmann

Ottobrunn, 20.08.2006.

Contents

List of Figures

List of Tables

Part I

Introduction

1

Circuit Simulation and DAEs

The numerical simulation of electrical circuits in the time domain requires the derivation of a mathematical model. In computer-aided electronics design a network approach is adopted that combines physical laws such as energy or charge conservation with the characteristic equations of the network elements. Most frequently the modified nodal analysis (MNA) is used [47, 53, 66]. The resulting model consists of a system of differential algebraic equations (DAEs).

In the following section the MNA is introduced. We show how to derive the model equations that are used to describe electrical circuits. The next section is then devoted to studying basic properties of the resulting differential algebraic equations.

1.1 Basic Circuit Modelling

Before discussing the modified nodal analysis in the general setup, an introductory example shall be studied first.

Example 1.1. The Miller Integrator depicted in Figure 1.1 is an electrical realisation of the mathematical concept of integration. Given a time-varying input signal v at node u_1 the circuit delivers the node potential $u_3 = -\frac{G_1}{C_1} \int v \, \mathrm{d}t$ as output. This relation will be confirmed in Section 3.2, Example 3.8.

The particular input-output relation of the Miller Integrator can be interpreted as an inductor in admittance form. Thus the Miller Integrator is heavily used for integrated filter circuits in order to substitute inductors, which are expensive to obtain using integrated technologies. This example is taken from [66] where the Miller Integrator is discussed in detail.

In order to apply nodal analysis, we consider the network as a directed graph, pick the node potentials u_1, u_2, u_3 as variables and write down the elements characteristic equations in admittance. In particular we seek relations expressing the branch current i explicitly in terms of the branch voltage u,

$$i_R = G_1 (u_1 - u_2), \qquad i_C = C_1 (u_3 - u_2)'. \tag{1.1a}$$

The equations (1.1a) are Ohm's law for a linear resistor with conductance G_1 and Faraday's law for a linear capacitor with capacitance C_1, respectively. The prime in the expression for i_C denotes the time derivative.

As in [66] an ideal operational amplifier with limited amplification a is used for simplicity, such that the operational amplifier can be modelled as a voltage-controlled voltage source ,

$$u_3 = a \cdot u_2. \tag{1.1b}$$

Similarly, the node potential u_1 is given explicitly by the independent source,

$$u_1(t) = v(t). \tag{1.1c}$$

The electrical interaction of the elements described by (1.1) is accomplished by Kirchhoff's laws which take into account the circuit's topology:

(a) Kirchhoff's Voltage Law (KVL): At every instant of time the algebraic sum of voltages along each loop of the network is equal to zero.

(b) Kirchhoff's Current Law (KCL): At every instant of time the algebraic sum of currents traversing each cutset of the network is equal to zero.

A cutset is a minimal subgraph such that removing this subgraph decomposes the network into two separate connected units. In particular for every node the set of branches, that are connected to this node, forms a cutset. Hence, applying KCL to every node leads to

$$0 = \quad i_{V_1} + i_R = \quad i_{V_1} + G_1 (u_1 - u_2), \tag{1.2a}$$
$$0 = -i_R - i_C = -G_1 (u_1 - u_2) - C_1 (u_3 - u_2)', \tag{1.2b}$$
$$0 = \quad i_C + i_{V_2} = \quad C_1 (u_3 - u_2)' + i_{V_2}, \tag{1.2c}$$

where the element relations (1.1a) have already been inserted.

In contrast to the classical nodal analysis, branch currents through voltage sources and inductors, if present, are considered as additional unknowns. Thus

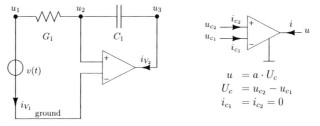

Figure 1.1: Miller Integrator circuit. The operational amplifier is modelled as a voltage-controlled voltage source with amplification factor a.

the equations of the modified nodal analysis are obtained by combining (1.1), (1.2) and $u_1 = v(t)$, but more structure is revealed using incidence matrices.

Notice that the topology of the circuit can be conveniently described using the element related matrices

$$A_C = \begin{bmatrix} 0 \\ -1 \\ 1 \end{bmatrix}, \qquad A_R = \begin{bmatrix} 1 \\ -1 \\ 0 \end{bmatrix}, \qquad A_V = \begin{bmatrix} 1 & 0 \\ 0 & 0 \\ 0 & 1 \end{bmatrix}.$$

Each column represents a branch and the entries ± 1 indicate which nodes this branch is connected to and in which direction the current is flowing (with the sign being a matter of convention). The MNA equations (1.1), (1.2) for the Miller Integrator can thus be written as

$$A_C C_1 A_C^\top u' + A_R G_1 A_R^\top u + A_V i_V = 0, \qquad A_V^\top u = \begin{bmatrix} v(t) \\ a \cdot u_2 \end{bmatrix}. \qquad (1.3)$$

However, in case of nonlinear capacitors and/or resistors (1.3) has to be replaced by

$$A_C \dot{q}\big(A_C^\top u(t), t\big) + A_R g\big(A_R^\top u(t), t\big) + A_V i_V(t) = 0,$$
$$A_V^\top u(t) - \begin{bmatrix} v(t) \\ a \cdot u_2(t) \end{bmatrix} = 0. \qquad (1.3')$$

As before u and i_V are the vectors of node potentials and branch currents through voltage sources, respectively. Additionally we need to take the vector $q(A_C^\top u, \cdot)$ of charges into account. More details on the treatment of nonlinear capacitors and inductors can be found in [66, 78, 80]. □

With the Miller Integrator in mind we now turn to the investigation of the charge oriented modified nodal analysis for arbitrary circuits.

The first step in deriving the MNA equations is to interpret the circuit as a directed graph. The vertices are referred to as nodes and each branch represents a basic element such as resistors, capacitors, inductors, voltage sources and current sources. The characteristic device equations are given in Table 1.1. Basic

device	linear	nonlinear
resistor	$i_R = G\, u_R$	$i_R = g(u_R, t)$
capacitor	$i_C = C\, u_C'$	$i_C = \dot{q}_C(u_C, t)$
inductor	$u_L = L\, i_L'$	$u_L = \dot{\phi}_L(i_R, t)$

device	independent	controlled
voltage source	$u_V = v(t)$	$u_V = v(u_{ctrl}, i_{ctrl}, t)$
current source	$i_I = i(t)$	$i_I = i(u_{ctrl}, i_{ctrl}, t)$

Table 1.1: Characteristic equations for basic elements

elements are idealised descriptions of the corresponding physical devices. Real
devices, and in particular more complex elements such as diodes, transistors
etc., are replaced by so-called companion models. A companion model is an
equivalent circuit consisting of basic elements that approximate the behaviour
of a real, physical device.

The topology of the directed graph is then described using element related
incidence matrices A_R, A_C, A_L, A_V, A_I, just as we did in case of the Miller
Integrator. The three essential steps in setting up the equations of the charge
oriented modified nodal analysis are to

(a) apply KCL to every node except ground,

(b) insert the admittance form representation for the branch current of res-
    istors, capacitors and current sources,

(c) add the impedance form representation for inductors and voltage sources
    explicitly to the system.

Hence we arrive at the following system of equations

$$A_C \, \dot{q}(A_C^\top u, t) + A_R \, g(A_R^\top u, t) + A_L \, i_L$$
$$+ A_V \, i_V + A_I \, i_I(A^\top u, \dot{q}, i_L, i_V, t) \; = 0, \qquad (1.4\text{a})$$
$$\dot{\phi}(i_L, t) - A_L^\top \, u = 0, \qquad (1.4\text{b})$$
$$v_V(A^\top u, \dot{q}, i_L, i_V, t) - A_V^\top \, u = 0. \qquad (1.4\text{c})$$

The vector $q(A_C^\top u, t)$ represents charges while $\phi(i_L, t)$ is the vector of fluxes.
Thus, (1.4) is referred to as the charge/flux oriented formulation of the network
equations.

A dot indicates time derivatives such that $\dot{q}(A_C^\top u, t) = \frac{\mathrm{d}}{\mathrm{d}t}\big[q(A_C^\top u(t), t)\big]$ is the
vector of branch currents through the circuit's capacitors. Similarly, the vector
$\dot{\phi}(i_L, t) = \frac{\mathrm{d}}{\mathrm{d}t}\big[\phi(i_L(t), t)\big]$ denotes currents through inductors.

For many analog circuits such as switched capacitor filters or charge pumps
charge conservation is a crucial property. In fact, the original intent of the
charge oriented formulation (1.4) was to ensure charge conservation. By construction (1.4) guarantees that for each charge storing element one terminal
charge is given as the negative sum of all other terminal charges. Expanding
the charge vector around the exact solution u_*,

$$q(A_C^\top u(t), t) = q(A_C^\top u_*(t), t) + \frac{\partial q(A_C^\top u_*(t), t)}{\partial u} \cdot \Delta u + \mathcal{O}(\Delta u^2),$$

shows that with increasing numerical accuracy, $\Delta u \to 0$, the exact charges are
approximated.

It is shown in [66] that a similar situation cannot be guaranteed for the con-
ventional formulation, where generalised conductance and inductance matrices
$\frac{\partial q(A_C^\top u, t)}{\partial u}$, $\frac{\partial \phi(i_L)}{\partial i_l}$ are used. See also [78, 80] for more details.

The charge/flux oriented formulation (1.4) exhibits the form

$$A\left[d(x,t)\right]' + b(x,t) = 0,\tag{1.5}$$

with

$$x = \begin{bmatrix} u \\ i_L \\ i_V \end{bmatrix}, \qquad A = \begin{bmatrix} A_C & 0 \\ 0 & I \\ 0 & 0 \end{bmatrix}, \qquad d(x,t) = \begin{bmatrix} q(A_C^\top u, t) \\ \phi(i_L, t) \end{bmatrix}$$

and the obvious definition of b. Typically the (constant) matrix A is rectangular having $n_C + n_L + n_V$ rows but only $n_C + n_L$ columns. The prime $[d(x,t)]' = \frac{\mathrm{d}}{\mathrm{dt}}[d(x(t),t)]$ denotes differentiation with respect to time.

The time derivative x' of the dependent variable is not given explicitly as a function of x and t. Since the matrix $A\frac{\partial d(x,t)}{\partial x}$ might be singular, (1.5) is not an ordinary differential equation but of differential algebraic type.

This thesis is devoted to studying numerical methods for differential algebraic equations (DAEs) of the form (1.5). We will start by investigating some basic properties of these equations.

1.2 Differential Algebraic Equations

In the previous section we saw that the charge oriented modified nodal analysis leads to differential algebraic equations of the form

$$A\left[d(x,t)\right]' + b(x,t) = 0.\tag{1.5}$$

These equations are different from ordinary differential equations in many ways. Some of the more prominent differences will be introduced by considering two simple examples.

Example 1.2. The circuit depicted in Figure 1.2 (a) represents a simple VRC loop. Using the modified nodal analysis the corresponding circuit DAE reads

$$\begin{bmatrix} 0 \\ 1 \\ 0 \end{bmatrix} (C_1 \cdot u_2)' + \begin{bmatrix} G_1 & -G_1 & 1 \\ -G_1 & G_1 & 0 \\ 1 & 0 & 0 \end{bmatrix} \begin{bmatrix} u_1 \\ u_2 \\ i_V \end{bmatrix} = \begin{bmatrix} 0 \\ 0 \\ v(t) \end{bmatrix},\tag{1.6}$$

where u_1, u_2 are again the node potentials and i_V represents the current though the voltage source.

This DAE can be solved analytically by noting that the third equation yields $u_1(t) = v(t)$ such that the second equation turns out to be the ordinary differential equation

$$u_2'(t) = C_1^{-1}G_1\big(v(t) - u_2(t)\big).$$

Once a solution for this ODE is found, the current i_V through the voltage source is found from the first equation,

$$i_V(t) = G_1\big(u_2(t) - u_1(t)\big).$$ □

Example 1.2 clearly shows that not all components of a DAE solution need to be differentiable. For u_1 and i_V it suffices to require continuity. The special formulation of the leading term in (1.6) figures out precisely which derivatives are involved.

Similarly, not all components can be assigned initial values. Here we may choose an arbitrary initial condition for u_2, but as u_1 and i_V are fixed in terms of u_2 and the problem data, initial conditions for these components need to be consistent, i.e. $u_1(t_0) = v(t_0)$ and $i_V(t_0) = G_1\big(u_2(t_0) - u_1(t_0)\big)$.

Example 1.3. In Figure 1.2 (b) the position of the resistor was changed and the MNA equations read

$$\begin{bmatrix} 1 \\ 0 \end{bmatrix} (C_1 \cdot u_1)' + \begin{bmatrix} G_1 & 1 \\ 1 & 0 \end{bmatrix} \begin{bmatrix} u_1 \\ i_V \end{bmatrix} = \begin{bmatrix} 0 \\ v(t) \end{bmatrix}.$$

In contrast to the previous example all components are fixed by the problem data and there is no freedom to pose any initial condition:

$$u_1(t) = v(t), \qquad\qquad i_V(t) = -C_1 u_1'(t) - G_1 u_1(t).$$

Even though the Examples 1.2 and 1.3 seem quite similar, Example 1.3 nevertheless introduces a new level of complexity. The crucial point is that in order to compute i_V one needs to differentiate u_1 such that input function v needs to be smooth.

Since differentiation is an unstable operation, this situation may lead to severe problems. As in [155] consider e.g. an input signal $v(t) = 5 - \delta(t)$ with a small perturbation $\delta(t) = \varepsilon \sin(\omega t)$. Independently of ω the perturbation is bounded in magnitude by $|\varepsilon|$. However, since the current through the voltage source is given by

$$i_V(t) = C_1 \varepsilon\, \omega \cos(\omega t) - 5G_1 + G_1 \varepsilon \sin(\omega t),$$

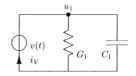

Figure 1.2 (a): *VRC* loop   Figure 1.2 (b): *VRC* in parallel

Figure 1.2: Two circuits consisting of a voltage source, resistor and capacitor.

the error resulting from the perturbation δ is found to be $C\varepsilon\,\omega\cos(\omega t) + G\varepsilon\sin(\omega t)$. For large values of ω this quantity can become arbitrarily large as is illustrated in Figure 1.3. $\qquad\qquad\qquad\qquad\qquad\qquad\qquad\qquad$ □

The solution of differential algebraic equations may require differentiation of certain parts of the input functions. Hence it is not enough, as for ODEs, to require continuity of the right-hand side, but some components may require additional smoothness.

The question whether differentiation is involved in solving a given DAE and, if so, up to what degree, distinguishes different levels of complexity both in the analytical and numerical treatment of DAEs. As a means to measure these difficulties, each DAE is assigned an integer number – its index.

Many different index concepts are available in the literature. Some of them will be briefly reviewed in Section 3.1. Roughly speaking, for an index-μ equation inherent differentiations up to order $\mu - 1$ are involved. Consequently, the circuit in Figure 1.2 (a) leads to an index-1 DAE while Figure 1.2 (b) comprises an index-2 equation. Ordinary differential equations are assigned index 0.

For a large number of circuits the index can be read directly from the circuit's topology. This is possible given that

(a) the capacitance, inductance and conductance matrices

$$C(w,t) = \tfrac{\partial q_C(w,t)}{\partial w}, \qquad L(w,t) = \tfrac{\partial \phi_L(w,t)}{\partial w}, \qquad G(w,t) = \tfrac{\partial g(w,t)}{\partial w}$$

are positive-definite[1] and

(b) the controlled sources satisfy certain topological conditions (specified in [63]).

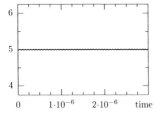
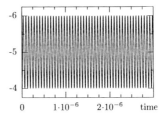

Figure 1.3 (a): Input signal $v(t)$ with perturbation $\delta(t) = \varepsilon\sin(\omega t)$.

Figure 1.3 (b): The current $i(t)$ with error of size $\varepsilon\,\omega\,C_1$.

Figure 1.3: Analytical solution of the circuit from Figure 1.2 (b) with $C_1 = 1\,\mu\mathrm{F}$, $R_1 = \frac{1}{G_1} = 1\,\Omega$ for a perturbed input signal ($\varepsilon = 10^{-2}$, $\omega = 10^8$).

---

[1]A matrix M is said to be positive definite if $x^\top M x > 0$ for every $x \neq 0$. This definition does *not* require symmetry.

The condition (a) corresponds to nonlinear capacitors, inductors and resistors that are strictly locally passive. On the other hand, (b) requires a careful analysis of controlled voltage and current sources since special circuit configurations may impede a topological index test.

Topological index tests were first investigated in [136]. The precise set of allowed controlled sources is given in [61, 63, 154]. There the proof of the next theorem can be found as well.

Theorem 1.4. *Consider a lumped electrical circuit consisting of capacitors, inductors, resistors, voltage sources and current sources. Let the conditions (a) and (b) (in the sense of [63]) be satisfied. If the circuit contains neither a loop of voltage sources nor a cutset of current sources, then the charge oriented modified nodal analysis leads to a DAE with index $\mu \leq 2$.*

The MNA equations have

- *index 0, if the circuit contains no voltage source and the network graph is spanned by a tree consisting exclusively of capacitive branches,*

- *index 1, if the circuit contains a voltage source or there is no tree of capacitive branches, and, in addition, the following two topological conditions are satisfied*

    T_1 :   *there is no CV loop containing at least one voltage source,*

    T_2 :   *there is no LI cutset,*

- *index 2, otherwise.* $\square$

The numerically unstable index-2 components are given by the branch currents through voltage sources of CV loops but also by the branch voltages of the inductors and current sources of LI cutsets [62]. In particular these branch voltages do not necessarily coincide with the Cartesian components of the vector x but are linear combinations of these.

In view of Theorem 1.4 the loop consisting of capacitor and voltage source in Figure 1.2 (b) is the reason for the index being 2. In Figure 1.2 (a) this loop is broken by the resistor such that the index is only 1.

The inclusion of controlled sources in Theorem 1.4 is crucial as controlled sources are implicitly included in every companion model for semiconductor devices. Fortunately the conditions stated in [63] guarantee that these source can be treated within the framework of Theorem 1.4. Hence topological index tests proofed to be a very powerful tool for modern circuit diagnosis.

Nevertheless, both assumptions (a) and (b) are vital for Theorem 1.4 to be applicable. The positive definiteness may be violated by independent charge and flux sources used e.g. to model α radiation or external magnetic fields [66]. On the other hand, controlled sources such as the operational amplifier in the context of the Miller Integrator from Example 1.1 are not covered by (b). In fact, the index of the Miller Integrator depends not only on topological properties but also on device parameters. A summary is given in Figure 1.4.

In particular the inclusion of a further capacitor C_2 between node 2 and ground gives rise to higher index cases as this capacitor closes a CV loop together with C_1 and the controlled voltage source. See [66] for more details.

Although the index may depend on device parameters, most circuits are covered by Theorem 1.4 and topological methods are a well established and powerful tool for dealing with the differential algebraic equations arising in modern circuit simulation.

DAEs are the subject of current research [100, 117, 118, 138]. They do present challenges both for mathematical analysis and scientific computing. We will return to investigating linear and nonlinear DAEs in much more detail in Chapter 3 and 4, respectively.

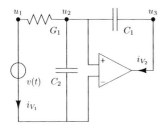

Technical parameters		Index
$C_2 = 0$	$a \neq 1$	1
	$a = 1$	2
$C_2 \neq 0$	$a \neq 1 + \frac{C_1}{C_2}$	2
	$a = 1 + \frac{C_1}{C_2}$	3

Figure 1.4: Impact of topology and technical parameters on the index of the Miller Integrator circuit.

2

Numerical integration schemes

In the previous chapter we derived a mathematical model for electrical circuits. The charge oriented modified nodal analysis leads to differential algebraic equations

$$A\big[d\big(x(t),t\big)\big]' + b\big(x(t),t\big) = 0. \tag{2.1}$$

Each DAE is assigned an index in order to measure its complexity concerning analytical and numerical treatment. Already for index-1 equations initial values can not be chosen arbitrarily but they have to be consistent with the problem data[1]. DAEs with index ≥ 2 involve inherent differentiations such that these equations are ill posed and certain parts of the input functions need to be differentiated.

Nevertheless, as Theorem 1.4 shows, index-1 and index-2 equations appear most frequently for electrical circuits. In fact, the index μ typically satisfies $1 \leq \mu \leq 2$. Even though certain controlled sources may lead to an index $\mu > 2$, these configurations can be detected using topological index tests, such that a regularisation[2] becomes possible. Thus, in electrical circuit simulation one is confronted with solving DAEs (2.1) having index $\mu = 1$ or $\mu = 2$.

It was mentioned earlier that modern semiconductor devices are treated using companion models. Regarding the switching tasks of these devices it is clear that the model equations will suffer from poor smoothness properties. Consequently, the coefficients of (2.1) are continuous but usually non-smooth.

Modern circuits comprise timescales of several orders of magnitude, such that the resulting circuit DAEs are usually stiff. In addition to this, the large number of devices leads to MNA models consisting of up to 10^5 or 10^6 equations and computational time is a major issue.

In summary, DAEs (2.1) in integrated circuit design present challenging problems for numerical computation. Low order methods need to be used due to

---

[1]An initial value x_0 is consistent if there is a solution passing through x_0.

[2]Additional capacitors to ground may often regularise a given circuit.

the model's poor smoothness properties. Since the equations are mainly stiff, A-stable methods are preferable.

Gear [71] proposed BDF[3] methods already in 1971. Their robustness and reliability, in particular in combination with the trapezoidal rule, have made these methods to become the standard in circuit simulation.

Unfortunately, linear multistep methods such as the BDF or the trapezoidal rule often suffer from undesired stability properties. For instance there is no A-stable method of order higher than 2. One step methods such as Runge-Kutta methods with improved stability properties may prove an alternative. However, the computational costs for these (fully implicit) methods is prohibitive in circuit simulation. Diagonally implicit methods, on the other hand, will suffer from order reduction.

In the next two sections we will briefly review both linear multistep methods and Runge-Kutta methods. Each class of integration schemes has its own advantages but also suffers from certain shortcomings and potential problems will be indicated. In Chapter 2.3 general linear methods will then be introduced as a means to overcome these difficulties.

2.1 Linear Multistep Methods

Given the DAE (2.1) together with a consistent initial condition $x(t_0) = x_0$, we want to determine the solutions x on some time interval $[t_0, T] \subset \mathbb{R}$, i.e. we seek approximations x_k to the exact solution on a (usually non-uniform) grid of discrete time points $\{ t_k \,|\, k = 0, 1, \dots, N \}$.

In order to proceed from t_n to t_{n+1} using a stepsize $h = t_{n+1} - t_n$ linear multistep methods make use of existing approximations $x_i \approx x(t_{n-i})$ and $d_i' \approx [d(x(t_{n-i}), t_{n-1})]'$ at previous time points, $i = 0, \dots, k-1$. These quantities are used in order to compute x_{n+1} and d_{n+1}' such that $A\, d_{n+1}' + b(x_{n+1}, t_{n+1}) = 0$ and

$$\sum_{i=0}^{k} \alpha_i \, d(x_{n+1-i}, t_{n+1-i}) = h \sum_{i=0}^{k} \beta_i \, d_{n+1-i}'.$$

For an ordinary differential equation $y' = f(y, t)$ this scheme simplifies to

$$\sum_{i=0}^{k} \alpha_i \, y_{n+1-i} = h \sum_{i=0}^{k} \beta_i \, f(y_{n+1-i}, t_{n+1-i}). \tag{2.2}$$

Since (2.2) has to be solved for x_{n+1} we need to require $\alpha_0 \neq 0$.

Linear multistep methods were used by Adams and Bashforth as early as in 1883 [6]. The so-called Adams-Bashforth methods are characterised by the

---

[3]Backward Difference Formulae

parameters $\alpha_0 = \alpha_1 = 1$ and $\alpha_2 = \cdots = \alpha_k = \beta_0 = 0$. For $\beta_0 \neq 1$ one obtains the Adams-Moulton methods [126].

Different choices of the coefficients α, β lead to a large variety of methods. Among the most prominent are the BDF methods first proposed by Curtiss and Hirschfelder [48]. In order to derive these methods from (2.2), set $\beta_1 = \cdots = \beta_k = 0$ and choose the coefficients α_i such that

$$\alpha(z) = \sum_{i=0}^{k} \alpha_i z^i = \beta_0 \sum_{i=1}^{k} \frac{1}{k}(1-z)^k.$$

The remaining parameter β_0 is fixed by the condition $\alpha(0) = \alpha_0 = 1$. The special choice of $\alpha(z)$ ensures that the polynomial p which interpolates the values $\{(t_{n+1-i}, y_{n+1-i}) \mid i = 0, \ldots, k\}$ satisfies $p'(t_{n+1}) = f(x_{n+1}, t_{n+1})$.

The coefficients of the methods BDF_k for $k = 1, \ldots, 6$ are given in Table 2.1. For $k \geq 7$ the BDF formulae are not stable and only the methods with $1 \leq k \leq 6$ are useful for numerical computation [25].

Since the work of Gear [70, 71] BDF methods are most successfully used for solving stiff equations. The application to DAEs has been investigated in [10, 11] and the code DASSL, written by Petzold [128], is still one of the most competitive solvers for low-index DAEs.

One of the main reasons for the BDFs being so popular also in circuit simulation is the fact that only one function evaluation is required per step. Hence BDF methods can be implemented with relatively low costs even for very large systems. However, the stability properties are not always desirable.

Example 2.1. Consider the circuit in Figure 2.1 (a), where a voltage source, capacitor, inductor and resistor are connected in series. As this circuit contains two energy storing elements, the circuit's complete response can be calculated by solving the second order differential equation [56]

$$0 = U'' + \frac{1}{LG}U' + \frac{1}{LC}U - \frac{1}{LC}v(t) \tag{2.3}$$

k	β_0	α_1	α_2	α_3	α_4	α_5	α_6
1	1	-1					
2	$\frac{2}{3}$	$-\frac{4}{3}$	$\frac{1}{3}$				
3	$\frac{6}{11}$	$-\frac{18}{11}$	$\frac{9}{11}$	$-\frac{2}{11}$			
4	$\frac{12}{25}$	$-\frac{48}{25}$	$\frac{36}{25}$	$-\frac{16}{25}$	$\frac{3}{25}$		
5	$\frac{60}{137}$	$-\frac{300}{137}$	$\frac{300}{137}$	$-\frac{200}{137}$	$\frac{75}{137}$	$-\frac{12}{137}$	
6	$\frac{20}{49}$	$-\frac{120}{49}$	$\frac{150}{49}$	$-\frac{400}{147}$	$\frac{75}{49}$	$-\frac{24}{49}$	$\frac{10}{147}$

Table 2.1: Coefficients of the BDF methods ($\alpha_0 = 1$)

for the capacitor voltage $U = u_1 - u_2$. For simplicity assume that $C = 1\,\mathrm{F}$, $L = 1\,\mathrm{H}$ and that $v \equiv 1\,\mathrm{V}$ is a DC source delivering a constant voltage drop between node 1 and ground. The initial conditions

$$U(0) = 2\,\mathrm{V} \qquad \text{and} \quad i_L(0) = -\frac{C}{2\,L\,G}$$

fix the capacitor voltage to be

$$U(t) = e^{-\alpha t}\cos(\omega\,t) + 1, \quad \text{with} \qquad \alpha = \tfrac{1}{2\,L\,G}, \quad \omega^2 = \tfrac{1}{L\,C} - \alpha^2, \quad (2.4)$$

provided that $G > \frac{1}{2\Omega}$. Since $u_1 \equiv 1\,\mathrm{V}$, the potential at the second node is given by $u_2(t) = -e^{-\alpha t}\cos(\omega\,t)$. For large values of G this is practically a pure cosine function, but for small values G, say $R = \frac{1}{G} = 100\,\mathrm{m\Omega}$, the node potential u_2 decays quickly (see Figure 2.1 (b) and Figure 2.1 (c)).

Stability problems of the BDF formulae become apparent in Figure 2.1 (d). There the BDF$_2$ was used to solve the undamped circuit using a constant stepsize $h = \frac{\pi}{4}$ which corresponds to a sampling rate of about 8 timesteps per period. Although the *undamped* circuit was solved, the solution for u_2 shows a striking similarity to the exact solution of the *damped* circuit.   □

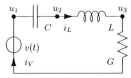

Figure 2.1 (a): An RLC series circuit.

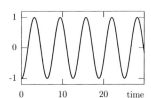

Figure 2.1 (b): Potential u_2 for $R = \frac{1}{G} = 1\,\mathrm{m\Omega}$ (exact values).

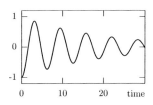

Figure 2.1 (c): Potential u_2 for $R = \frac{1}{G} = 100\,\mathrm{m\Omega}$ (exact values).

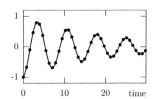

Figure 2.1 (d): Potential u_2 for $R = \frac{1}{G} = 1\,\mathrm{m\Omega}$ (BDF solution).

Figure 2.1: An RLC series circuit ($C = 1\,\mathrm{F}$, $L = 1\,\mathrm{H}$, $v \equiv 1\,\mathrm{V}$): For small values of $R = \frac{1}{G}$ the node potential u_2 is nearly undamped but it decays quickly for a large resistance. The BDF$_2$ method, when applied to the un-damped circuit, produces results as if the circuit was damped.

As indicated in the previous example, the BDF methods, in particular for order 1 and 2 show strong numerical damping. This is a most disadvantageous situation for the simulation of electrical circuits[4] as it is not clear from the simulation results whether the damping is inherent in the circuit or just a numerical artefact. Even though a stepsize control algorithm may avoid these kind of problems, the steps taken will be excessively small since the stepsize sequence will be governed by stability rather than accuracy.

There are methods that do preserves both amplitude and phase of any given oscillation. One example is the trapezoidal rule, that is obtained from (2.2) by setting $k = 1$, $\alpha_0 = -\alpha_1 = 1$, $\beta_0 = \beta_1 = \frac{1}{2}$, such that

$$y_{n+1} = y_n + \frac{h}{2}\big(f(y_n, t_n) + f(y_{n+1}, t_{n+1})\big).$$

Example 2.2. As an example the *VRC* circuit from Figure 1.2 (b) on page 20 will be solved using the trapezoidal rule. For simplicity choose $R_1 = \frac{1}{G_1} = 1\,\Omega$, $C_1 = 1\,\mathrm{F}$ and the input signal

$$v(t) = \left\{ \begin{array}{ll} 1 & , t \leq \frac{\pi}{2}, \\ \cos\left(t - \frac{\pi}{2}\right) & , t > \frac{\pi}{2}. \end{array} \right.$$

This function models an independent voltage source that delivers a DC signal of $1\,\mathrm{V}$ until the breakpoint $t_* = \frac{\pi}{2}$ is reached. There the input switches to a sinusoidal signal.

The trapezoidal rule is applied with constant stepsize $h = \frac{1}{4}$ on the interval $[0, 3\pi]$ using consistent initial values. Figure 2.2 shows that the numerical result for the node potential $u(t) = v(t)$ is in good agreement with the exact value. Nevertheless, at the breakpoint errors are introduced. These perturbations lead to a highly oscillatory behaviour of the numerical approximation to $i_V(t) = -C_1 u'(t) - G_1 u(t)$. □

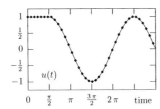 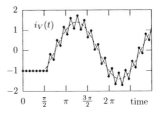

Figure 2.2: The trapezoidal rule applied to the *VRC* circuit from Figure 1.2 (b). For the node potential u the numerical result agrees with the exact solution but i_V oscillates around the exact solution (plotted as a dashed line in both pictures).

---

[4]For other applications such as e.g. the simulation of multibody systems with friction, the BDF method's numerical damping may well be a desired feature.

The previous two examples indicate that, depending on the type of oscillation, different behaviour of an integrator is required:

(a) Oscillations of physical significance, as in Example 2.1, should be preserved.

(b) High frequent oscillations caused by numerical noise, perturbations, inconsistent initial values or errors from the Newton iteration should be damped out quickly.

In order to analyse the damping properties of a numerical method, (2.3) can be studied in more detail. Rewriting (2.3) as a first order ordinary differential equation,

$$
\begin{bmatrix} U \\ i_L \end{bmatrix}' = \begin{bmatrix} 0 & \frac{1}{C} \\ -\frac{1}{L} & -\frac{1}{LG} \end{bmatrix} \begin{bmatrix} U \\ i_L \end{bmatrix} + \begin{bmatrix} 0 \\ \frac{1}{L} v(t) \end{bmatrix}
$$

the eigenvalues of the coefficient matrix are found to be $\lambda_{1/2} = -\alpha \pm \omega\, i$, where $\alpha = \frac{1}{2LG}$, $\omega^2 = \frac{1}{LC} - \alpha^2$ and i is the imaginary unit. ω is real number for $G^2 > \frac{C}{4L}$ and the corresponding homogeneous system has oscillatory solutions with frequency ω. These oscillations are stable for $\alpha \geq 0$, i.e. when λ_i has negative real part. See (2.4) for an example. If $|\lambda_i|$ is large compared to the time scale of the problem, then the differential equation is called *stiff*.

In order to assess a methods linear stability properties we need to study the linear scalar test equation

$$
y'(t) = \lambda\, y(t), \tag{2.5}
$$

where $\lambda = -\alpha + \omega\, i$ is a complex number having negative real part. Thus the exact solution is of the form $e^{-\alpha t}\big(A_1 \cos(\omega\, t) + A_2 \sin(\omega\, t)\big)$, such that it decays for $t \geq 0$.

The trapezoidal rule, when applied to (2.5) with constant stepsize, yields the recursion

$$
y_{n+1} = y_n + \frac{\lambda\, h}{2}(y_n + y_{n+1}) \qquad \Leftrightarrow \qquad y_{n+1} = \frac{2 + \lambda\, h}{2 - \lambda\, h}\, y_n.
$$

The mapping $R : \mathbb{C} \rightarrow \mathbb{C}$, $z \mapsto \frac{2+z}{2-z}$ is known as the method's stability function. Stable solutions are obtained provided that the product $z = \lambda\, h$ of eigenvalue and stepsize satisfies $|R(\lambda\, h)| \leq 1$. Thus the set $\{ z \in \mathbb{C} \,|\, 1 \geq |R(z)| \}$ is called the stability region. Observe that $|R(\omega i)| = |\frac{2+\omega i}{2-\omega i}| = 1$ along the imaginary axis. Thus the stability region for the trapezoidal rule coincides exactly with the left half of the complex plane.

On the other hand, the BDF$_1$ method, often referred to as the implicit Euler scheme, reflects (2.5) by means of the recursion

$$
y_{n+1} = y_n + \lambda\, h\, y_{n+1} \qquad \Leftrightarrow \qquad y_{n+1} = \frac{1}{1 - \lambda\, h}\, y_{n+1},
$$

such that the stability region is given by the set $\{\,z \in \mathbb{C} \,|\, 1 \leq |1-z|\,\}$ (Figure 2.3 (a)). For higher order BDF methods, write $Y_n = \begin{bmatrix} y_n & y_{n-1} & \cdots & y_{n+1-k} \end{bmatrix}^{\mathsf{T}}$ such that the k-step BDF applied to (2.5) reads

$$
Y_{n+1} = \begin{bmatrix}
\frac{-\alpha_1}{1-z\beta_0} & \frac{-\alpha_2}{1-z\beta_0} & \cdots & \frac{-\alpha_{k-1}}{1-z\beta_0} & \frac{-\alpha_k}{1-z\beta_0} \\
1 & 0 & \cdots & 0 & 0 \\
0 & 1 & \cdots & 0 & 0 \\
\vdots & \vdots & \ddots & \vdots & \vdots \\
0 & 0 & \cdots & 1 & 0
\end{bmatrix} Y_n = M_k(z)\,Y_n,
$$

where λh was again replaced by the complex variable z. In contrast to one step methods (e.g. the trapezoidal rule or the implicit Euler scheme) and their stability function $R(z)$, linear stability is governed by a stability *matrix* $M(z)$. The stability region is characterised by those values $z \in \mathbb{C}$, where $M(z)$ is power bounded. For the BDF_k formulae with $1 \leq k \leq 5$ the stability regions are plotted in Figure 2.3 (a).

For $k \leq 2$ the left half of the complex plane is included in the stability region. Thus, whenever the exact solution is stable, the numerical approximation will be stable as well. This statement is not true for $k \geq 3$ as the boundary of the stability region crosses the imaginary axis.

Methods showing this behaviour are in general not suited for solving stiff equations. In the stiff case, where eigenvalues $\lambda = -\alpha + \omega\,\mathrm{i}$ with large α are present, it may be necessary to reduce the stepsize drastically until $z = \lambda h$ again belongs to the stability region. Hence, for solving stiff equations a property called A-stability is desirable.

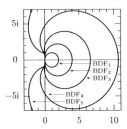

Figure 2.3 (a): Stability regions of the BDF_k methods.

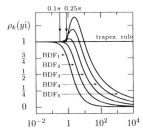

Figure 2.3 (b): Stability behaviour along the imaginary axis.

Figure 2.3: Stability properties of the BDF_k methods and the trapezoidal rule. For each $1 \leq k \leq 5$ the stable area in Figure 2.3 (a) is given by the outside of the kidney shaped region. For Figure 2.3 (b) the modulus $\rho_k(z)$ of the largest eigenvalue of $M_k(z)$ is plotted along the imaginary axis. For a sampling rate of about $8 - 20$ steps per period $\rho(y\mathrm{i})$ is required to stay close to 1 for $y \in [0.1\pi, 0.25\pi]$.

Definition 2.3. *A numerical method is called A-stable, if the left half of the complex plane is included in the method's the stability region.*

This definition is due to Dahlquist [49]. Dahlquist was also able to show that there is no A-stable linear multistep method with order $p > 2$. This fact is known as the second Dahlquist barrier, which is confirmed in Figure 2.3 (a).

Even though the BDF_k methods are not A-stable for $k \geq 3$, the sector

$$\{\, r\, e^{(\pi-\theta)\mathrm{i}} \in \mathbb{C} \,|\, r \in \mathbb{R},\, \theta \in [-\alpha, \alpha] \,\}$$

is contained in the stability region for some suitable value of α (see Table 2.2).

Definition 2.4. *A numerical method is called $A(\alpha)$-stable if the stability region includes the sector* $\{\, r\, e^{(\pi-\theta)\mathrm{i}} \in \mathbb{C} \,|\, r \in \mathbb{R},\, r \geq 0,\, \theta \in [-\alpha, \alpha] \,\}$.

Apart from looking at A- and $A(\alpha)$-stability it is interesting to study the stability matrix along the imaginary axis, since purely imaginary eigenvalues $\lambda = \omega\,\mathrm{i}$ give rise to undamped oscillations. Let $\rho(z)$ denote the modulus of the largest eigenvalue of $M(z)$. Assuming a sampling rate of about $8 - 20$ steps for oscillations of physical significance, we require that $\rho(\omega\,\mathrm{i})$ stays close to 1 for $\omega \in [0.1\pi, 0.25\pi]$. This range is marked in Figure 2.3 (b) using vertical lines.

It becomes clear from Figure 2.3 (b) that both the implicit Euler scheme and the BDF_2 method will damp out oscillations of physical significance as $\rho(\omega\,\mathrm{i}) < 1$ holds in the desired range. The BDF methods with $k \geq 3$ will amplify these oscillations due to $\rho(\omega\,\mathrm{i}) > 1$. The behaviour in both cases is far from optimal.

For the trapezoidal rule, on the other hand, we already found $|R(\omega\,\mathrm{i})| = \rho(\omega\,\mathrm{i}) = 1$, such that the amplitude of any oscillation will be preserved. This is true not only for oscillations of physical significance but also for high frequent numerical noise, a fact that explains the behaviour from Example 2.2.

More details on this type of analysis can be found in [66, 81].

Considering the damping behaviour as discussed above, a commonly used strategy in order to simulate electrical circuits both efficiently and reliably is to start the integration with the BDF method but to continue with the trapezoidal rule after a few successful steps. If convergence problems arise or a breakpoint (switching point) is reached, the BDF method is used for a few steps followed by the trapezoidal rule etc. Obviously, the frequent change of the method causes problems for an efficient error estimation and stepsize selection. This cyclic approach cannot be generalised to order 3 computations where the BDF_3 is used exclusively. This method is not A-stable and due to Dahlquist's second barrier, there is no A-stable linear multistep method of this order at all.

BDF_1	BDF_2	BDF_3	BDF_4	BDF_5	BDF_6
$90°$	$90°$	$86.03°$	$73.35°$	$51.84°$	$17.84°$

Table 2.2: $A(\alpha)$-stability of the BDF methods (taken from [85])

Apart from stability constraints, further complications are associated with changing the stepsize and/or the order. Both aspects are essential for a competitive code, but require considerable overhead for linear multistep methods.

It is far simpler to change the stepsize and order for a Runge-Kutta method. More importantly, there are Runge-Kutta methods with much improved stability properties. Hence this class of methods shall be investigated next.

2.2 Runge-Kutta Methods

In contrast to linear multistep methods Runge-Kutta methods do not use past information but confine themselves to using only one initial approximation $y_n \approx y(t_n)$ at the beginning of a step. In order to obtain a highly accurate approximation $y_{n+1} \approx y(t_n + h)$ at the end of this step, intermediate stages $Y_i \approx y(t_n + c_i\, h)$ are calculated. These ideas were first proposed by Runge [142] and Heun [87], but it was Kutta [103] who completely characterised the set of order 4 methods.

Since the fundamental work of Butcher [15, 18, 16] Runge-Kutta methods received much attention in the literature. Particular methods are constructed in [1, 17, 57] while implementation issues are dealt with in [14, 86] and many other references. The monograph [22] provides in-depth information.

The theory of Runge-Kutta methods for the solution of differential algebraic equations is developed in [84, 85, 104, 140]. Unfortunately, these results do not cover all aspects of scientific computing in electrical circuit design. [84, 85] focus on systems in Hessenberg form such that the charge oriented formulation is not covered. On the other hand, [104] restricts attention to index-1 equations.

Given an ordinary differential equation $y' = f(y,t)$ and an initial approximation y_n, a Runge-Kutta method is of the form

$$Y_i = y_n + h\sum_{j=1}^{s} a_{ij}\, f(Y_j, t_n + c_j\, h), \qquad y_{n+1} = y_n + h\sum_{i=1}^{s} b_i\, f(Y_i, t_n + c_i\, h),$$

where s intermediate stages Y_i are calculated, $i = 1, \ldots, s$. The position of these internal stages is indicated by the vector $c = \begin{bmatrix} c_1 & \cdots & c_s \end{bmatrix}^\top$, i.e. we intend to have $Y_i \approx y(t_n + c_i\, h)$. The method's parameters can conveniently be represented in a Butcher-tableau

$$
\begin{array}{c|ccc}
c_1 & a_{11} & \cdots & a_{1s} \\
\vdots & \vdots & \ddots & \vdots \\
c_s & a_{s1} & \cdots & a_{ss} \\
\hline
 & b_1 & \cdots & b_s
\end{array}
\qquad \text{or} \qquad
\begin{array}{c|c}
c & \mathcal{A} \\
\hline
 & b^\top
\end{array}.
$$

Some well known examples are given in Table 2.3. Two of these methods are stiffly accurate, i.e the vector $b^\top$ coincides with the last row of $\mathcal{A}$. In case of stiffly accurate methods the numerical result $y_{n+1} = Y_s$ is given by the last stage. This property is of particular importance when studying stiff equations and DAEs [84, 86, 90].

The coefficients of a Runge-Kutta method need to satisfy certain order conditions that guarantee that the error committed in one step can be estimated in terms of $\mathcal{O}(h^{p+1})$ for an order p method. In Chapter 8 order conditions will be discussed in considerable detail within the framework of general linear methods for DAEs. Here, it suffices to note that Runge-Kutta methods are often constructed using the simplifying assumptions

$$B(\eta): \quad \sum_{i=1}^{s} b_i\, c_i^{k-1} = \frac{1}{k}, \qquad k = 1, \ldots, \eta,$$

$$C(\xi): \quad \sum_{j=1}^{s} a_{ij}\, c_j^{k-1} = \frac{1}{k}\, c_i^{k}, \qquad k = 1, \ldots, \xi, \quad i = 1, \ldots, s,$$

$$D(\zeta): \quad \sum_{i=1}^{s} b_i\, c_i^{k-1}\, a_{ij} = \frac{1}{k}\, b_j\, (1 - c_j^{k}), \quad k = 1, \ldots, \zeta, \quad j = 1, \ldots, s.$$

For a method of order p it is necessary that $B(p)$ holds. If the condition $C(q)$ is satisfied, then the method has stage order q, i.e. the intermediate stages are calculated with accuracy $Y_i = y(t_n + c_i\, h) + \mathcal{O}(h^{q+1})$. More details are given in [22, 25]. There the following fundamental result is proved.

Theorem 2.5. *If the coefficients of a Runge-Kutta method satisfy $B(p)$, $C(\xi)$, $D(\zeta)$ with $\xi + \zeta + 1 \geq p$ and $2\xi + 2 \geq p$, then the method is of order p, i.e. $y_{n+1} = y(t_{n+1}) + \mathcal{O}(h^{p+1})$ if y_{n+1} is the numerical result calculated from the exact initial value $y(t_n)$ using stepsize h.* $\qquad\square$

The simplifying assumptions can also be used for the construction of Runge-Kutta methods. For example the Gauss methods are obtained by choosing

$\begin{array}{c\|cc} 0 & & \\ 1 & \frac{1}{2} & \frac{1}{2} \\ \hline & \frac{1}{2} & \frac{1}{2} \end{array}$	$\begin{array}{c\|cc} 1-\frac{\sqrt{2}}{2} & 1-\frac{\sqrt{2}}{2} & \\ 1 & \frac{\sqrt{2}}{2} & 1-\frac{\sqrt{2}}{2} \\ \hline & \frac{\sqrt{2}}{2} & 1-\frac{\sqrt{2}}{2} \end{array}$	$\begin{array}{c\|cccc} 0 & & & & \\ \frac{1}{2} & \frac{1}{2} & & & \\ \frac{1}{2} & 0 & \frac{1}{2} & & \\ 1 & 0 & 0 & 1 & \\ \hline & \frac{1}{6} & \frac{1}{3} & \frac{1}{3} & \frac{1}{6} \end{array}$	$\begin{array}{c\|cc} \frac{1}{2}-\frac{\sqrt{3}}{6} & \frac{1}{4} & \frac{1}{4}-\frac{\sqrt{3}}{6} \\ \frac{1}{2}+\frac{\sqrt{3}}{6} & \frac{1}{4}+\frac{\sqrt{3}}{6} & \frac{1}{4} \\ \hline & \frac{1}{2} & \frac{1}{2} \end{array}$
trapezoidal rule (order 2)	SDIRK method[5] (order 2)	'classical' RK method (order 4)	Gauss method with 2 stages (order 4)

Table 2.3: Some examples of Runge-Kutta methods

---

[5]SDIRK is an abbreviation for "Singly Diagonally Implicit Runge-Kutta method". More details will be given later in this section.

$c_1, \ldots, c_s$ to be the zeros of the shifted Legendre polynomials of degree s,

$$P_s(t) = \frac{1}{s!}\frac{d^s}{dt^s}\left(t^s\,(t-1)^s\right),$$

and then satisfying $B(s)$ and $C(s)$. These methods have stage order s and order $2s$, which is the highest possible order for an s-stage Runge-Kutta method.

Another important family of Runge-Kutta methods is based on the Radau quadrature formulae, i.e. $c_1, \ldots, c_s$ are the zeros of $P_s - P_{s-1}$ with $c_s = 1$. Choosing b and $\mathcal{A}$ such that $B(s)$ and $C(s)$ are satisfied, yields the RadauIIA methods [25, 58, 85], the first few being listed in Table 2.4. The order of a RadauIIA method is $p = 2s - 1$ and the stage order is $q = s$.

The code RADAU written by Hairer and Wanner [85, 86] is based on the RadauIIA methods. RADAU became one of the standard solvers for ODEs and DAEs up to index 3. One of the main reasons for RADAU's popularity are the excellent stability properties of the RadauIIA method.

A Runge-Kutta method, when applied to (2.5), reads

$$Y_i = y_n + \lambda\,h\sum_{j=1}^{s} a_{ij}\,Y_j, \qquad\qquad y_{n+1} = y_n + \lambda\,h\sum_{i=1}^{s} b_i\,Y_i \qquad (2.7)$$

such that the numerical approximations y_n satisfy the recurrence

$$y_{n+1} = \left(1 + z\,b^\top(I - z\,\mathcal{A})^{-1}e\right)y_n, \qquad z = \lambda\,h, \qquad e = \begin{bmatrix}1\\ \vdots \\ 1\end{bmatrix} \in \mathbb{R}^s.$$

The stability function $R(z) = 1 + z\,b^\top(I - z\,\mathcal{A})^{-1}e$ determines the stability region $\{\, z \in \mathbb{C} \,|\, 1 \ge |R(z)| \,\}$ for a given Runge-Kutta method.

Figure 2.4 shows that all RadauIIA methods are A-stable – a fact that is proved e.g. in [85]. The SDIRK method from Table 2.3 is A-stable as well. Even though the stability regions of the Gauss methods are not plotted here, it can be shown that these methods are all A-stable and their stability region coincides exactly with left half of the complex plane. Hence they are high order generalisations of the trapezoidal rule. This statement is confirmed in Figure 2.4 (b), where $|R(z)|$ is plotted along the imaginary axis.

					$\frac{2}{5} - \frac{\sqrt{6}}{10}$	$\frac{11}{45} - \frac{7\sqrt{6}}{360}$	$\frac{37}{225} - \frac{169\sqrt{6}}{1800}$	$-\frac{2}{225} + \frac{3\sqrt{6}}{225}$
			$\frac{1}{3}$	$\frac{5}{12}$	$-\frac{1}{12}$	$\frac{2}{5} + \frac{\sqrt{6}}{10}$	$\frac{37}{225} + \frac{169\sqrt{6}}{1800}$	$\frac{11}{45} + \frac{7\sqrt{6}}{360}$	$-\frac{2}{225} - \frac{3\sqrt{6}}{225}$
1	1		1	$\frac{3}{4}$	$\frac{1}{4}$	1	$\frac{4}{9} - \frac{\sqrt{6}}{36}$	$\frac{4}{9} + \frac{\sqrt{6}}{36}$	$\frac{1}{9}$
	1			$\frac{3}{4}$	$\frac{1}{4}$		$\frac{4}{9} - \frac{\sqrt{6}}{36}$	$\frac{4}{9} + \frac{\sqrt{6}}{36}$	$\frac{1}{9}$

Table 2.4: The RadauIIA methods of order $p = 1, 3, 5$

This plot indeed shows the superior stability properties of the RadauIIA methods: High frequent oscillations are damped out quickly while oscillations of physical significance are preserved. In particular, $|R(yi)|$ stays close to 1 for $y \in [0.1\,\pi, 0.25\,\pi]$.

For the SDIRK method Figure 2.4 (b) suggests a good preservation of low frequent oscillations as well.

Nevertheless, Runge-Kutta methods are not popular in circuit simulation. The main reason are their high computational costs. Recall from (2.6) that the stages Y_i satisfy

$$Y = h(\mathcal{A} \otimes I_m)\, F(Y) + (e \otimes I_m)\, y_n, \quad Y = \begin{bmatrix} Y_1 \\ \vdots \\ Y_s \end{bmatrix}, \quad F(Y) = \begin{bmatrix} f(Y_1, t_n + c_1\,h) \\ \vdots \\ f(Y_s, t_n + c_s\,h) \end{bmatrix},$$

where $A \otimes B$ denotes the Kronecker product for matrices [91] and m is the problem size. In general, this is a nonlinear system of dimension $s\cdot m$ and an iterative procedure such as Newton's method has to be used [86, 110].

Given an initial guess Y^0 the costs for the stage evaluation comprise, for each iteration,

(a) the evaluation of the residual $\delta^k = Y^k - h(\mathcal{A} \otimes I_m)\, F(Y^k) - (e \otimes I_m)\, y_n$,

(b) evaluating the Jacobian $J = \left(\frac{\partial f_i}{\partial Y_j} \right)$,

(c) computing the LU factors of the iteration matrix $LU = I_s \otimes I_m - h(\mathcal{A} \otimes J)$,

(d) solving the linear system $LU\,\Delta Y^{k+1} = -\delta^k$.

In particular the costs for (c) and (d) depend heavily on the structure of the Runge-Kutta matrix $\mathcal{A}$. If $\mathcal{A}$ is a full matrix, then (c) requires $\mathcal{O}(s^3 m^3)$ operations, and for (d) another $\mathcal{O}(s^2 m^2)$ operations are necessary [93, 130].

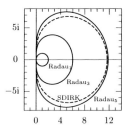

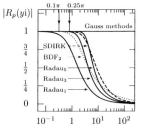

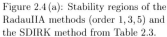

Figure 2.4 (a): Stability regions of the RadauIIA methods (order $1, 3, 5$) and the SDIRK method from Table 2.3.

Figure 2.4 (b): Stability behaviour along the imaginary axis. The BDF$_2$ is included for comparison.

Figure 2.4: Stability properties of some Runge-Kutta methods. For a sampling rate of $8 - 20$ steps per period Radau$_5$ shows almost no damping of physical oscillations as $|R_5(yi)|$ stays close to 1 for $y \in [0.1\pi, 0.25\pi]$.

Thus, for large m the linear algebra routines for Runge-Kutta methods ($s \geq 2$) require much more work as compared to linear multistep methods ($s = 1$). Since Runge-Kutta methods use more functions evaluations per step, the computational costs can be prohibitive for the application in electrical circuit design as function calls are most expensive due to the evaluation of complex transistor models.

Obviously this is only a rough estimate and it is not fair to judge methods based only on these crude computations. Runge-Kutta methods such as the RadauIIA formulae can often be implemented much more efficiently [85]. On the other hand keeping track of the backward information requires some overhead in linear multistep methods. Thus, in order to compare the computational costs of Runge-Kutta and linear multistep methods, we consider a real world application.

Example 2.6. The Ring Modulator circuit from Figure 2.5 is an electrical device that mixes a low-frequent input signal U_{in_1} with a high-frequent input signal U_{in_2}. The corresponding mathematical model was introduced by Horneber [92]. It is a widely used benchmark circuit described in detail in the Test Set for Initial Value Problem Solvers [124] of Bari University (formerly maintained by CWI Amsterdam). There artificial parasitic capacitances C_s are used in order to regularise the circuit and an ODE model is obtained [65].

This regularisation comes at the price of high frequent parasitic oscillations being introduced. Omitting the artificial capacitances removes these parasitic effects, but the circuit turns into an index-2 system. As in [84, 132] this index-2 formulation shall be investigated here.

For the MNA, the diodes are replaced by nonlinear resistors described by the voltage-current relation $I = g(U) = \gamma \left(e^{\delta U} - 1 \right)$. The various constants can be

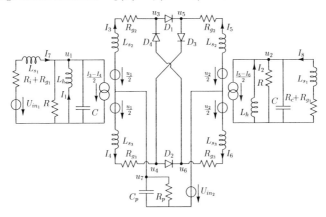

Figure 2.5: Circuit diagram for the Ring Modulator

found in Table 2.5. Writing down the equations of the charge oriented modified nodal analysis as described in Section 1.1 yields the DAE[6]

$$A\big[d\big(x(t)\big)\big]' + b\big(x(t), t\big) = 0 \quad \text{where} \quad x = \begin{bmatrix} u_1 & \cdots & u_7 & I_1 & \cdots & I_8 \end{bmatrix}^\top$$

comprises potentials for the node 1–7 and currents $I_1, \ldots, I_8$ through the eight inductors. The coefficients A, d and b are given by

$$A = \begin{bmatrix} I_2 & 0 \\ 0 & 0 \\ 0 & I_9 \end{bmatrix} \!\begin{matrix} \}2 \\ \}4 \\ \}9 \end{matrix}, \quad d(x) = \begin{bmatrix} C\,u_1 \\ C\,u_2 \\ C_p\,u_7 \\ L_h\,I_1 \\ L_h\,I_2 \\ L_{s_2}\,I_3 \\ L_{s_3}\,I_4 \\ L_{s_2}\,I_5 \\ L_{s_3}\,I_6 \\ L_{s_1}\,I_7 \\ L_{s_1}\,I_8 \end{bmatrix}, \quad b(x,t) = \begin{bmatrix} \frac{1}{R}u_1 - I_1 + \frac{1}{2}I_3 - \frac{1}{2}I_4 - I_7 \\ \frac{1}{R}u_2 - I_2 + \frac{1}{2}I_5 - \frac{1}{2}I_6 - I_8 \\ I_3 + g(u_6 - u_3) - g(u_3 - u_5) \\ I_4 + g(u_5 - u_4) - g(u_4 - u_6) \\ I_5 + g(u_3 - u_5) - g(u_5 - u_4) \\ I_6 + g(u_4 - u_6) - g(u_6 - u_3) \\ \frac{1}{R_p}u_7 - I_3 - I_4 \\ u_1 \\ u_2 \\ -\frac{u_1}{2} + u_3 - u_7 + R_{g_2}I_3 \\ \frac{u_1}{2} + u_4 - u_7 + R_{g_3}I_4 \\ -\frac{u_2}{2} + u_5 + R_{g_2}I_5 - U_{in_2}(t) \\ \frac{u_2}{2} + u_6 + R_{g_3}I_6 - U_{in_2}(t) \\ u_1 + (R_i + R_{g_1})I_7 - U_{in_1}(t) \\ u_2 + (R_c + R_{g_1})I_8 \end{bmatrix}.$$

As an example consider node 3. Kirchhoff's current law shows that

$$0 = i_{R_{g_2}} + i_{D_4} - i_{D_1} = I_3 + g(u_6 - u_3) - g(u_3 - u_5).$$

On the other hand, each inductor yields a further differential equation, e.g.

$$L_{s_2}\,\dot{I}_3 = u_{L_{s_2}} = \tfrac{u_1}{2} + u_7 - u_{R_{g_2}} - u_3 = \tfrac{u_1}{2} + u_7 - R_{g_2}I_3 - u_3.$$

The input signals $U_{in_1}(t) = \frac{1}{2}\sin(2\,\pi\,t \cdot 10^3)$, $U_{in_2}(t) = 2\sin(2\,\pi\,t \cdot 10^4)$ produce an output signal u_2 at node 2 (see Figure 2.6).

As the capacitors and inductors are linear, it is straightforward to reformulate the equations in the form $M\,y' = f(y, t)$ that is given in the Test Set [124]. This formulation can be dealt with by DASSL and RADAU, two of the most successful linear multistep and Runge-Kutta solvers, respectively. Although

C	$=$	$1.6 \cdot 10^{-8}$	R	$= 25000$
C_p	$=$	10^{-8}	R_p	$= 50$
L_h	$=$	4.45	R_{g_1}	$= 36.3$
L_{s_1}	$=$	0.002	R_{g_2}	$= 17.3$
L_{s_2}	$=$	$5 \cdot 10^{-4}$	R_{g_3}	$= 17.3$
L_{s_3}	$=$	$5 \cdot 10^{-4}$	R_i	$= 50$
γ	$= 40.67286402 \cdot 10^{-9}$		R_c	$= 600$
δ	$=$	17.7493332		

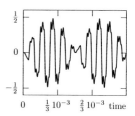

Table 2.5: Technical parameters for the Ring Modulator

Figure 2.6: The Ring Modulator's output signal u_2 at node 2.

[6]The formulation given in [124] uses the voltages $U_1 = u_1$, $U_2 = u_2$, $U_3 = u_3 - u_7$, $U_4 = u_7 - u_4$, $U_5 = u_5 - U_{in_2}$, $U_6 = U_{in_2} - u_6$, $U_7 = u_7$ instead of the node potentials.

DASSL is intended to solve DAEs of index $\mu \leq 1$, it is still a tough competitor for RADAU on this index-2 test problem.

In order to generate problems of arbitrary size, N instances of the Ring Modulator were solved simultaneously. Hence, systems of dimension $N \cdot 15$ are treated here.

From the work-precision diagram in Figure 2.7 (a) we see that for $N = 1$ the Runge-Kutta code RADAU is clearly superior, as higher accuracy is achieved in shorter time. However, already for $N = 50$, i.e. when solving a system of dimension 750, this situation is turned upside down. Now the BDF-solver DASSL outperforms RADAU, even though DASSL was not designed for solving index-2 problems (see Figure 2.7 (b)). □

We saw that Runge-Kutta methods with a full matrix $\mathcal{A}$ in general require $\mathcal{O}(s^3m^3) + \mathcal{O}(s^2m^2)$ operations for evaluating the stages. When the problem is large, Runge-Kutta methods are therefore expected to be computationally more expensive than linear multistep methods. This effect became clearly visible in the previous example.

However, the costs for solving the nonlinear equations in a Runge-Kutta scheme depend heavily on the structure of $\mathcal{A}$. For an explicit method, i.e. when $a_{ij} = 0$ for $j \geq i$, there is no nonlinear system to be solved at all.

In case of differential algebraic equations $A[d(x,t)]' + b(x,t) = 0$ and the corresponding numerical scheme

$$A Y_i' + b(X_i, t_n + c_i h) = 0, \qquad\qquad i = 1, \ldots, s, \qquad (2.8\text{a})$$

$$d(X_i, t_n + c_i h) = y_n + h \sum_{j=1}^{s} a_{ij} Y_j', \qquad y_{n+1} = y_n + h \sum_{i=1}^{s} b_i Y_i' \qquad (2.8\text{b})$$

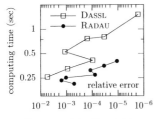

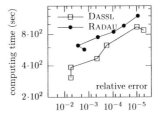

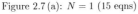

Figure 2.7 (a): $N = 1$ (15 eqns)   Figure 2.7 (b): $N = 50$ (750 eqns)

Figure 2.7: Computing time vs. the achieved accuracy for different problem sizes. For the smaller problem problem RADAU is superior, while DASSL is more efficient for the larger problem. The computations were performed on an Intel® Pentium® M processor with 1.4 GHz, 512 MB RAM, running the software from [124].

the coefficient matrix A in (2.8a) is singular. Thus certain parts of the stage derivatives Y_i' need to be calculated from (2.8b) rather than from (2.8a). As a consequence implicit methods with a nonsingular matrix $\mathcal{A}$ have to be used. Different levels of implicitness can be considered:

(a) *singly diagonally implicit methods*: $\mathcal{A}$ has lower triangular structure and the diagonal elements $a_{ii} = \lambda \neq 0$ are all equal.

$$\begin{bmatrix} \lambda & & \\ \vdots & \ddots & \\ a_{s1} & \cdots & \lambda \end{bmatrix}$$

(b) *diagonally implicit methods*: The matrix $\mathcal{A}$ has still lower triangular structure but the diagonal elements may be different (some, not all, may even be zero).

$$\begin{bmatrix} a_{11} & & \\ \vdots & \ddots & \\ a_{s1} & \cdots & a_{ss} \end{bmatrix}$$

(c) *fully implicit methods*: $\mathcal{A}$ is a full matrix.

(d) *singly implicit methods*: The matrix $\mathcal{A}$ is a full matrix but it has a one-point spectrum $\sigma(\mathcal{A}) = \lambda$.

$$\begin{bmatrix} a_{11} & \cdots & a_{1s} \\ \vdots & \ddots & \vdots \\ a_{s1} & \cdots & a_{ss} \end{bmatrix}$$

In case of diagonally implicit methods (b), the stages can be evaluated sequentially. Thus the computational work is considerably lowered to about $s\,m^3 + s\,m^2$ operations. Recall that the iteration matrix for stage number i is given by $I - h\,a_{ii}\,\frac{\partial f}{\partial y}$. If all diagonal elements are equal, as is the case for singly diagonally implicit methods, there is a good chance that the same Jacobian can be used for every stage such that the costs are further reduced. Finally, singly implicit methods can be implemented as cheaply as singly diagonally implicit methods (at least for large problems) [14], but they require an additional transformation to be carried out.

As a consequence, singly diagonally implicit methods such as the SDIRK method given in Table 2.3 may be the way to go. SDIRK methods were introduced by Butcher [16]. Further contributions were made by Nørsett [141] and Alexander [1]. Unfortunately, SDIRK methods suffer from the order reduction phenomenon, which is closely related to the method's stage order.

Example 2.7. Prothero and Robinson [131] introduced the test problem

$$y'(t) = L\big(y(t) - g(t)\big) + g'(t), \qquad\qquad y(t_0) = g(t_0), \qquad\qquad (2.9)$$

in order to study the behaviour of numerical methods for stiff problems. The exact solution is given by the smooth function $y(t) = g(t)$ which is assumed to be slowly varying. For this example, assume $g(t) = \cos(t)$. If L is negative and large in magnitude, then the problem can be arbitrarily stiff.

For stable methods of order p the order conditions ensure that the global error satisfies $y_n - g(t_n) = \mathcal{O}(h^p)$ as h tends to zero, but this relation may not hold for larger stepsizes. The order can be estimated numerically by solving (2.9) using different stepsizes. If the global error is plotted against the stepsize on a doubly logarithmic scale, then the slope indicates the order of convergence.

These calculations were carried out for the two-stage Gauss and SDIRK methods from Table 2.3 with order $p = 4$ and $p = 2$, respectively. Figure 2.8 shows that for moderate values of L the methods do enjoy order p behaviour. If L is chosen to be large in magnitude, say $L = -10^7$, then the order being numerically observed degenerates to the stage order q.

This effect can be understood by noting that the global error is given by

$$g(t_n) - y_n = R(hL)\big(g(t_{n-1}) - y_{n-1}\big) + \epsilon_0 + hLb^\top(I - hL\mathcal{A})^{-1}\begin{bmatrix}\epsilon_1 \\ \vdots \\ \epsilon_s\end{bmatrix},$$

where $\epsilon_0 = \frac{h^{p+1}}{p!}\big(\frac{1}{p+1} - \sum_{i=1}^s b_i c_i^p\big) + \mathcal{O}(h^{p+2})$, $\epsilon_i = \frac{h^{q_i+1}}{q_i!}\big(\frac{1}{q_i+1} - \sum_{j=1}^s a_{ij}c_j^{q_i}\big) + \mathcal{O}(h^{q_i+2})$, $i = 1, \ldots, s$, are quadrature errors for the numerical result and the stages, respectively. R is the stability function. Due to the expansion

$$hLb^\top(I - hL\mathcal{A})^{-1}\epsilon = -b^\top\mathcal{A}^{-1}\epsilon - (hL)^{-1}b^\top\mathcal{A}^{-1}\mathcal{A}^{-1}\epsilon - \cdots,$$

which is relevant for large hL, we find

$$g(t_n) - y_n = R(hL)\big(g(t_{n-1}) - y_{n-1}\big) + \frac{h^3}{36}g^{(3)}(t_{n-1}) + \mathcal{O}(h^4) = \mathcal{O}(h^2)$$

for the Gauss method with $q = \min(q_1, q_2) = \min(2, 2) = 2$ and

$$g(t_n) - y_n = R(hL)\big(g(t_{n-1}) - y_{n-1}\big) - \frac{h\sqrt{2}}{4L}g^{(2)}(t_{n-1}) + \mathcal{O}(h^2) \underset{\underset{\text{L-stability}}{\uparrow}}{=} \mathcal{O}(h)$$

for the SDIRK method where $q = \min(q_1, q_2) = \min(1, 2) = 1$. Owing to its L-stability, i.e. $\lim_{z\to\infty} R(z) = 0$, the SDIRK method exhibits order 1 convergence since for large hL the global error is essentially given by the local one. See also [25] for an in-depth explanation of this order reduction phenomenon.   □

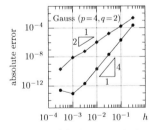

Figure 2.8 (a): Gauss method, $s = 2$.

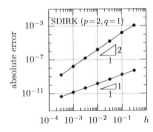

Figure 2.8 (b): SDIRK from Table 2.3.

Figure 2.8: Order of convergence for a Gauss and SDIRK method. In the non-stiff case $L = -10$ (—•—) the methods show order p behaviour. For $L = -10^7$ (—•—), i.e. in the stiff regime, the numerically observed order is governed by the stage order q.

For stiff ODEs and even more so for DAEs, the order of convergence is governed not only by the method's order but also by its stage order. In the stiff case, as in the previous example, the order of convergence being numerically observed is determined by the stage order q. In particular, when the stage order is significantly lower than the order, this means an unacceptable degeneration in performance. Unfortunately, SDIRK methods are restricted to stage order 1 [52]. If an explicit first stage is used [105], then the stage order may be 2, but if a higher stage order is desired, some components of the vector c will lie outside the unit interval $[0, 1]$.

While studying Runge-Kutta methods in this section, it turned out that some of these methods have excellent stability properties, but there are severe drawbacks as well. Implementations of (fully-implicit) Runge-Kutta methods are too expensive for large problems due to the nonlinear system that needs to be solved. In order to reduce the costs, singly diagonally implicit methods have been considered. Unfortunately this class of methods is restricted to a low stage order and order reduction will occur for stiff ODEs and DAEs.

2.3 General Linear Methods

Traditionally integration schemes are classified as being either a linear multistep or a one-step method. Both classes have been considered in the previous two sections. It turned out that linear multistep methods, and in particular the BDF schemes, can be implemented efficiently for large problems, but the stability properties are not always satisfactory for integrated circuit design. One-step methods, with Runge-Kutta methods being the most prominent, have superior stability properties but they do suffer from high computational costs. Singly diagonally implicit methods (SDIRK) are affected by the order reduction phenomenon.

The implicit Euler method and the trapezoidal rule do belong to both classes. Hence linear multistep methods as well as Runge-Kutta methods can be regarded as generalisations of these integration schemes. While linear multistep methods make use of information from the past, Runge-Kutta methods invest more work per step in order to calculate highly accurate approximations to the exact solution. As a consequence, mathematical analysis such as order conditions, stability concepts or the derivation of methods is often different for these two classes of methods.

Several attempts have been made in order to overcome difficulties associated with each class of methods while keeping its advantages. Hybrid methods allow more than one function evaluation in a linear multistep scheme [69]. Using cyclic compositions of multistep methods it became possible to break Dahlquist's barriers [7]. On the other hand, Rosenbrock methods aim at reducing the costs for a Runge-Kutta scheme by linearising the nonlinear system and incorporating the Jacobian into the numerical scheme [85, 96]. Since only linear systems need to be solved, these methods are of particular interest for simulating electrical circuits of huge dimension. The investigations in [79, 81] lead to the development of CHORAL, a charge-oriented ROW method. In [129] two-step-W-methods are considered, where a general matrix W may be used instead of the Jacobian.

In order to cover both linear multistep and Runge-Kutta methods in one unifying framework, Butcher [18] introduced general linear methods (GLMs) for the solution of ordinary differential equations

$$y'(t) = f\big(y(t), t\big), \qquad\qquad y(t_0) = y_0.$$

A general linear methods uses r pieces of input information $y_1^{[n]}, \ldots, y_r^{[n]}$, when proceeding from t_n to $t_{n+1} = t_n + h$ with a stepsize h. Similar to Runge-Kutta methods s internal stages

$$Y_i = h \sum_{j=1}^{s} a_{ij} f(Y_j, t_n + c_j h) + \sum_{j=1}^{r} u_{ij} y_j^{[n]}, \qquad i = 1, \ldots, s, \qquad (2.10a)$$

are calculated and another r quantities

$$y_i^{[n+1]} = h \sum_{j=1}^{s} b_{ij} f(Y_j, t_n + c_j h) + \sum_{j=1}^{r} v_{ij} y_j^{[n]}, \qquad i = 1, \ldots, r, \qquad (2.10b)$$

are passed on to the next step. Occasionally Y_i and $y_i^{[n]}$ are referred to as internal and external stages, respectively. Using the more compact notation

$$Y = \begin{bmatrix} Y_1 \\ \vdots \\ Y_s \end{bmatrix}, \qquad F(Y) = \begin{bmatrix} f(Y_1, t_n + c_1 h) \\ \vdots \\ f(Y_s, t_n + c_s h) \end{bmatrix}, \qquad y^{[n]} = \begin{bmatrix} y_1^{[n]} \\ \vdots \\ y_r^{[n]} \end{bmatrix}$$

(2.10) can be written as

$$Y = (\mathcal{A} \otimes I_m)\, h\, F(Y) + (\mathcal{U} \otimes I_m)\, y^{[n]},$$
$$y^{[n+1]} = (\mathcal{B} \otimes I_m)\, h\, F(Y) + (\mathcal{V} \otimes I_m)\, y^{[n]}.$$

The integer m denotes the problem size and $\otimes$ represents the Kronecker product for matrices [91]. It is only a slight abuse of notation when the Kronecker product is often omitted, i.e.

$$\left[\begin{array}{c} Y \\ \hline y^{[n+1]} \end{array} \right] = \left[\begin{array}{c|c} \mathcal{A} & \mathcal{U} \\ \hline \mathcal{B} & \mathcal{V} \end{array} \right] \left[\begin{array}{c} h\, F(Y) \\ \hline y^{[n]} \end{array} \right].$$

This formulation of a GLM is due to Burrage and Butcher [13]. The matrices $\mathcal{A}$, $\mathcal{U}$, $\mathcal{B}$ and $\mathcal{V}$ represent the method's coefficients. Additionally, each method is characterised by four integers:

s: the number of internal stages,
r: the number of external stages,

p: the order of the method,

q: the stage order of the method.

The internal stages are approximations $Y_i \approx y(t_n + c_i h)$ to the exact solution, but the external stages are fairly general. Commonly adopted choices are

$$y^{[n]} \approx \begin{bmatrix} y(t_n) \\ y(t_n - h) \\ \vdots \\ y(t_n - (r-1)h) \end{bmatrix} \quad \text{or} \quad y^{[n]} \approx \begin{bmatrix} y(t_n) \\ h\,y'(t_n) \\ \vdots \\ h^{r-1}y^{(r-1)}(t_n) \end{bmatrix}, \quad (2.11)$$

the former representing a method of multistep type while the latter is a Nordsieck vector. Notice that compared to Nordsieck's original formulation [127] factorials have been omitted for convenience. Methods in Nordsieck form became popular through the work of Gear [70]. General linear methods of this type are considered, among other references, in [29, 40, 158].

Different choices of the vector $y^{[n]}$ are often related by linear transformations. In this sense the representation of a method using the matrices $\mathcal{A}$, $\mathcal{U}$, $\mathcal{B}$ and $\mathcal{V}$ is not unique as two different methods may be equivalent owing to such a transformation [25].

Many classical integration schemes can be cast into general linear form. For simplicity we will restrict attention to Runge-Kutta and BDF methods but more exotic schemes such as the cyclic composition methods mentioned earlier belong to this class as well. Many examples are given in [26] but also in the extensive monograph [25].

Example 2.8. Consider a Runge-Kutta scheme given by the Butcher tableau

$$\frac{c \;\;\; \mathcal{A}}{\;\;\; b^\top} \;=\; \begin{array}{c|ccc} c_1 & a_{11} & \cdots & a_{1s} \\ \vdots & \vdots & \ddots & \vdots \\ c_s & a_{s1} & \cdots & a_{ss} \\ \hline & b_1 & \cdots & b_s \end{array} \,.$$

A Runge-Kutta method uses only the initial values y_n as input such that $r = 1$. Thus the scheme (2.7), written in general linear form, reads

$$\begin{bmatrix} Y \\ y^{[n+1]} \end{bmatrix} = \begin{bmatrix} \mathcal{A} & e \\ \hline b^\top & 1 \end{bmatrix} \begin{bmatrix} f(Y) \\ y^{[n]} \end{bmatrix} \qquad\qquad \square$$

Example 2.9. The BDF scheme

$$\sum_{i=0}^{k} \alpha_i\, y_{n+1-i} = h\,\beta_0\, f(y_{n+1}, t_{n+1})$$

from Section 2.1 with $\alpha_0 = 1$ can be written as

$$y_{n+1} = h\,\beta_0\, f(y_{n+1}, t_{n+1}) - \sum_{i=1}^{k} \alpha_i\, y_{n+1-i}.$$

This quantity $Y = y_{n+1}$ can be interpreted as the single internal stage of step number n. The approximations y_{n+1-i} to the exact solution at previous timepoints are collected in the vector $y^{[n]}$. Hence a general linear formulation of the BDF scheme is given by

$$
\begin{bmatrix} y_{n+1} \\ \hline y_{n+1} \\ y_n \\ \vdots \\ y_{n+2-k} \end{bmatrix} = \left[\begin{array}{c|cccc} \beta_0 & -\alpha_1 & \cdots & -\alpha_{k-1} & -\alpha_k \\ \hline \beta_0 & -\alpha_1 & \cdots & -\alpha_{k-1} & -\alpha_k \\ 0 & 1 & & & 0 \\ \vdots & & \ddots & & \vdots \\ 0 & & & 1 & 0 \end{array} \right] \cdot \begin{bmatrix} h\, f(y_{n+1}) \\ \hline y_n \\ y_{n-1} \\ \vdots \\ y_{n+1-k} \end{bmatrix} . \qquad \square
$$

General linear methods were introduced already in 1966. Even though these methods are investigated in [13, 20, 22], it was the derivation of DIMSIMs[7] in 1993 [23] that started a renewed interest in using GLMs for practical computations. The subclass of DIMSIM methods is characterised by $s = r = p = q$. Often more general methods are included by only requiring that the quantities s, r, p, and q are all approximately equal [25].

DIMSIMs for non-stiff and stiff equations were constructed mainly by Butcher and Jackiewicz [32, 33, 34]. In [35, 36, 94] implementation issues are addressed. The potential of these methods in a parallel computing environment is investigated in [28, 60]. DIMSIMs are also closely related to singly implicit Runge-Kutta methods with an explicit first stage [105] as well as to the two-stage-W-methods of [129].

Recently it turned out that allowing one more stage, i.e. $s = r = p + 1 = q + 1$, has advantages both for the construction and analysis of GLMs [25, 158]. Methods with inherent Runge-Kutta stability can be constructed using only linear operations. These methods allow a very accurate estimation of the local truncation error [40].

The reason for general linear methods becoming so popular is the fact that within this class diagonally implicit methods exist that have high stage order and a stability behaviour similar to Runge-Kutta methods.

Example 2.10. Consider the general linear method

$$
\mathcal{M} = \left[\begin{array}{cc|cc} 1 - \frac{\sqrt{2}}{2} & 0 & 1 & 1 - \frac{\sqrt{2}}{2} \\ \frac{\sqrt{2}}{4} & 1 - \frac{\sqrt{2}}{2} & 1 & \frac{\sqrt{2}}{4} \\ \hline \frac{\sqrt{2}}{4} & 1 - \frac{\sqrt{2}}{2} & 1 & \frac{\sqrt{2}}{4} \\ 0 & 1 & 0 & 0 \end{array} \right] \quad \text{with} \quad c = \begin{bmatrix} 2 - \sqrt{2} \\ 1 \end{bmatrix} . \quad (2.12)
$$

It is assumed that the two quantities passed from step to step are approximations to the exact solution and the first scaled derivative. Since $y^{[n]} \approx \begin{bmatrix} y(t_n) \\ h\, y'(t_n) \end{bmatrix}$, this method is therefore in Nordsieck form.

---

[7]DIMSIM is short for Diagonally Implicit Multistage Integration Method.

Note that the matrix $\mathcal{A}$ has diagonally implicit structure such that the stages can be calculated sequentially. Thus the implementation costs are similar to the SDIRK method from Table 2.3. In fact, these two methods even share the same linear stability behaviour. This can be seen by applying $\mathcal{M}$ to the linear scalar test equation (2.5),

$$\left.\begin{array}{l} Y = z\mathcal{A}Y + \mathcal{U}y^{[n]} \\ y^{[n+1]} = z\mathcal{B}Y + \mathcal{V}y^{[n]} \end{array}\right\} \quad \Rightarrow \quad y^{[n+1]} = \big(\mathcal{V} + z\mathcal{B}(I - z\mathcal{A})^{-1}\mathcal{U}\big)y^{[n]}.$$

The stability matrix $M(z) = \mathcal{V} + z\mathcal{B}(I - z\mathcal{A})^{-1}\mathcal{U}$ has only one nonzero eigenvalue $R(z) = \frac{2+(2-4\lambda)z+(1-4\lambda+2\lambda^2)z^2}{2(\lambda z-1)^2}$ and $R(z)$ agrees with the stability function of the SDIRK method. Hence their stability regions coincide (see Figure 2.9 (a) and Figure 2.4 (a)).

However, the general linear method is superior when solving stiff equations. The general linear structure allows the stage order to be 2 such that there is no order reduction. In Figure 2.9 (b) this statement is confirmed numerically for the Prothero-Robinson problem (2.9).   □

For any order p the framework of general linear methods allows the construction of diagonally implicit schemes having high stage order $q = p$ (see [158] but also Section 10.1). This makes GLMs very attractive for solving not only stiff ODE but also DAEs. Nevertheless, there are only a few references available in the literature addressing general linear methods in the context of DAEs.

In a technical report Chartier [46] considered GLMs for DAEs having index 1 or 2. This work is similar to the Runge-Kutta approach [84], in particular in its restriction to DAEs in Hessenberg form. It was mentioned earlier that this class does not cover the equations of the charge oriented modified nodal analysis. GLMs for index-3 Hessenberg systems are studied by Schneider in [143].

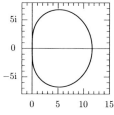

Figure 2.9 (a): Stability region (outside of the encircled area).

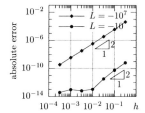

Figure 2.9 (b): Order of convergence.

Figure 2.9: Stability behaviour of the GLM (2.12) and the numerically observed order for the Prothero-Robinson example (2.9). $\mathcal{M}$ has the same stability region as the SDIRK method with order 2. Due to the stage order being 2 there is no order reduction for stiff problems.

There is also a technical report by Butcher and Chartier [27] were methods are constructed for stiff ODEs and DAEs.

In this thesis DIMSIMs as well as methods with $s = r = p + 1 = q + 1$ are studied for nonlinear differential algebraic equations having index 1 or 2. The DAEs studied here originate mainly from the charge oriented modified nodal analysis (MNA) and the emphasis is on solving DAEs of the form

$$A\big[d\big(x(t)\big)\big]' + b\big(x(t), t\big) = 0, \tag{2.13}$$

where the leading term is properly stated. Other applications such as chemical reaction dynamics or the simulation of mechanical multibody systems (after a suitable regularisation) often fall into this class as well [59].

Studying numerical methods for DAEs of this type requires a refined analysis of structural properties. Part II is therefore devoted to studying (2.13) in considerable detail. In particular the concept of the tractability index will lead to a decoupling procedure for nonlinear DAEs. This new decoupling procedure makes only mild assumptions on the smoothness of the coefficients and is thus suited for dealing with the MNA equations.

In Part III the decoupling procedure is used to study general linear methods for DAEs of the form (2.13). After reviewing briefly ODE theory for GLMs, these methods are applied to implicit index-1 equations. The order conditions and convergence results obtained in Chapter 8 are then transferred to properly stated DAEs (2.13) with index 2. Here, the decoupling procedure will be the central tool. The details of this construction are presented in Chapter 9.

After deriving all requirements a GLMs has to fulfil in order to be applicable for solving DAEs, practical aspects are addressed in Part IV. Here, in Chapter 10, general linear methods with $s = r = p + 1 = q + 1$ but also DIMSIMs with $s = r = p = q$ are constructed. Implementation issues such as error estimation and order control are addressed for each class of methods such that a variable-stepsize variable-order implementation is possible in each case.

Numerical experiments in Chapter 12 are used to compare these two implementations. The application to realistic circuits in integrated circuit design gives strong evidence that general linear methods, in particular the DIMSIM implementation, are competitive if not superior to classical codes.

Part II

Differential Algebraic Equations

3

Linear Differential Algebraic Equations

In Chapter 1 differences between differential and differential algebraic equations have been addressed. In particular it may happen that parts of the original data need to be differentiated in order to obtain a solution. As we are aiming at solving DAEs numerically, this differentiation process has to be performed numerically as well. This in turn, is far from trivial since small perturbations in the original data may lead to errors of arbitrary size – an effect already demonstrated in Example 1.3. The question whether there are inherent differentiations involved and, if so, up to what degree, distinguishes different classes of numerical complexity for DAEs. These different levels are usually measured by some kind of index concept.

In the next section some of the most frequently used index notions will be reviewed. Then, in Section 3.2, the tractability index will be treated in more detail. DAEs with a properly stated leading term are introduced and analysed using the framework provided by this index concept. In particular the decoupling of general index-μ equations is briefly addressed for linear DAEs. The main focus, however, is on DAEs with index 1 or 2.

3.1 Index Concepts

There is a wide variety of different index-concepts available in the literature. For linear DAEs with constant coefficients,

$$E\,x'(t) + F\,x(t) = 0, \tag{3.1}$$

the Kronecker index is most natural and will therefore be addressed first. Given that the pair (E, F) forms a regular matrix pencil, i.e. there is $\lambda \in \mathbb{C}$ such that $\det(\lambda E + F) \neq 0$, there are nonsingular matrices U, V that transform E and F simultaneously into Kronecker-Weierstraß normal form given by

$$UEV = \begin{bmatrix} I & 0 \\ 0 & N \end{bmatrix}, \qquad\qquad UFV = \begin{bmatrix} C & 0 \\ 0 & I \end{bmatrix}.$$

The matrix $N = \text{diag}[N_1, \ldots, N_k]$ has block-diagonal form, where N_i is a Jordan block to the eigenvalue 0. Proofs of this result are given in [68, 85, 97]. Using the transformed variables $\begin{bmatrix} u \\ v \end{bmatrix} = V^{-1}x$ and $\begin{bmatrix} a \\ b \end{bmatrix} = U q$ equation (3.1) can be written as the decoupled system

$$u'(t) + Cu(t) = a(t), \qquad\qquad v(t) = \sum_{i=0}^{\mu-1}(-N)^i b^{(i)}(t). \qquad (3.2)$$

The number μ is the index of nilpotency for the matrix N, i.e. $N^{\mu-1} \neq 0$ but $N^\mu = 0$. From (3.2) it is clear that the inherent dynamics of the DAE (3.1) is described by the ordinary differential equation $u'(t) = -Cu(t) + a(t)$. Only for this component initial conditions may be prescribed. The remaining part of the solution, $v(t)$, is already fixed by the right hand side. However, in order to calculate $v(t)$, parts of the right hand side need to be differentiated up to $\mu - 1$ times. If $\mu = 1$, and therefore $N = 0$, no differentiation is necessary. But if $\mu = 2$, then b' appears in the representation for v. Therefore the DAE (3.1) is said to have index μ.

In the case where E is non-singular, i.e. (3.1) is in fact an ordinary differential equation, the block N does not appear at all in the Kronecker-Weierstraß normal form. Hence this special case is assigned the index $\mu = 0$.

Unfortunately there is no immediate generalisation of the Kronecker index to DAEs with time-dependent coefficients. The case of nonlinear DAEs is even further beyond the scope of the Kronecker index. The reason is that important invariants of constant coefficient systems may be changed under time-dependent transformations. In the ongoing strive to find suitable index concepts for large classes of DAEs many different index-concepts where developed, each of them emphasising different aspects of the Kronecker index.

It was remarked earlier that for equations with higher index small perturbations in the data my lead to large errors in the solution. The perturbation index introduced in [84] is defined as a measure for this sensitivity of solutions with respect to perturbations of the problem data. The perturbation index is most relevant for numerical computations. Unfortunately, it can be quite difficult to calculate the index analytically. Another drawback is that for some problems the perturbation index may differ significantly from other indices attributed to the same problem based on the index concepts described below [43, 72].

It is conceptually much easier to count, as for the Kronecker index, the number of times that the equations need to be differentiated in order to derive an ODE representation. This idea was applied to general nonlinear DAEs by Campbell, Gear and Petzold. They introduced the concept of the differentiation index [10, 43, 73]. The following definition is taken from [85].

Definition 3.1. *Let the nonlinear DAE*

$$f\big(x'(t), x(t), t\big) = 0$$

*be sufficiently smooth. The DAE is said to have (differentiation) index μ, if μ
is the minimal number of differentiations*

$$f\big(x'(t), x(t), t\big) = 0,$$

$$\frac{\mathrm{d}}{\mathrm{d}t}\, f\big(x'(t), x(t), t\big) = 0, \quad \ldots \quad , \quad \frac{\mathrm{d}^{\mu}}{\mathrm{d}t^{\mu}} f\big(x'(t), x(t), t\big) = 0 \tag{3.3}$$

*such that the equations (3.3) allow to extract an explicit ordinary differential
system $x'(t) = \varphi\big(x(t), t\big)$ using only algebraic manipulations.*

The differentiation index received much attention and it is widely used. But
for practical computations the setup of the derivative array (3.3) may be prohibitive due to high computational costs. Also, Definition 3.1 tacitly assumes
that $f \in C^{\mu}$ is sufficiently smooth to calculate (3.3), which often does not
hold for applications such as circuit simulation or multi-body dynamics. We
will see later that weaker smoothness requirements are often sufficient. One
particular disadvantage of the differentiation index is the fact that it can't be
applied to over- and underdetermined systems as the solvability concept of the
differentiation index requires unique solvability.

The strangeness index introduced by Kunkel and Mehrmann [98, 99] generalises
the differentiation index. Its main advantage is not only the possibility to
include over- and underdetermined systems, but also to provide normal forms
for time-dependent differential algebraic equations. We will briefly review some
of the key points related to the strangeness index. More details are given in the
papers by Kunkel and Mehrmann and in particular in the monograph [101].

Within the context of the Kronecker index two matrix pairs (E_1, F_1) and
(E_2, F_2) are considered to be equivalent if they are related by a similarity
transformation, i.e. $E_2 = UE_1V$ and $F_2 = UF_1V$ with nonsingular matrices
U, V. The equation is then written in terms of the transformed variable
$y = V^{-1}x$. If V depends on time, then $x = Vy$ needs to be differentiated
in order to obtain the transformed equation. Thus, due to the product rule,
$x' = Vy' + V'y$ holds and the additional term $V'y$ has to be taken into considera-
tion. Consequently, Kunkel and Mehrmann consider two matrix pairs (E_i, F_i),
$E_i, F_i \in C(\mathcal{I}, \mathbb{C}^{m \times n})$, $i = 1, 2$, equivalent, if there are pointwise nonsingular
matrix functions $U \in C(\mathcal{I}, \mathbb{C}^{m \times m})$ and $V \in C(\mathcal{I}, \mathbb{C}^{n \times n})$ such that

$$E_2 = UE_1V, \qquad\qquad F_2 = UF_1V + UE_1V'.$$

This indeed defines an equivalence relation $(E_1, F_1) \sim (E_2, F_2)$. Under additional constant rank assumptions it is possible to derive the (global) normal
form

$$(E_1, F_1) \sim (E_2, F_2) = \left(\begin{bmatrix} I_s & 0 & 0 & 0 \\ 0 & I_d & 0 & 0 \\ 0 & 0 & 0 & 0 \\ 0 & 0 & 0 & 0 \\ 0 & 0 & 0 & 0 \end{bmatrix}, \begin{bmatrix} 0 & A_{12} & 0 & A_{14} \\ 0 & 0 & 0 & A_{24} \\ 0 & 0 & I_a & 0 \\ I_s & 0 & 0 & 0 \\ 0 & 0 & 0 & 0 \end{bmatrix} \right) \begin{matrix} s \\ d \\ a \\ s \\ m-d-a \end{matrix}$$

where the blocks A_{ij} are again matrix functions on $\mathcal{I}$ and the numbers s, d and a are invariants of the equivalence relation [101]. Writing down the corresponding linear differential algebraic equation, (3.1) is found to be equivalent to the system

$$x_1' + A_{12}x_2 + A_{14}x_4 = q_1 \tag{3.4a}$$
$$x_2' + A_{24}x_4 = q_2 \qquad \text{(dynamic part)} \tag{3.4b}$$
$$x_3 = q_3 \qquad \text{(algebraic part)} \tag{3.4c}$$
$$x_1 = q_4 \tag{3.4d}$$
$$0 = q_5 \qquad \text{(consistency condition)} \tag{3.4e}$$

The interpretation of (3.4b), (3.4c) and (3.4e) is straightforward. However, there is a "strange" coupling between (3.4a) and (3.4d) such that the number s is called "strangeness". Differentiating (3.4d) and inserting the result into (3.4a), the latter equation becomes purely algebraic. The resulting modified matrix pair is again denoted as (E_2, F_2) for simplicity. The process of finding the global normal form and eliminating the strangeness part can be repeated iteratively to obtain a sequence of characteristic values (s_i, d_i, a_i) for (E_i, F_i). A pair (E_i, F_i) is called strangeness free if $s_i = 0$. If this is the case, then the sequence becomes stationary. The smallest number $i \in \mathbb{N}_0$, such that (E_i, F_i) is strangeness free, is defined to be the strangeness index.

If the strangeness index μ is well-defined, i.e. the constant rank assumptions hold for every step of the construction, then the normal form of (E_μ, F_μ) reads

$$\begin{bmatrix} I_{d_\mu} & & \\ & 0 & \\ & & 0 \end{bmatrix} \begin{bmatrix} x_1 \\ x_2 \\ x_3 \end{bmatrix}' + \begin{bmatrix} 0 & 0 & A_{13} \\ & I_{a_\mu} & 0 \\ & & 0 \end{bmatrix} \begin{bmatrix} x_1 \\ x_2 \\ x_3 \end{bmatrix} = \begin{bmatrix} f_1 \\ f_2 \\ f_3 \end{bmatrix}.$$

For the original DAE (3.1) the following properties hold:

- The problem (3.1) is solvable if and only if $f_3 = 0$.
- An initial condition $x(t_0) = x_0$ is consistent if and only if $x(t_0) = x_0$ implies $x_2(t_0) = f_2(t_0)$.
- The problem (3.1) is uniquely solvable if and only if $m - d_\mu - a_\mu = 0$.

Obviously the strangeness index is a powerful tool for analysing differential algebraic equations. Over- and underdetermined systems are included most naturally. In particular the resulting normal forms provide much inside into the structure of a given DAE.

On the other hand even for simple examples it may be difficult to calculate the normal forms due to the complexity of the linear algebra computations involved [146]. This is true in particular for nonlinear problems. Thus numerical software developed in conjunction with the strangeness index requires the user to provide the derivative array (3.3) as input [102].

In this thesis we are mainly concerned with the analytical and numerical investigation of differential algebraic equations arising from modelling electrical circuits. Input signals but also the equations modelling electrical devices such as MOSFETs or BJTs[1] will typically have only low smoothness properties. Thus for our investigations we have to avoid using the derivative array not only due to these smoothness considerations but also because of the problem size. If a circuit contains hundred-thousands of devices, the calculation of the derivative array is not recommended in practice.

Hence we will focus on the tractability index originally introduced by März [113]. It does not require the calculation of the derivative array – neither for analytical nor for computational purposes. Much of the smoothness requirements on the data will be replaced by assumptions on certain subspaces to be spanned by continuously differentiable functions. These subspaces can be conveniently analysed in the context of electrical circuit simulation.

This index concept aims at determining the exact smoothness requirements necessary for solving a given DAE [117, 121]. The equation's structure is analysed in detail by determining an inherent ordinary differential equation that describes the dynamics of the system [115, 116]. Recently over- and underdetermined systems have been included into the general theory as well [55, 119]. Normal forms are considered in [118].

As the tractability index is seminal for the work presented in this thesis, we will investigate this concept in considerable detail. The next section is devoted to defining the index and investigating its consequences for linear differential algebraic equations. We will determine the so-called inherent regular ordinary differential equation for linear DAEs. A decoupling procedure will be the main tool leading not only to the inherent regular ODE but also to a refined characterisation of solvability, consistent initialisation and smoothness requirements.

Other index concepts such as the the geometric index [133] will not be investigated further. All index concepts exist in their own right. Each has its own advantages and disadvantages, but for integrated circuit design the tractability index seems to be most appropriate.

3.2 Linear DAEs with Properly Stated Leading Terms

The tractability index is defined for differential algebraic equations of the form

$$A(t)\big[D(t)x(t)\big]' + B(t)x(t) = q(t), \qquad t \in \mathcal{I}. \tag{3.5}$$

Two matrices $A(t) \in \mathbb{R}^{m \times n}$ and $D(t) \in \mathbb{R}^{n \times m}$ are used in order to formulate the leading term. In general both matrices will be rectangular. The leading term in (3.5) figures out precisely which derivatives are actually involved.

---

[1]MOSFET: Metal Oxide Semiconductor Field Effect Transistor and BJT: Bipolar Junction Transistor

Notice that DAEs in standard form (3.1) can be cast easily into the form (3.5), using e.g. a projector P_E along the kernel of E [2, 88]. Also, recall from Section 1.1 that the equations of the modified nodal analysis are exactly of this type, provided that the circuit is linear. In the MNA equations (1.5) the vector $d(x(t), t)$ represented charges and fluxes of the network. The leading term of (3.5) represents the linear analogy to this situation.

However, A and D may not be chosen completely arbitrary. We have to make sure that there is no overlap between A and D but also no gap in between them. More precisely, A and D have to be well-matched in the following sense.

Definition 3.2. *The leading term of* (3.5) *is properly stated if*

$$\ker A(t) \oplus \operatorname{im} D(t) = \mathbb{R}^n, \qquad\qquad t \in \mathcal{I},$$

and there is a continuously differentiable projector function $R \in C^1\big(\mathcal{I}, L(\mathbb{R}^n)\big)$ such that

$$\operatorname{im} R(t) = \operatorname{im} D(t), \qquad \ker R(t) = \ker A(t) \qquad t \in \mathcal{I}.$$

We require that the matrix functions A, D and B are continuous. If the leading term is properly stated then, by definition, A and D have a common constant rank [116]. A continuous function $x : \mathcal{I} \to \mathbb{R}^m$ is a solution of (3.5) if it satisfies the equation pointwise. In order for this to hold, the D-part of x, i.e. $t \mapsto D(t)x(t)$, needs to be differentiable. Consequently the appropriate solution space is given by

$$C_D^1(\mathcal{I}, \mathbb{R}^m) = \big\{ x \in C(\mathcal{I}, \mathbb{R}^m) \,|\, Dx \in C^1(\mathcal{I}, \mathbb{R}^n) \big\}.$$

The linear DAE (3.5) will now be investigated in more detail.

Let Q_0 be a continuous projector function onto the kernel of AD. If $P_0 = I - Q_0$ denotes the complementary projector, we have $x = P_0 x + Q_0 x$ and (3.5) can be written as

$$A(Dx)' + Bx = q \qquad \Leftrightarrow \qquad A(Dx)' + BP_0 x + BQ_0 x = q. \qquad (3.6)$$

The t arguments have been omitted for simplicity. For a further refined reformulation a generalised reflexive inverse D^- is introduced for D such that

$$DD^- D = D, \qquad D^- DD^- = D^-, \qquad DD^- = R, \qquad D^- D = P_0. \qquad (3.7)$$

Generalised matrix inverses are studied in [160]. Due to the first two properties, the products DD^- and $D^- D$ are projector functions. By requiring $DD^- = R$ and $D^- D = P_0$ the inverse D^- becomes uniquely defined. However, it still depends on P_0 and thus on the choice of Q_0.

Obviously we have $DP_0 = D$ and $DQ_0 = Q_0 D^- = 0$. Using Definition 3.2 it turns out that $A = AR = ADD^-$. Hence (3.6) can be rewritten as

$$(AD + BQ_0)[D^-(Dx)' + Q_0 x] + BP_0 x = q. \tag{3.6'}$$

What have we gained by introducing the reformulation (3.6')? To see the benefit of the above calculations let us assume, for the moment, that the matrix function $G_1 = AD + BQ_0$ remains nonsingular on $\mathcal{I}$. Then (3.5) is equivalent to

$$D^-(Dx)' + Q_0 x + G_1^{-1} BD^- Dx = G_1^{-1} q, \tag{3.6''}$$

which is found by scaling with G_1^{-1}. Multiplication with the projector functions P_0 and Q_0, respectively, transforms this equation into the decoupled system

$$P_0 D^- u' + P_0 G_1^{-1} BD^- u = P_0 G_1^{-1} q \tag{3.8a}$$
$$z_0 + Q_0 G_1^{-1} BD^- u = Q_0 G_1^{-1} q. \tag{3.8b}$$

The abbreviations $u = Dx$ and $z_0 = Q_0 x$ were introduced in order to make the decoupled structure clearer.

Similar to the approach taken when defining the Kronecker index, the reformulation (3.8) leads to a decoupled system of two equations. (3.8a) is completely independent of (3.8b). But once a solution of (3.8a) is found, a solution of the original DAE can be constructed as

$$x = P_0 x + Q_0 x = D^- u + z_0 = (I - Q_0 G_1^{-1} B) D^- u + Q_0 G_1^{-1} q.$$

Observe that in order to calculate $Q_0 x$ no differentiation is necessary. Therefore we will later refer to this case, where G_1 is nonsingular, as index 1.

Even though (3.8) provides a decoupling of the original DAE, this structure is not quite satisfying. Matters would be much clearer if (3.8a) was an ordinary differential equation. Using the product rule this goal is easily achieved.

Notice that $P_0 = D^- D$ and $\ker P_0 = \ker D$ imply that (3.8a) is equivalent to

$$Ru' + DG_1^{-1} BD^- u = DG_1^{-1} q$$

and since R is a smooth function, we find

$$u' = (Ru)' = R'u - DG_1^{-1} BD^- u + DG_1^{-1} q \tag{3.8a'}$$

Thus the component $u = Dx$ is determined by an inherent regular ordinary differential equation. For u initial conditions may be freely chosen. The remaining part $z_0 = Q_0 x$ of the solution is explicitly given in terms of u.

It is obvious that this decoupling hinges on the fact that $G_1 = AD + BQ_0$ is nonsingular. However, the procedure can be generalised for higher index cases. This is done by considering $G_0 = AD$, $B_0 = B$ and the following sequence of matrices and subspaces for $i \geq 0$. All expressions in this sequence are meant pointwise for $t \in \mathcal{I}$.

$$N_i = \ker G_i, \tag{3.9a}$$
$$S_i = \{\, z \in \mathbb{R}^m \mid B_i z \in \operatorname{im} G_i \,\} = \{\, z \in \mathbb{R}^m \mid B z \in \operatorname{im} G_i \,\}, \tag{3.9b}$$
$$Q_i = Q_i^2, \qquad \operatorname{im} Q_i = N_i, \qquad P_i = I - Q_i, \tag{3.9c}$$
$$G_{i+1} = G_i + B_i Q_i, \tag{3.9d}$$
$$B_{i+1} = B_i P_i - G_{i+1} D^- (DP_0 \cdots P_{i+1} D^-)' DP_0 \cdots P_i. \tag{3.9e}$$

The expression (3.9c) means that a projector function Q_i onto N_i is introduced and that P_i is its complementary projector. When defining the sequence (3.9) we have to make sure that the involved derivatives exist.

Definition 3.3. *The sequence* (3.9) *is said to be admissible up to* $k \in \mathbb{N}$, *if the following properties hold for* $i = 0, 1, \ldots, k$:

(a) $\operatorname{rank} G_i(t) = r_i$ *is constant on* $\mathcal{I}$,

(b) $N_0 \oplus \cdots \oplus N_{i-1} \subset \ker Q_i$ *for* $i \geq 1$,

(c) $Q_i \in C(\mathcal{I}, \mathbb{R}^{m \times m})$ *and* $DP_0 \cdots P_i D^- \in C^1(\mathcal{I}, \mathbb{R}^{n \times n})$.

The condition (b) ensures that certain products of projector functions are again projectors. Property (c) states the required smoothness properties.

We saw in the above example that $G_0 = AD$ was singular, but the nonsingular matrix function G_1 enabled a decoupling of the original DAE. This situation is generalised to higher indices in the following definition. See [3, 75, 116, 120] for more details.

Definition 3.4. *Equation* (3.5) *is called a regular DAE with tractability index* μ *if there is a sequence* (3.9) *that is admissible up to* μ *with*

$$0 \leq r_0 \leq \cdots \leq r_{\mu-1} < m \qquad and \qquad r_\mu = m.$$

It was pointed out earlier that D^- depends on the choice of Q_0. Similarly the matrix functions G_i depend on the particular choice of the projector functions Q_j for $j = 0, \ldots, i-1$. In spite of this, the tractability index itself is well-defined. In [116] it is proved that the index is independent of the special choice of the admissible sequence (3.9). There it is also shown that the index remains invariant under regular transformations and refactorisations.

Remark 3.5. In [3] the index is defined in terms of the subspaces N_i and S_i appearing in the sequence (3.9). In fact, both approaches are equivalent (see [116]) and the index can be defined as follows:

- A regular index-1 DAE can be equivalently characterised by the condition $N_0(t) \cap S_0(t) = \{0\}$ for $t \in \mathcal{I}$.

- A regular index-2 DAE can be equivalently characterised by requiring $\dim\big(N_0(t) \cap S_0(t)\big) = \operatorname{const} > 0$ and $N_1(t) \cap S_1(t) = \{0\}$ for $t \in \mathcal{I}$.

In particular the space $N_0(t) \cap S_0(t)$ will later become of vital importance for studying nonlinear index-2 equations.

Finally, recall from [75, Theorem A.13] that $N_i(t) \cap S_i(t) = \{0\}$ is equivalent to $N_i(t) \oplus S_i(t) = \mathbb{R}^m$. Thus for index-$\mu$ equations the canonical projector $\bar{Q}_{\mu-1}$ onto $N_{\mu-1}$ along $S_{\mu-1}$ can always be defined.   □

Similarly to the case of index-1 equations discussed above, the sequence (3.9) allows to decouple index-μ equations into their characteristic parts. A careful rearrangement of (3.5) and rescaling with the matrix G_μ^{-1} shows that (3.5) can be written as

$$P_{\mu-1} \cdots P_0 D^- (Dx)' + G_\mu^{-1} B P_0 \cdots P_{\mu-1} x + \sum_{j=0}^{\mu-1} Q_j x$$

$$+ \sum_{i=1}^{\mu-1} \sum_{j=1}^{i} P_{\mu-1} \cdots P_j D^- \left(D P_0 \cdots P_j D^- \right)' D P_0 \cdots P_{i-1} Q_i x = G_\mu^{-1} q.$$

For $\mu = 1$ this equation exactly coincides with (3.6") since $P_0 D^- = D^-$. The derivatives $(D P_0 \cdots P_j D^-)'$ start to appear for $\mu \geq 2$. In any case the decomposition

$$I = P_0 \cdots P_{\mu-1} + Q_0 P_1 \cdots P_{\mu-1} + \cdots + Q_{\mu-2} P_{\mu-1} + Q_{\mu-1} \qquad (3.10)$$

is used to split this equation into $\mu + 1$ separate parts. Notice that each term of (3.10) is a projector function due to the requirement (b) in Definition 3.3. Thus, after quite long and technical calculations as performed e.g. in [116, 120] it turns out that a regular index-μ DAE (3.5) is equivalent to the decoupled system

$$u' - \left(D P_0 \cdots P_{\mu-1} D^- \right)' u + D P_0 \cdots P_{\mu-1} G_\mu^{-1} B D^- u \qquad (3.11a)$$
$$= D P_0 \cdots P_{\mu-1} G_\mu^{-1} q,$$

$$z_k = \mathcal{L}_k q - \mathcal{K}_k D^- u + \sum_{j=k+1}^{\mu-1} \mathcal{N}_{kj} (D z_j)' + \sum_{j=k+2}^{\mu-1} \mathcal{M}_{kj} z_j \qquad (3.11b)$$

where $k = 0, \ldots, \mu - 1$. The components u and z_k are given by

$$u = D P_0 \cdots P_{\mu-1} x, \quad z_0 = Q_0 x,$$
$$z_k = P_0 \cdots P_{k-1} Q_k x, \quad k = 1, \ldots, \mu - 1. \qquad (3.12)$$

The continuous coefficients $\mathcal{L}_k$, $\mathcal{K}_k$, $\mathcal{N}_{kj}$ and $\mathcal{M}_{kj}$ are defined in terms of D, D^- and the projector functions Q_i, P_i. Detailed expressions are given in [120].

For our investigations it is important to note that (3.11a) is an ordinary differential equation for the component u which does not depend on any of the other variables z_k. This equation is called the *inherent regular ordinary differential equation* of the index-μ DAE (3.5).

Once a solution u of the inherent regular ODE is found, $z_{\mu-1}$ is calculated as $z_{\mu-1} = \mathcal{L}_{\mu-1}q - \mathcal{K}_{\mu-1}D^- u$. In order to calculate $z_{\mu-2}$ the D part of this component needs to be differentiated according to

$$z_{\mu-2} = \mathcal{L}_{\mu-2}q - \mathcal{K}_{\mu-2}D^- u + \mathcal{N}_{\mu-2,\mu-1}(Dz_{\mu-1})'.$$

Successively the remaining parts $z_{\mu-3}, \dots, z_0$ are calculated and additional derivatives may be involved. Finally the solution x of the original DAE is obtained as

$$x = D^- u + z_0 + \cdots + z_{\mu-1}.$$

This situation is summarised in the following theorem.

Theorem 3.6. *Let (3.5) be a regular DAE with tractability index μ. Then*

(a) *$\operatorname{im} DP_0 \cdots P_{\mu-1}$ is a (time-varying) invariant subspace of the inherent regular ODE (3.11a).*

(b) *If $x \in C_D^1(\mathcal{I}, \mathbb{R}^m)$ is a solution of the DAE (3.5), then the components $u \in C^1(\mathcal{I}, \mathbb{R}^n)$, $z_0 \in C(\mathcal{I}, \mathbb{R}^m)$ and $z_1, \dots, z_{\mu-1} \in C^1(\mathcal{I}, \mathbb{R}^m)$ defined in (3.12) form a solution of the decoupled system (3.11).*

(c) *If $u \in C^1(\mathcal{I}, \mathbb{R}^n)$, $z_0 \in C(\mathcal{I}, \mathbb{R}^m)$ and $z_1, \dots, z_{\mu-1} \in C^1(\mathcal{I}, \mathbb{R}^m)$ satisfy (3.11) with $u(t_0) \in \operatorname{im}\big(DP_0 \cdots P_{\mu-1}\big)(t_0)$, then $x = D^- u + z_0 + \cdots + z_{\mu-1}$ is a solution of the original DAE (3.5).*   □

Proofs of the above result can be found in [116, 120].

The following statement about the existence and uniqueness of solutions for initial value problems in regular index-μ DAEs is a direct consequence of the preceding theorem.

Corollary 3.7. *Let (3.5) be a regular DAE with tractability index μ. Assume that the coefficients and the right-hand side q are sufficiently smooth. Then for each $x^0 \in \mathbb{R}^m$ the initial value problem*

$$A(Dx)' + Bx = q, \qquad x(t_0) - x^0 \in \big(N_0 \oplus \cdots \oplus N_{\mu-1}\big)(t_0)$$

is uniquely solvable in $C_D^1(\mathcal{I}, \mathbb{R}^m)$.

Proof. Define $u_0 = \big(DP_0 \cdots P_{\mu-1}\big)(t_0)\, x^0$ and let u be the solution of the inherent regular ODE (3.11a) satisfying $u(t_0) = u_0$. As $\operatorname{im} DP_0 \cdots P_{\mu-1}$ is an invariant subspace of this equation, $u = DP_0 \cdots P_{\mu-1}u \in \operatorname{im} DP_0 \cdots P_{\mu-1}$ follows for $t \in \mathcal{I}$.

Let z_k be the vector from (3.11b) and consider $x = D^- u + z_0 + \cdots + z_{\mu-1}$. By Theorem 3.6 this is a solution satisfying $\big(DP_0 \cdots P_{\mu-1}\big)(t_0)\, x(t_0) = u(t_0) = u_0 = \big(DP_0 \cdots P_{\mu-1}\big)(t_0)\, x^0$. Therefore $x(t_0) - x^0 \in \big(N_0 \oplus \cdots \oplus N_{\mu-1}\big)(t_0)$ is satisfied as well.

If there was another solution of the IVP, say $\bar{x}$, then Theorem 3.6 shows that the corresponding parts $\bar{u}$ and $\bar{z}_k$ satisfy (3.11b) as well. Thus $\bar{u} = u$, $\bar{z}_k = z_k$ and the two solutions coincide.   □

Observe that in Corollary 3.7 special care was taken when formulating the initial condition. Given an arbitrary vector $x^0 \in \mathbb{R}^m$, the exact solution has to satisfy

$$x(t_0) - x^0 \in \big(N_0 \oplus \cdots \oplus N_{\mu-1}\big)(t_0).$$

Then $x(t_0)$ is the consistent initial value corresponding to x^0. Nevertheless, computing $x(t_0)$ from x^0 often poses serious difficulties for realistic applications [61].

The results of Theorem 3.6 and Corollary 3.7 were obtained using arbitrary admissible sequences (3.9). If we restrict ourselves to choosing special *canonical* projector functions Q_i, the structure of (3.11b) can be simplified considerably. In fact, in [118] fine and complete decouplings are studied. Choosing canonical projector functions it is possible to obtain $\mathcal{K}_k = 0$, such that the coupling coefficients disappear. Thus u and the components z_k can be calculated independently of each other.

Consider again the case of index-1 equations. It was shown earlier that the inherent regular ODE is given by (3.8a'), (3.8b),

$$u' = (DD^-)'u - DG_1^{-1}BD^-u + DG_1^{-1}q. \tag{3.11a'}$$

$$z_0 = Q_0 x = Q_0 G_1^{-1} q - Q_0 G_1^{-1} B D^- u. \tag{3.11b'}$$

Thus we are lead to the solution representation

$$x = D^- u + z_0 = (I - Q_0 G_1^{-1} B)D^- u + Q_0 G_1^{-1} q. \tag{3.13}$$

Observe that $I - Q_0 G_1^{-1} B$ is precisely the canonical projector onto S_0 along the subspace N_0. This is shown e.g. in [75]. If the choice $\bar{Q}_0 = Q_0 G_1^{-1} B$ leads to an admissible sequence, then (3.11) turns out to be the completely decoupled system

$$u' = R'u - D\bar{G}_1^{-1}B\bar{D}^-u + D\bar{G}_1^{-1}q,$$
$$z_0 = \bar{Q}_0 \bar{G}_1^{-1} q, \qquad\qquad x = \bar{D}^- u + \bar{Q}_0 \bar{G}_1^{-1} q.$$

This can be seen by noting that the term $\bar{Q}_0 \bar{G}_1^{-1} B\bar{D}^- = 0$ vanishes. Thus the decoupled system is in standard canonical form (SCF) in the sense of [42].

The index-2 case is more complicated. However, working carefully through the expressions given in [120] we find

$$u' = (DP_1 D^-)'u - DP_1 G_2^{-1}BD^-u + DP_1 G_2^{-1}q, \tag{3.11a''}$$

$$z_1 = P_0 Q_1 x = P_0 Q_1 G_2^{-1} q - \mathcal{K}_1 D^- u, \tag{3.11b''}$$

$$z_0 = \quad Q_0 x = Q_0 P_1 G_2^{-1} q - \mathcal{K}_0 D^- u + Q_0 Q_1 D^- (Dz_1)'.$$

The coupling coefficients are given by (cf. [118])

$$\mathcal{K}_1 = P_0 Q_1 G_2^{-1} B P_0 P_1, \tag{3.14a}$$

$$\mathcal{K}_0 = Q_0 P_1 G_2^{-1} \big(B + G_0 D^- (D P_1 D^-)' D \big) P_0 P_1. \tag{3.14b}$$

Again, $\bar{Q}_1 = Q_1 G_2^{-1} B P_0$ is the canonical projector onto N_1 along S_1. If the sequence based on $\bar{Q}_0 = Q_0$ and $\bar{Q}_1$ is admissible, then $\bar{Q}_1 = \bar{Q}_1 \bar{G}_2^{-1} B \bar{P}_0$ implies $\mathcal{K}_1 = \bar{P}_0 \bar{Q}_1 \bar{P}_1 = 0$ such that (3.11) reads

$$u' = (D \bar{P}_1 \bar{D}^-)' u - D \bar{P}_1 \bar{G}_2^{-1} B \bar{D}^- u + D \bar{P}_1 \bar{G}_2^{-1} q, \tag{3.15a}$$

$$z_1 = \bar{P}_0 \bar{Q}_1 \bar{G}_2^{-1} q, \tag{3.15b}$$

$$z_0 = \bar{Q}_0 \bar{P}_1 \bar{G}_2^{-1} q - \mathcal{K}_0 \bar{D}^- u + \bar{Q}_0 \bar{Q}_1 \bar{D}^- (D z_1)'. \tag{3.15c}$$

In the case of index-2 equations the exact solution can therefore be written as

$$
\begin{aligned}
x &= D^- u + z_0 + z_1 \\
&= \big(I - \bar{\mathcal{K}}_0 \big) D^- u + \big(\bar{Q}_0 \bar{P}_1 + \bar{P}_0 \bar{Q}_1 \big) \bar{G}_2^{-1} q + \bar{Q}_0 \bar{Q}_1 \bar{D}^- \big(D \bar{Q}_1 \bar{G}_2^{-1} q \big)' \quad (3.16)
\end{aligned}
$$

This is called a *fine* decoupling. In contrast to a *complete* decoupling the coefficient $\bar{\mathcal{K}}_0$ is still present. Fine decouplings such as the above system were studied in considerable detail in [90, 115, 120, 145]. However, in [3] it is shown that

$$\bar{Q}_0 = Q_0 P_1 G_2^{-1} B + Q_0 Q_1 D^- (D Q_1 D^-)', \qquad \bar{Q}_1 = (I - \bar{Q}_0 P_0) Q_1 = Q_1$$

yield a complete decoupling with $\bar{\mathcal{K}}_0 = 0$, provided that $\bar{Q}_0$, $\bar{Q}_1$ give rise to an admissible sequence (3.9).

Example 3.8. The Miller Integrator already introduced in Example 1.1 is described by the linear DAE

$$
\begin{bmatrix} 0 \\ -1 \\ 1 \\ 0 \\ 0 \end{bmatrix} \left(\begin{bmatrix} 0 & -C & C & 0 & 0 \end{bmatrix} \begin{bmatrix} u_1 \\ u_2 \\ u_3 \\ i_{V_1} \\ i_{V_2} \end{bmatrix} \right)' + \begin{bmatrix} G & -G & 0 & 1 & 0 \\ -G & G & 0 & 0 & 0 \\ 0 & 0 & 0 & 0 & 1 \\ 0 & -a & 1 & 0 & 0 \\ 1 & 0 & 0 & 0 & 0 \end{bmatrix} \begin{bmatrix} u_1 \\ u_2 \\ u_3 \\ i_{V_1} \\ i_{V_2} \end{bmatrix} = \begin{bmatrix} 0 \\ 0 \\ 0 \\ 0 \\ v(t) \end{bmatrix}.
$$

The amplification factor of the operational amplifier is given by the real number a. Using the formulation above, the leading term is properly stated with $R = 1$. Calculating the matrix sequence (4.6) we find

$$
G_0 = \begin{bmatrix} 0 & 0 & 0 & 0 & 0 \\ 0 & C & -C & 0 & 0 \\ 0 & -C & C & 0 & 0 \\ 0 & 0 & 0 & 0 & 0 \\ 0 & 0 & 0 & 0 & 0 \end{bmatrix}, \quad Q_0 = \begin{bmatrix} 1 & 0 & 0 & 0 & 0 \\ 0 & 1 & 0 & 0 & 0 \\ 0 & 1 & 0 & 0 & 0 \\ 0 & 0 & 0 & 1 & 0 \\ 0 & 0 & 0 & 0 & 1 \end{bmatrix}, \quad G_1 = \begin{bmatrix} G & -G & 0 & 1 & 0 \\ -G & C+G & -C & 0 & 0 \\ 0 & -C & C & 0 & 1 \\ 0 & 1-a & 0 & 0 & 0 \\ 1 & 0 & 0 & 0 & 0 \end{bmatrix}.
$$

Due to $\det G_1 = (a-1)C$ the Miller Integrator represents an index-1 problem for $a \neq 1$. The particular choice of the projector function Q_0 fixes the generalised inverse $D^- = \begin{bmatrix} 0 & 0 & \frac{1}{C} & 0 & 0 \end{bmatrix}^\top$ and the decoupled system (3.11) reads

$$
u' = \frac{G}{(a-1)C} u - G\,v(t), \qquad z_0 = v(t) \begin{bmatrix} 1 \\ 0 \\ 0 \\ -G \\ G \end{bmatrix} + \frac{1}{(1-a)C} \begin{bmatrix} 0 \\ 1 \\ 1 \\ G \\ -G \end{bmatrix}.
$$

If the amplification factor a tends towards infinity, the inherent regular ODE reads $u' = -G\,v(t)$ such that the potential at node 3 is $u_3 = -\frac{G}{C}\int v(t)\,dt$ such that the circuit indeed performs an integration.

For $a = 1$ the matrix G_1 is singular and the matrix sequence needs to be extended in order to determine the index:

$$Q_1 = \frac{1}{G}\begin{bmatrix} 0 & 0 & 0 & 0 & 0 \\ 0 & -C & C & 0 & 0 \\ 0 & -C-G & C+G & 0 & 0 \\ 0 & -CG & CG & 0 & 0 \\ 0 & CG & -CG & 0 & 0 \end{bmatrix}, \qquad G_2 = \begin{bmatrix} G & -G & 0 & 1 & 0 \\ -G & C+G & -C & 0 & 0 \\ 0 & -C & C & 0 & 1 \\ 0 & -1 & 1 & 0 & 0 \\ 1 & 0 & 0 & 0 & 0 \end{bmatrix}.$$

Notice that $DP_1D^- = 0$ such that $G_2 = G_1 + BP_0Q_1$. The projector Q_1 chosen here is already the canonical one, i.e. $\operatorname{im} Q_1 = N_1$ and $\ker Q_1 = S_1$. This can be seen by checking the relation $Q_1 = Q_1 G_2^{-1} BP_0$. Calculating $\det G_2 = -G$ shows that the index is 2 and we obtain the decoupled system

$$u' = 0, \qquad z_1 = 0, \qquad z_0 = v(t) \cdot \begin{bmatrix} 1 & 1 & 1 & 0 & 0 \end{bmatrix}^\top. \qquad \square$$

Since our main focus is on dealing with equations having index 1 or 2, we will not dwell on higher index DAEs any further. We content ourselves with remarking that studying fine and complete decouplings naturally leads to the investigation of normal forms for linear DAEs with tractability index μ. Under suitable regularity assumptions it can be shown that each regular index-μ equation is equivalent to a DAE in standard canonical form (SCF),

$$\begin{aligned} y' + My &= L_y q, \\ Nw' + w &= L_w q, \end{aligned}$$

where N is strictly upper triangular. See [42, 118] for more details.

Currently there is much interest in studying the relationship of the strangeness and the tractability index using the corresponding normal forms. It has been shown that both index concepts essentially agree for $\mu = \{1, 2, \ldots, 5\}$ (using the counting of the tractability index). It is conjectured that the relationship "$\mu_{\text{tractability}} = \mu_{\text{strangeness}} + 1$" holds for the general case as well. More details can be found in [108].

3.3 Examples

The framework of the tractability index might not be as easily accessible as the differentiation or the perturbation index but it is most convenient for studying DAEs in electrical circuit simulation. The matrix sequence and the decoupling procedure introduced above provide a detailed insight into the structure of a given DAE. For electrical networks the index $\mu = 0, 1, 2$ can be checked using fast and reliable algorithms based on graph theory [61, 63].

However, the tractability index does not only offer a refined structural analysis, but it often leads to significant improvements for numerical computations. In particular the properly stated leading term plays a crucial role here.

Example 3.9. Consider the well-known problem of Gear and Petzold [73],

$$\begin{bmatrix} 0 & 0 \\ 1 & \eta\, t \end{bmatrix} x'(t) + \begin{bmatrix} 1 & \eta\, t \\ 0 & 1+\eta \end{bmatrix} x(t) = \begin{bmatrix} e^{-t} \\ 0 \end{bmatrix}, \tag{3.17a}$$

which constitutes a linear index-2 DAE. The parameter η is a real number. It is shown in [73] that BDF methods fail completely for $\eta = -1$. For other parameter values $-1 < \eta < -0.5$ the numerical solution is exponentially unstable. In [84] the DAE (3.17a) is said to pose difficulties to every numerical method.

These statements are confirmed by the numerical results given in Figure 3.1. There (3.17a) was solved on the interval $[0,3]$ using the BDF_2 formula and the RadauIIA method with two stages. A constant stepsize $h = 0.05$ was used. The exact solution is given by $x_1(t) = (1 - \eta\, t)e^{-t}$ and $x_2(t) = e^{-t}$, such that $x^0 = [1,1]^{\top}$ is a consistent initial value. Obviously both numerical methods fail even for moderate values of η due to the exponential instability.

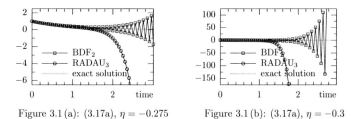

Figure 3.1 (a): (3.17a), $\eta = -0.275$ Figure 3.1 (b): (3.17a), $\eta = -0.3$

Figure 3.1: Numerical solution (2nd component). Using the standard formulation (3.17a) the solution explodes even for moderate values of η.

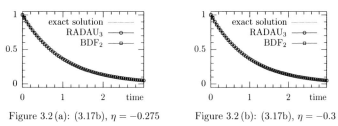

Figure 3.2 (a): (3.17b), $\eta = -0.275$ Figure 3.2 (b): (3.17b), $\eta = -0.3$

Figure 3.2: Numerical solution (2nd component). The properly stated DAE (3.17b) yields correct results and no difficulties arise.

For comparison consider the following equivalent reformulation using a properly stated leading term,

$$\begin{bmatrix} 0 \\ 1 \end{bmatrix} \left(\begin{bmatrix} 1 & \eta\,t \end{bmatrix} x(t) \right)' + \begin{bmatrix} 1 & \eta\,t \\ 0 & 1 \end{bmatrix} x(t) = \begin{bmatrix} e^{-t} \\ 0 \end{bmatrix}. \tag{3.17b}$$

Solving the reformulated problem yields correct numerical results as can be seen in Figure 3.2. In particular the accuracy of the numerical result does not depend on the parameter η. We will see later, in Section 9.1, that this behaviour is ensured by the two subspaces $DN_1 = \mathbb{R}$, $DS_1 = \{0\}$ being constant, a fact that was first observed in [90]. $\square$

The numerical results for Example 3.9 indicate that an appropriate formulation of the problem ensures a correct behaviour of the numerical solution. The standard formulation $Ex' + Fx = q$, by contrast, inevitably leads to serious difficulties. This effect can be studied in detail for the following example.

Example 3.10. Consider the simple linear DAE

$$\lambda t\, u' + (\lambda - 1) z' = 0, \qquad\qquad (\lambda t - 1) u + (\lambda - 1) z = 0 \tag{3.18a}$$

depending on the parameter $\lambda \in \mathbb{R}$. J.C. Butcher received this example from Caren Tischendorf. It is documented in [24].

Differentiating the second equation of (3.18a) and inserting the result into the first one shows that

$$u' = \lambda\, u, \qquad\qquad z = \tfrac{\lambda t - 1}{1 - \lambda}\, u. \tag{3.18b}$$

Hence the dynamics is given by the inherent ordinary differential equation $u' = \lambda\, u$. The remaining component z is fixed in terms of u. A numerical method, when applied to (3.18a), should refect this situation properly, i.e. we expect the decoupled system (3.18b) to be discretised correctly.

For simplicity consider the implicit Euler scheme. Discretising (3.18a) yields

$$\tfrac{\lambda t_{n+1}}{h} \left(u_{n+1} - u_n \right) + \tfrac{\lambda - 1}{h} \left(z_{n+1} - z_n \right) = 0,$$
$$(\lambda t_n - 1) u_n + (\lambda - 1) z_n = 0.$$

Solving the second equation for z_n, the first equation simplifies to

$$u_{n+1} = u_n + h\lambda u_n.$$

Thus the inherent ODE is in fact discretised using the *explicit* Euler scheme. This will have severe consequences such as strong stepsize restrictions due to stability requirements.

Similar to the previous example, the DAE

$$\begin{bmatrix} 1 \\ 0 \end{bmatrix} \left(\begin{bmatrix} \lambda - 1 & \lambda t \end{bmatrix} \begin{bmatrix} z(t) \\ u(t) \end{bmatrix} \right)' + \begin{bmatrix} 0 & -\lambda \\ \lambda - 1 & \lambda t - 1 \end{bmatrix} \begin{bmatrix} z(t) \\ u(t) \end{bmatrix} = 0$$

is an equivalent reformulation of (3.18a). Due to the properly stated leading term it is possible to ensure that the implicit Euler scheme leads to

$$u_{n+1} = u_n + h\lambda u_{n+1}, \qquad\qquad (1 - \lambda)z_n = (\lambda t_n - 1)\, u_n,$$

such that the inherent ODE is indeed discretised by the implicit scheme. In order to guarantee this behaviour, one has to ensure that $\operatorname{im} D$ remains constant [89]. □

4

Nonlinear Differential Algebraic Equations

In the previous chapter linear differential algebraic equations

$$A(t)\big[D(t)x(t)\big]' + B(t)x(t) = q(t) \tag{4.1}$$

were studied. This class is of key relevance for analysing DAEs since important features such as the properly stated leading term and the decoupling procedure can be introduced. However, when modelling complex processes in real applications one most often is confronted to deal with nonlinear equations.

Example 4.1. Logical elements such as the NAND gate in Figure 4.1 are the central building blocks for complex adding circuits and registers. The NAND gate consists of two n-channel enhancement MOSFETs[1] (ME), one n-channel depletion MOSFET (MD) and a load capacitance C. The transistors can be produced in CMOS technology leading to low power consumption, high density and a fast switching behaviour [78].

The gate voltages of the two MEs are controlled by input signals V_1 and V_2. The response at node 1 is *low* (false) only in the case of both input signals being *high* (true). The NAND gate thus represents the logical expression $\neg(V_1 \wedge V_2)$.

For the numerical simulation of the NAND gate the MOSFETs are replaced by suitable equivalent circuits. Using the companion model from Figure 4.1 (b) the modified nodal analysis leads to a DAE of the form

$$A\,\dot{q}\big(x(t)\big) + b\big(x(t), t\big) = 0 \tag{4.2}$$

where $x = \begin{bmatrix} u_1 & \cdots & u_{12} & I_1 & I_2 & I_{BB} & I_{DD} \end{bmatrix}^\top$ comprises all node potentials and the branch currents of the four voltage sources. $A \in \mathbb{R}^{16 \times 13}$ is a constant matrix and the functions q and b are given in Table 4.1.

The nonlinear functions q_{gd}, q_{gs}, q_{db}, q_{sb} and i_{bd}, i_{bs}, i_{ds} as well as all parameter values are given in [151]. Notice that both the leading term and the function b contain nonlinearities. □

---

[1] MOSFET: Metal Oxide Semiconductor Field Effect Transistor.

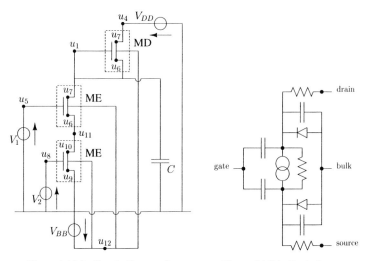

Figure 4.1 (a): Circuit diagram for the NAND gate.

Figure 4.1 (b): Equivalent circuit replacing the MOSFETs.

Figure 4.1: The NAND gate model taken from [151].

$$
q(x) =
\begin{bmatrix}
C\,u_1 \\
q_{gd}(u_1 - u_3) \\
q_{gs}(u_1 - u_2) \\
q_{db}(u_3 - u_{12}) \\
q_{sb}(u_2 - u_{12}) \\
q_{gd}(u_5 - u_7) \\
q_{gs}(u_5 - u_6) \\
q_{db}(u_7 - u_{12}) \\
q_{sb}(u_6 - u_{12}) \\
q_{gd}(u_8 - u_{10}) \\
q_{gs}(u_8 - u_9) \\
q_{db}(u_{10} - u_{12}) \\
q_{sb}(u_9 - u_{12})
\end{bmatrix}
\qquad
b(x,t) =
\begin{bmatrix}
\frac{1}{R_s}(u_1 - u_2) - \frac{1}{R_d}(u_7 - u_1) \\
\frac{1}{R_s}(u_2 - u_1) + \frac{1}{R_{sd}}(u_2 - u_3) + i_{bs}(u_{12} - u_2) \ldots \\
\quad + i_{ds}(u_3 - u_2, u_1 - u_2, u_{12} - u_2) \\
\frac{1}{R_d}(u_3 - u_4) - \frac{1}{R_{sd}}(u_2 - u_3) + i_{bd}(u_{12} - u_3) \ldots \\
\quad - i_{ds}(u_3 - u_2, u_1 - u_2, u_{12} - u_2) \\
\frac{1}{R_d}(u_4 - u_3) + I_{DD} \\
I_1 \\
\frac{1}{R_s}(u_6 - u_{11}) + \frac{1}{R_{sd}}(u_6 - u_7) + i_{bs}(u_{12} - u_6) \ldots \\
\quad + i_{ds}(u_7 - u_6, u_5 - u_6, u_{12} - u_6) \\
\frac{1}{R_d}(u_7 - u_1) - \frac{1}{R_{sd}}(u_6 - u_7) + i_{bd}(u_{12} - u_7) \ldots \\
\quad - i_{ds}(u_7 - u_6, u_5 - u_6, u_{12} - u_6) \\
I_2 \\
\frac{1}{R_{sd}}(u_9 - u_{10}) + \frac{u_9}{R_s} + i_{bs}(u_{12} - u_9) \ldots \\
\quad + i_{ds}(u_{10} - u_9, u_8 - u_9, u_{12} - u_9) \\
\frac{1}{R_d}(u_{10} - u_{11}) - \frac{1}{R_{sd}}(u_9 - u_{10}) + i_{bd}(u_{12} - u_{10}) \ldots \\
\quad - i_{ds}(u_{10} - u_9, u_8 - u_9, u_{12} - u_9) \\
\frac{1}{R_s}(u_{11} - u_6) - \frac{1}{R_d}(u_{10} - u_{11}) \\
-i_{bs}(u_{12} - u_2) - i_{bd}(u_{12} - u_3) - i_{bs}(u_{12} - u_6) \ldots \\
\quad - i_{bd}(u_{12} - u_7) - i_{bs}(u_{12} - u_9) \ldots \\
\quad - i_{bd}(u_{12} - u_{10}) + I_{BB} \\
u_4 - V_{DD} \\
u_{12} - V_{BB} \\
u_5 - V_1(t) \\
u_8 - V_2(t)
\end{bmatrix}
$$

Table 4.1: The functions q and b for the NAND gate model

In this chapter nonlinear equations such as (4.2) are addressed. In Section 4.1 the concept of regular DAEs with tractability index μ is extended to the nonlinear case. For index-1 equations a decoupling procedure similar to the results of Section 3.2 is used to prove existence and uniqueness of solutions in Section 4.2. The case of nonlinear index-2 equations is considerably harder to analyse. Hence the analysis is split into two parts:

Many electrical circuits, when modelled using the modified nodal analysis, lead to a special structure where the so-called index-2 components enter the model equations linearly. Equations of this type are treated in Chapter 5. The case of general index-2 equations with a properly stated leading term is addressed in Chapter 6. There it is shown how a careful analysis of index-2 equations within the framework of the tractability index leads to criteria for the existence and uniqueness of solutions that are easily accessible in practical applications.

4.1 The Index of Nonlinear DAEs

As a nonlinear version of (4.1) we consider differential algebraic equations

$$A(t)\big[d(x(t),t)\big]' + b\big(x(t),t\big) = 0, \qquad\qquad t \in \mathcal{I}. \qquad (4.3)$$

$A : \mathcal{I} \to \mathbb{R}^{m \times n}$ is assumed to be a continuous matrix function defined on the interval $\mathcal{I} \subset \mathbb{R}$. As seen in Example 4.1 the mappings $d : \mathcal{D} \times \mathcal{I} \to \mathbb{R}^n$ and $b : \mathcal{D} \times \mathcal{I} \to \mathbb{R}^m$ represent nonlinearities of the model. d and b are defined on $\mathcal{D} \times \mathcal{I} \subset \mathbb{R}^{m+1}$ where $\mathcal{D}$ represents some domain in $\mathbb{R}^m$. We assume both functions to be continuous. Furthermore it is assumed that continuous partial derivatives $d_x = \frac{\partial d}{\partial x}$ and $b_x = \frac{\partial b}{\partial x}$ exist.

Similar to Definition 3.2 the notion of a properly stated leading term is needed. Observe that in case of linear DAEs with $d(x,t) = D(t)x(t)$ the matrices A and $D = d_x$ were used to define properly stated DAEs. Thus we arrive at this straightforward generalisation:

Definition 4.2. *The leading term of* (4.3) *is properly stated if*

$$\ker A(t) \oplus \operatorname{im} d_x(x,t) = \mathbb{R}^n \qquad \forall \ (x,t) \in \mathcal{D} \times \mathcal{I},$$

and if there is a smooth projector function $R \in C^1\big(\mathcal{I}, \mathbb{R}^{n \times n})\big)$ *such that*

$$\ker R(t) = \ker A(t), \quad \operatorname{im} R(t) = \operatorname{im} d_x(x,t) \quad and \quad d(x,t) = R(t)d(x,t)$$

for every $(x,t) \in \mathcal{D} \times \mathcal{I}$.

This definition was given in [114]. In particular, $\operatorname{im} d_x$ does not depend on x for properly stated DAEs.

If x satisfies (4.3) pointwise, then it is sufficient for x to be continuous, but the mapping $t \mapsto d\big(x(t),t\big)$ needs to be differentiable. Hence a function $x \in C(\mathcal{I}_x, \mathbb{R}^m)$, $\mathcal{I}_x \subset \mathcal{I}$, is said to be a solution of (4.3) provided that

- for every $t \in \mathcal{I}_x$ the mapping $x(t)$ takes values in $\mathcal{D}$,
- the mapping $t \mapsto d\big(x(t), t\big)$ is continuously differentiable
- and x satisfies the DAE (4.3) pointwise for $t \in \mathcal{I}_x$.

Unfortunately this concept of a solution does not lead to a linear function space. Thus we consider DAEs with a linear leading term,

$$A(t)\big[D(t)x(t)\big]' + b\big(x(t), t\big) = 0. \tag{4.4}$$

DAEs of this type are sometimes referred to as quasi-linear. Obviously solutions lie in the linear space

$$C_D^1(\mathcal{I}, \mathbb{R}^m) := \big\{ z \in C(\mathcal{I}, \mathbb{R}^m) \mid Dz \in C^1(\mathcal{I}, \mathbb{R}^n) \big\}.$$

From the point of view of mathematical analysis (4.4) is much easier to handle than (4.3), but the quasi-linear form (4.4) does not seem to be general enough to include circuits such as the NAND gate from Example 4.1. Luckily this is *not* the case as the general formulation (4.3) can be transformed into (4.4) by considering the enlarged system

$$A(t)\big[R(t)y(t)\big]' + b\big(x(t), t\big) = 0, \tag{4.5a}$$
$$y(t) - d(x(t), t) = 0. \tag{4.5b}$$

A new variable $y(t) = d\big(x(t), t\big)$ is introduced that allows to artificially split the equation in two. R is the projector function from Definition 4.2 characterising the properly stated leading term.

With $\hat{x} = \big(\begin{smallmatrix} x \\ y \end{smallmatrix}\big)$, $\hat{A} = \big(\begin{smallmatrix} A \\ 0 \end{smallmatrix}\big)$, $\hat{D} = \big(\begin{smallmatrix} 0 & R \end{smallmatrix}\big)$ and $\hat{b}(\hat{x}, t) = \big(\begin{smallmatrix} b(x,t) \\ y - d(x,t) \end{smallmatrix}\big)$ the enlarged system (4.5) is seen to be of type (4.4). Even though (4.5) may have twice the dimension of (4.3), this does not pose a problem for practical computations. Whenever the variable y is referenced, an explicit evaluation of (4.5b) is possible. Hence the computational effort for solving (4.3) and (4.5) will be the same. In fact, implementations do rely on (4.3), but the mathematical analysis is carried out using the enlarged system (4.5) in the formulation (4.4).

This approach is justified in [88, 114]. There it is shown that (4.3) and (4.5) are equivalent in the sense that there is a one-to-one correspondence of solutions, subspaces, the index and so on. Thus it is no restriction to concentrate on DAEs of the form (4.4).

Remark 4.3. In the literature [46, 84, 85] nonlinear DAEs are most often assumed to be given in Hessenberg form

$$y' = f(y, z), \qquad 0 = g(y).$$

For index 2 one has to ensure that $g_y f_z$ has a bounded inverse in the neighbourhood of a solution.

In general the MNA equations and in particular the split system (4.5) are not of Hessenberg type. Conversely, the class of Hessenberg DAEs is included in the formulation (4.4) by considering

$$ x = \begin{bmatrix} y \\ z \end{bmatrix}, \quad A = \begin{bmatrix} I \\ 0 \end{bmatrix}, \quad D = \begin{bmatrix} I & 0 \end{bmatrix}, \quad b(x,t) = \begin{bmatrix} -f(y,z) \\ g(y) \end{bmatrix}. \qquad \square $$

As in the case of linear DAEs one introduces admissible sequences of matrix functions and subspaces in order to analyse nonlinear DAEs with a properly stated leading term. The details of this construction can be found in [116, 117, 146]. Here we will content ourselves with summarising the key results necessary for our later investigations.

Pointwise for $t \in \mathcal{I}$, $x \in \mathcal{D}$ we introduce

$$ \left. \begin{aligned} G_0(t) &= A(t)D(t), \quad B(x,t) = b_x(x,t) = \frac{\partial}{\partial x} b(x,t), \\ N_0(t) &= \ker G_0(t), \quad S_0(x,t) = \left\{ z \in \mathbb{R}^m \mid B(x,t)z \in \operatorname{im} G_0(t) \right\}. \end{aligned} \right\} \quad (4.6a) $$

Observe that due to the nonlinearity b the matrix $B(x,t)$ as well as the subspace $S_0(x,t)$ are state-dependent in general. If $Q_0 \in C(\mathcal{I}, \mathbb{R}^{m \times m})$ is again a projector function onto N_0, then G_1, N_1 and S_1 depend on x as well,

$$ \left. \begin{aligned} G_1(x,t) &= G_0(t) + B(x,t)Q_0(t), \\ N_1(x,t) &= \ker G_1(x,t), \\ S_1(x,t) &= \left\{ z \in \mathbb{R}^m \mid B(x,t)z \in \operatorname{im} G_1(x,t) \right\}. \end{aligned} \right\} \quad (4.6b) $$

Let $Q_1(x,t)$ be a projector onto $N_1(x,t)$ and define $P_0 = I - Q_0$, $P_1 = I - Q_1$. It is possible to extend this sequence by introducing

$$ \left. \begin{aligned} C(x^1,x,t) &= \left(DP_1D^-\right)_x(x,t)x^1 + \left(DP_1D^-\right)_t(x,t), \\ B_1(x^1,x,t) &= B(x,t)P_0(t) - G_1(x,t)D^-(t)C(x^1,x,t)D(t), \\ G_2(x^1,x,t) &= G_1(x,t) + B_1(x^1,x,t)Q_1(x,t) \end{aligned} \right\} \quad (4.6c) $$

for $t \in \mathcal{I}$, $x \in \mathcal{D}$. As in the previous chapter $D^-(t)$ is the generalised reflexive inverse of $D(t)$ defined by

$$ DD^-D = D, \quad D^-DD^- = D^-, \quad D^-D = P_0, \quad DD^- = R. $$

The newly introduced variable $x^1 \in \mathbb{R}^m$ is necessary for a correct definition of the matrix C. To better understand this construction recall that for linear DAEs the matrix B_1 is defined as

$$ B_1(t) = B(t)P_0(t) - G_1(t)D^-(t)\left(DP_1D^-\right)'(t)D(t). $$

For nonlinear DAEs, P_1 depends on x and so does DP_1D^- in general. If $\bar{x}$ is a smooth mapping that takes values in $\mathcal{D}$, then

$$\frac{\mathrm{d}}{\mathrm{d}t}\left(DP_1D^-\right)(\bar{x}(t),t) = \left(DP_1D^-\right)_x(\bar{x}(t),t)\,\bar{x}'(t) + \left(DP_1D^-\right)_t(\bar{x}(t),t)$$
$$= C\big(\bar{x}'(t),\bar{x}(t),t\big).$$

For a moment let us consider the linearisation of (4.4) along $\bar{x}$. Denote the coefficients of the corresponding linear DAE by

$$\bar{A}(t) = A(t), \qquad \bar{D}(t) = \frac{\partial}{\partial x}d\big(\bar{x}(t),t\big), \qquad \bar{B}(t) = \frac{\partial}{\partial x}b\big(\bar{x}(t),t\big).$$

Then the above construction ensures that $\bar{G}_1(t) = G_1\big(\bar{x}(t),t\big)$ as well as $\bar{G}_2(t) = G_2\big(\bar{x}'(t),\bar{x}(t),t\big)$, revealing a close relationship between the matrix sequence and the corresponding sequence for the linearised equation.

Definition 4.4. *The sequence* (4.6) *is said to be admissible if*

(a) $\operatorname{rank} G_0 = r_0$, $\operatorname{rank} G_1 = r_1$ *and* $\operatorname{rank} G_2 = r_2$ *are constant on* $\mathcal{I}$,

(b) $N_0 \subset \ker Q_1$,

(c) $Q_0 \in C(\mathcal{I}, \mathbb{R}^{m\times m})$, $Q_1 \in C(\mathcal{D}\times\mathcal{I}, \mathbb{R}^{m\times m})$, $DP_1D^- \in C^1(\mathcal{D}\times\mathcal{I}, \mathbb{R}^{n\times n})$.

Comparing with Definition 3.3 we have defined admissibility up to index 2. In [117, 121] these sequences are extended to higher index cases as well. Doing so the technical effort increases considerably as further additional variables x^i need to be introduced. For our investigations the case of index 1 and 2 is most relevant. Hence attention will be restricted this situation.

Definition 4.5. *The DAE* (4.4) *with a properly stated leading term is regular with tractability index* $\mu \in \{1,2\}$ *on* $\mathcal{D}\times\mathcal{I}$ *if there is an admissible sequence* (4.6) *such that* $r_{\mu-1} < r_\mu = m$.

For $\mu \in \{1,2\}$ this definition generalises the corresponding Definition 3.4 for linear DAEs.

Remark 4.6. A regular index-1 DAE can be equivalently characterised by the condition $N_0(t) \cap S_0(x,t) = \{0\}$ for $(x,t) \in \mathcal{D}\times\mathcal{I}$. On the other hand, a regular index-2 DAE can be equivalently characterised by requiring $\dim\big(N_0(t) \cap S_0(x,t)\big) = \operatorname{const} > 0$ and $N_1(x,t) \cap S_1(x,t) = \{0\}$ for $(x,t) \in \mathcal{D}\times\mathcal{I}$ (see also Remark 3.5). $\qquad\square$

An important consequence of Definition 4.5 concerns linearisations of regular index-μ equations.

Theorem 4.7. *Let* (4.4) *be a regular index-μ DAE in the sense of Definition 4.5. Then for every function* $\bar{x} \in C^1(\mathcal{I}, \mathbb{R}^m)$ *with* $\bar{x}(t) \in \mathcal{D}$ *the linearisation along* $\bar{x}$ *is regular with tractability index* μ.

Proof. This result follows at once from the above construction. More details are given in [117]. □

Example 4.8. We want to finish this section with determining the index of the nonlinear circuit from Figure 4.2. This example is taken from [61] but a slightly modified version will be used here. In particular the resistor is modelled using a nonlinear conductance such that $i_R = (2+t)u_1^2$. The current source is controlled by the current i_V trough the voltage source. These are rather artificial parameters and it is not claimed that the circuit is of any practical use. Nevertheless, the circuit from Figure 4.2 allows a better understanding of how to put the above construction into practice.

The charge oriented modified nodal analysis leads to the DAE

$$\begin{bmatrix} 1 & 0 \\ 0 & 1 \\ 0 & 0 \end{bmatrix} \left(\begin{bmatrix} 1 & 0 & 0 \\ 0 & 1 & 0 \end{bmatrix} \begin{bmatrix} u_1 \\ u_2 \\ i_V \end{bmatrix} \right)' + \begin{bmatrix} (2+t)u_1^2 + i_V + i(i_V, t) \\ -i_V \\ u_1 - u_2 - v(t) \end{bmatrix} = 0,$$

where linear capacitors with capacity $C_1 = C_2 = 1\,\mathrm{F}$ are used. Constructing the matrix sequence (4.6) yields

$$G_0 = \begin{bmatrix} 1 & 0 & 0 \\ 0 & 1 & 0 \\ 0 & 0 & 0 \end{bmatrix}, \quad Q_0 = \begin{bmatrix} 0 & 0 & 0 \\ 0 & 0 & 0 \\ 0 & 0 & 1 \end{bmatrix}, \quad D^- = \begin{bmatrix} 1 & 0 \\ 0 & 1 \\ 0 & 0 \end{bmatrix}, \quad B = \begin{bmatrix} 2(2+t)u_1 & 0 & \psi-1 \\ 0 & 0 & -1 \\ 1 & -1 & 0 \end{bmatrix}.$$

The function $\psi(i_V, t) = \frac{\partial i(i_V,t)}{\partial i_V} + 2$ was introduced in order to shorten notations. It is assumed that ψ remains nonzero for all arguments (i_V, t). Notice that B depends not only on (u_1, t) but also on i_V via ψ. Since the matrix

$$G_1 = AD + BQ_0 = \begin{bmatrix} 1 & 0 & \psi-1 \\ 0 & 1 & -1 \\ 0 & 0 & 0 \end{bmatrix}$$

is singular, the index of this circuit is $\mu \geq 2$. The matrix sequence can be extended by considering the projector

$$Q_1 = \frac{1}{\psi} \begin{bmatrix} \psi-1 & 1-\psi & 0 \\ -1 & 1 & 0 \\ -1 & 1 & 0 \end{bmatrix}$$

onto $\mathrm{im}\, G_1$. This particular choice of Q_1 leads to

$$C = (DP_1D^-)_x x^1 + (DP_1D^-)_t = \phi \begin{bmatrix} -1 & 1 \\ -1 & 1 \end{bmatrix}, \quad B_1 = \begin{bmatrix} 2(2+t)u_1 + \phi & -\phi & 0 \\ \phi & -\phi & 0 \\ 1 & -1 & 0 \end{bmatrix}$$

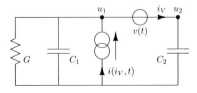

Figure 4.2: A nonlinear circuit with a current-controlled current source

with $\phi(i_V, t) = \frac{\partial \psi(i_V, t)}{\partial t} / \psi(i_V, t)^2$. Finally we arrive at

$$G_2 = G_1 + B_1 Q_1 = \begin{bmatrix} 1 + \phi + 2(2+t)u_1 \frac{\psi - 1}{\psi} & -\phi - 2(2+t)u_1 \frac{\psi - 1}{\psi} & \psi - 1 \\ \phi & 1 - \phi & -1 \\ 1 & -1 & 0 \end{bmatrix}$$

and $\det G_2 = -\psi$. As we assumed $\psi \neq 0$ the index is 2 due to G_2 being nonsingular. Observe that Q_1 is indeed the canonical projector onto N_1 along S_1 as can be seen by computing $Q_1 = Q_1 G_2^{-1} B P_0$.   □

4.2 Index-1 Equations

Nonlinear differential algebraic equations with index 1 are studied in [88]. This paper is an important step towards investigating nonlinear DAEs within the framework of the tractability index. Unfortunately this approach can not be easily generalised to index-2 equations.

Nevertheless similar ideas will mark the starting point for studying properly stated index-2 equations later on. As motivation and for later reference we will roughly sketch the main results of [88] in this section.

The central tool will be an appropriate decoupling procedure. In contrast to Section 3.2 the implicit function theorem [67] has to be used in order to treat nonlinear equations

$$A(t)\big[D(t)x(t)\big]' + b\big(x(t), t\big) = 0. \tag{4.4}$$

Recall that the index-1 case is precisely characterised by the fact that the matrix $G_1(x, t) = A(t)D(t) + B(x, t)Q_0(t)$ remains nonsingular on $\mathcal{D} \times \mathcal{I}$. In terms of the subspaces N_0 and S_0 this is equivalent to requiring $N_0(t) \cap S_0(x, t) = \{0\}$ for every point $(x, t) \in \mathcal{D} \times \mathcal{I}$ [3, 75]. Each solution x_* of (4.4) satisfies the obvious constraint

$$x_*(t) \in \mathcal{M}_0(t) = \{ z \in \mathcal{D} \,|\, b(z, t) \in \operatorname{im} A(t) \}.$$

In order to split x_* into its characteristic parts, it is advantageous to introduce

$$u_*(t) = D(t)x_*(t), \qquad\qquad w_*(t) = D^-(t)u_*'(t) + Q_0(t)x_*(t). \tag{4.7}$$

The definition of w_* and $DQ_0 = 0$ imply $Dw_* = DD^- u_*' = Ru_*'$. Thus the component u_* satisfies the ordinary differential equation

$$\begin{aligned} u_*'(t) &= (Ru_*)'(t) = R'(t)u_*(t) + R(t)u_*'(t) \\ &= R'(t)u_*(t) + D(t)w_*(t). \end{aligned}$$

Using the projector functions Q_0, P_0 and the decomposition

$$x_* = P_0 x_* + Q_0 x_* = D^- u_* + Q_0 w_*,$$

the index-1 DAE (4.4) can be equivalently written as[2]

$$0 = A(t)D(t)w_*(t) + b\Big(D^-(t)u_*(t) + Q_0(t)w_*(t), t\Big) =: F\big(w_*(t), u_*(t), t\big).$$

The mapping $F(w, u, \cdot) = ADw + b(D^-u + Q_0w, \cdot) = 0$ can be studied without assuming that x_* is a solution of (4.4). It turns out that due to the index-1 condition the equation $F(w, u, t) = 0$ can locally be solved for w.

Lemma 4.9. *Given* $(x_0, t_0) \in \mathcal{D} \times \mathcal{I}$ *and* $y_0 \in \operatorname{im} D(t_0)$ *such that*

$$x_0 \in \mathcal{M}_0(t_0), \qquad N_0(t_0) \cap S_0(x_0, t_0) = \{0\}, \qquad A(t_0)y_0 + b(x_0, t_0) = 0,$$

denote $u_0 = D(t_0)x_0$ *and* $w_0 = D(t_0)^- y_0 + Q_0(t_0)x_0$. *Then there exist* $\delta > 0$ *and a continuous mapping* $\mathsf{w} : B_\delta(u_0) \times \mathcal{I} \to \mathbb{R}^m$ *satisfying*

$$\mathsf{w}(u_0, t_0) = w_0, \qquad F\big(\mathsf{w}(u, t), u, t\big) = 0 \qquad \forall \, (u, t) \in B_\delta(u_0) \times \mathcal{I}.$$

Proof. It suffices to compute

$$F(w_0, u_0, t_0) = A(t_0)y_0 + b(x_0, t_0) = 0,$$
$$F_w(w_0, u_0, t_0) = A(t_0)D(t_0) + b_x(x_0, t_0)Q_0(t_0) = G_1(x_0, t_0).$$

Hence the assertion follows from the implicit function theorem as G_1 is nonsingular for index-1 equations. $\qquad\square$

Using this result we know that every solution x can be written as

$$x(t) = D^-(t)u(t) + Q_0(t)\mathsf{w}\big(u(t), t\big) \tag{4.8a}$$

where $u(t) = D(t)x(t)$ satisfies the ordinary differential equation

$$u'(t) = R'(t)u(t) + D(t)\mathsf{w}\big(u(t), t\big). \tag{4.8b}$$

This representation (4.8) provides deep insight into the structure if index-1 equations. As in Section 3.2 the equation (4.8b) is called the inherent regular ODE.

Theorem 4.10. *Let the assumptions of Lemma 4.9 be satisfied.*

(a) *The inherent regular ODE (4.8b) is uniquely determined by the DAE, i.e. it does not depend on the particular choice of* Q_0.

(b) $\operatorname{im} D$ *is a (time-varying) invariant subspace of (4.8b), i.e. the condition* $u(t_0) \in \operatorname{im} D(t_0)$ *implies* $u(t) \in \operatorname{im} D(t)$ *for all* $t \in \mathcal{I}$.

(c) *If* $\operatorname{im} D$ *is constant, then (4.8b) simplifies to* $u'(t) = D(t)\mathsf{w}\big(u(t), t\big)$.

(d) *Through each* $x_0 \in \mathcal{M}_0(t_0)$ *passes exactly one solution of (4.4).*

Proof. The proof can be found in [88]. $\qquad\square$

---

[2]To see this observe that $A = AR = ADD^-$.

5

Exploiting the Structure of DAEs in Circuit Simulation

The previous chapter gave a brief introduction to the tractability index for nonlinear DAEs

$$A(t)\big[d\big(x(t),t\big)\big]' + b\big(x(t),t\big) = 0, \qquad t \in \mathcal{I}. \tag{5.1}$$

The index-1 case was seen not to pose serious difficulties as the decoupling procedure from Section 4.2 is applicable. In particular, if the subspace $\operatorname{im} d_x(x,t)$ is constant, then numerical methods, when applied to (5.1), are known to discretise the inherent regular ODE correctly (compare Theorem 4.10 (c) and see [88] for more details).

In Chapter 1 the matrix A turned out to be constant for MNA equations. Given that the capacitance matrix $C = \frac{\partial q_C(x,t)}{\partial x}$ and the inductance matrix $L = \frac{\partial \phi_L(x,t)}{\partial x}$ are positive definite, the subspace $\operatorname{im} d_x(x,t)$ does not vary with x nor t. Index-1 equations can therefore be treated efficiently and qualitative properties of the inherent ODE are preserved by stiffly accurate Runge-Kutta schemes but also when using BDF methods [88, 89].

Given the structural assumptions from [61, 63] the index of (5.1) typically does not exceed 2. However, Theorem 1.4 shows that the index-2 case appears frequently in applications. In particular loops of capacitors and voltage sources (with at least one voltage source) lead to index-2 configurations. Similarly, cutsets of inductors and current sources give rise to index-2 equations as well.

It is unfortunate that the approach taken for index-1 equations cannot be generalised easily to index-2 DAEs of the general structure (5.1). It is not known how to define suitable variables such that just one application of the implicit function theorem yields a decoupled system such as (4.8). It is a mere supposition that such an approach is not possible for general index-2 equations (5.1) with a properly stated leading term.

However, if we restrict attention to DAEs of the special form

$$A(t)\big[D(t)x(t)\big]' + b\big(U(t)x(t),t\big) + \mathfrak{B}(t)T(t)x(t) = 0, \tag{5.2}$$

where

$$N_0(t) \cap S_0(x,t) \quad \text{does not depend on } x,$$

then it is possible to extend the decoupling procedure of Section 4.2 to (5.2). Roughly speaking, (5.2) ensures that the so-called index-2 components Tx enter the equations linearly. This situation and the additional projector functions U and T will be discussed in detail later on.

Equations of form (5.2) were already studied in [61]. There it is shown that, given some additional assumptions on the controlled sources of an electrical network, the charge oriented modified nodal analysis leads to DAEs having the special structure (5.2) (see e.g. [61, Corollary 3.2.8]).

Example 5.1. Consider once again the NAND gate from Example 4.1. If the companion model from Figure 4.1 (b) is inserted into the circuit diagram Figure 4.1 (a), it becomes obvious that the capacitors of the MOS-FET model lead to several CV-loops. In Figure 5.1 these loops are indicated using dashed lines. Hence the NAND gate comprises an index-2 system (see Theorem 1.4).

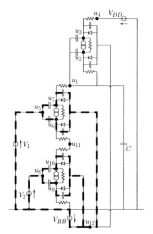

The numerically unstable index-2 components are given by the branch currents of voltage sources belonging to these loops, i.e. I_1, I_2 and I_{BB} for the NAND gate. Looking at the MNA equations presented in Example 4.1 on page 67 these components are found to enter the equations linearly.   □

Figure 5.1: The NAND gate with the equivalent circuit from Figure 4.1 (b)

We will explore equations of the type (5.2) in more detail. Recall that for index-2 DAEs $N_1(x,t) \oplus S_1(x,t) = \mathbb{R}^m$ holds for $(x,t) \in \mathcal{D} \times \mathcal{I}$. Thus Q_1 will always be chosen to be the canonical projector onto N_1 along S_1.

5.1 Decoupling Charge-Oriented MNA Equations

For regular DAEs (5.1) the characteristic subspaces

$$N_0(t) = \ker A(t)D(t), \qquad S_0(x,t) = \left\{ z \in \mathbb{R}^m \mid B(x,t)z \in \operatorname{im} A(t) \right\}.$$

were introduced in Section 4.2. Strictly speaking the enlarged system (4.5) was used instead of (5.1). Recall from (4.6) that $B(x,t) = b_x(x,t) = \frac{\partial}{\partial x}b(x,t)$.

The index is $\mu = 1$, if and only if these two subspaces intersect transversely, i.e. $N_0 \cap S_0 = \{0\}$. For $\mu = 2$ the intersection $N_1 \cap S_1 = 0$ is trivial, but the subspace $N_0 \cap S_0$ has constant nonzero dimension (see Remark 4.6). The latter subspace will be of vital importance for the subsequent analysis.

The following result is given in [61].

Lemma 5.2. *Let Q_0 and Q_1 be projector functions defined in the sequence[1]* *(4.6). Independently of the particular choice of Q_0 and Q_1 the relation*

$$\operatorname{im} Q_0(t) Q_1(x, t) = N_0(t) \cap S_0(x, t)$$

is satisfied for $(x, t) \in \mathcal{D} \times \mathcal{I}$.

Proof. Let $z = Q_0 Q_1 w \in \operatorname{im} Q_0 Q_1$. Obviously, $z \in \operatorname{im} Q_0 = N_0$. From

$$Bz = BQ_0 Q_1 w = (G_1 - AD)Q_1 w = -ADQ_1 w \in \operatorname{im} A$$

it is found that $z \in S_0$, and therefore $z \in N_0 \cap S_0$.

On the other hand, if $z \in N_0 \cap S_0$, then $z \in N_0 = \operatorname{im} Q_0$ implies $z = Q_0 z$. Since $z \in S_0$ there is a vector w such that $Bz = BQ_0 z = Aw$. Due to the properly stated leading term the matrix A satisfies $A = AR = ADD^-$ (see page 57) such that $Bz = ADD^- w$ follows.

Defining $u = D^- w - z$ yields $G_1 u = (AD + BQ_0)(D^- w - z) = 0$ and therefore $u \in \ker G_1 = \operatorname{im} Q_1$. Since $Q_0 D^- = Q_0 D^- DD^- = Q_0 P_0 D^- = 0$, we can conclude that $z = Q_0 z = Q_0(D^- w - u) = -Q_0 u = -Q_0 Q_1 u \in \operatorname{im} Q_0 Q_1$. $\qquad\square$

Recall from (3.11b″) that in the case of linear index-2 equations the component

$$z_0 = Q_0 x = Q_0 P_1 G_2^{-1} q - \mathcal{K}_0 D^- u + Q_0 Q_1 D^-(Dz_1)'$$

is calculated by differentiating Dz_1 (see page 61). It becomes clear that the derivative $(Dz_1)'$ is not affecting z_0 as a whole, but only the component that belongs to $\operatorname{im} Q_0 Q_1 = N_0 \cap S_0$.

Following [153] we introduce a projector function T onto $N_0 \cap S_0$ and define $U = I - T$. Calculating Uz_0 requires no differentiation at all. But in order to determine Tz_0 the component Dz_1 needs to be differentiated. The vector $Tx = Tz_0$ is therefore sometimes referred to as the index-2 variables.

Observe that $\operatorname{im} T = N_0 \cap S_0 \subset N_0 = \ker P_0$ implies $TP_0 = 0$, such that we have indeed

$$Tx = T(D^- u + z_1 + z_0) = T(P_0 P_1 x + P_0 Q_1 x + z_0) = Tz_0.$$

---

[1]Recall that the admissible sequences of Definition 4.4 leave some freedom for choosing the projector functions Q_i.

On the other hand $\operatorname{im} T \cap \operatorname{im} P_0 = \{0\}$ shows that without loss of generality the projector T can always be chosen such that $P_0 T = 0$. Taking into account that $Q_1 Q_0 = 0$ (cf. Definition 4.4) the following relations hold:

$$Q_0 T = T = T Q_0, \quad P_0 U = P_0 = U P_0, \quad Q_1 T = 0, \quad U Q_0 Q_1 = 0. \quad (5.3)$$

The benefit of introducing the additional projector functions U and T was realised in [61, Corollary 3.2.8]. Given the assumptions from [61, 63] on controlled sources, the structural condition

$$N_0(t) \cap S_0(x,t) \quad \text{does not depend on } x \qquad (5.4)$$

is found to be satisfied. Therefore T can be chosen to be independent of the state x and the equations of the charge oriented modified nodal analysis exhibit the particular structure

$$A(t)\big[D(t)x(t)\big]' + b\big(U(t)x(t),t\big) + \mathfrak{B}(t)T(t)x(t) = 0, \qquad (5.5a)$$

or, dropping t arguments,

$$A[Dx]' + b(Ux,\cdot) + \mathfrak{B}Tx = 0. \qquad (5.5b)$$

In fact, in [61] the subspace $N_0(t) \cap S_0(x,t)$ is shown to be constant and choosing a constant projector T becomes possible. Nevertheless we will use the slightly more general time-dependent version.

As in Section 4.2 let x_* be a solution of (5.5) and denote $x_0 = x_*(t_0)$. We introduce new variables

$$u_* = D\bar{P}_1 x_*, \qquad w_* = \bar{P}_1 D^-(Dx_*)' + (Q_0 + \bar{Q}_1)x_*, \qquad (5.6)$$

where $\bar{P}_1(t) = P_1(x_0,t)$ and $\bar{Q}_1(t) = Q_1(x_0,t)$. Here and in the sequel t arguments are generally omitted for better readability. The bar-notation for $\bar{P}_1$, $\bar{Q}_1$ is used to indicate that the x argument is replaced by the fixed initial value x_0.

By simple manipulations it turns out that

$$\bar{Q}_1 w_* = \bar{Q}_1 x_*, \qquad Q_0 w_* = -Q_0 \bar{Q}_1 D^-(Dx_*)' + Q_0 x_* + Q_0 \bar{Q}_1 x_*,$$
$$D\bar{P}_1 w_* = D\bar{P}_1 D^-(Dx_*)'.$$

Similar to the case of linear equations the solution x_* can be split as

$$x_* = P_0 \bar{P}_1 x_* + P_0 \bar{Q}_1 x_* + Q_0 x_* \qquad (5.7)$$
$$= D^- u_* + (P_0 \bar{Q}_1 + Q_0 \bar{P}_1)w_* + Q_0 \bar{Q}_1 D^-(Dx_*)'.$$

Keeping (5.3) in mind, the component $Ux_* = D^- u_* + (P_0 \bar{Q}_1 + U Q_0)w_*$ is found to be given in terms of u_* and w_* only. This is of particular importance, as we are aiming at rewriting (5.5) in terms of these new variables. In order to achieve this goal it remains to rewrite $A(Dx_*)' + \mathfrak{B}Tx_*$ using u_* and w_*.

As a first step recall that $\operatorname{im}Q_1 = \ker G_1$ and therefore

$$0 = G_1(x_0, \cdot)\bar{Q}_1 = (AD + \mathfrak{B}T)\bar{Q}_1 + B\big(Ux_0, \cdot\big)UQ_0Q_1 = (AD + \mathfrak{B}T)\bar{Q}_1.$$

Using this relation we can calculate

$$
\begin{aligned}
A(Dx_*)' + \mathfrak{B}Tx_* &= (AD + \mathfrak{B}T)D^-(Dx_*)' + \mathfrak{B}Tx_* \\
&= (AD + \mathfrak{B}T)\bar{P}_1 D^-(Dx_*)' + \mathfrak{B}Tx_* + (AD + \mathfrak{B}T)\bar{Q}_1 x_* \\
&= (AD + \mathfrak{B}T)\big[\, \bar{P}_1 D^-(Dx_*)' + TQ_0 x_* + UQ_0 x_* + \bar{Q}_1 x_* \,\big] \\
&= (AD + \mathfrak{B}T)w_*,
\end{aligned}
$$

such that the DAE (5.5) can be written equivalently as

$$F(u_*, w_*, \cdot) := \big(AD + \mathfrak{B}T\big)w_* + b\Big(D^- u_* + (P_0\bar{Q}_1 + UQ_0)w_*, \cdot\Big) = 0. \quad (5.8)$$

Lemma 5.3. *Let (5.5) be a regular DAE with index $\mu \in \{1,2\}$. Assume that $(x^0, t_0) \in \mathcal{D} \times \mathcal{I}$ and $y^0 \in \operatorname{im}D(t_0)$ satisfy*

$$A(t_0)y^0 + b\big(U(t_0)x^0, t_0\big) + \mathfrak{B}(t_0)T(t_0)x^0 = 0.$$

Denote

$$u_0 = D(t_0)\bar{P}_1(t_0)x^0, \qquad w_0 = \bar{P}_1(t_0)D^-(t_0)y^0 + \big(Q_0 + \bar{Q}_1\big)(t_0)x^0$$

and consider F from (5.8) being defined on a neighbourhood $\mathcal{N}_0 \subset \mathbb{R}^n \times \mathbb{R}^m \times \mathbb{R}$ of (u_0, w_0, t_0). Then there is a neighbourhood $\mathcal{N}_1 \subset \mathbb{R}^n \times \mathbb{R}$ of (u_0, t_0) and a continuous mapping $\mathsf{w} : \mathcal{N}_1 \to \mathbb{R}^m$ such that

$$\mathsf{w}(u_0, t_0) = w_0, \qquad F\big(u, \mathsf{w}(u,t), t\big) = 0 \qquad \forall\ (u,t) \in \mathcal{N}_1.$$

Proof. Due to (5.8) we have $F(u_0, w_0, t_0) = 0$ and

$$F_w(u, w, \cdot) = AD + \mathfrak{B}T + b_x\big(D^- u + (P_0\bar{Q}_1 + UQ_0)w, \cdot\big)(P_0\bar{Q}_1 + UQ_0).$$

We will show that $F_w(u_0, w_0, t_0)$ and $G_2(x^1, x^0, t_0)$ have common rank for every $x^1 \in \mathbb{R}^m$. Since $G_2(x^1, x^0, t_0)$ is nonsingular due to the index-2 condition, $F_w(u_0, w_0, t_0)$ will have the same property. Then the assertion follows from the implicit function theorem.

Recall that if (5.5) was an index-1 equation, the matrix sequence (4.6) implies that $G_2 = G_1$ is nonsingular. Hence index-1 DAEs are covered as well.

For the structure (5.5) the matrix sequence (4.6) yields

$$
\begin{aligned}
B(x, \cdot) &= b_x(Ux, \cdot)U + \mathfrak{B}T, \\
B_1(x^1, x, \cdot) &= B(x, \cdot)P_0 - G_1(x, \cdot)D^- C(x^1, x, \cdot)D.
\end{aligned}
$$

The matrix B_1 is used to define

$$G_2(x^1, x, \cdot) = G_1(x, \cdot) + B_1(x^1, x, \cdot)Q_1(x, \cdot).$$

Simple calculations show that this matrix can be factorised as

$$\Big[AD + B(x, \cdot)\big(Q_0 + P_0 Q_1(x, \cdot)\big)\Big]\Big[I - P_1(x, \cdot)D^-C(x^1, x, \cdot)DQ_1(x, \cdot)\Big].$$

The second factor is nonsingular with $I + P_1(x, \cdot)D^-C(x^1, x, \cdot)DQ_1(x, \cdot)$ being its inverse. The first factor, on the other hand, is independent of x^1. Evaluated at the point (x^0, t_0) this factor happens to agree with $F_w(u_0, w_0, t_0)$. Therefore $F_w(u_0, w_0, t_0)$ and $G_2(x^1, x^0, t_0)$ have indeed common rank.   □

Notice that the mapping w from the previous lemma is defined only locally around (u_0, t_0). For simplicity we assume that the definition domain is sufficiently large to cover the whole interval $\mathcal{I}$. If this is not the case, an appropriate smaller interval $\mathcal{I}$ has to be considered.

Finally, from (5.7) the solution representation

$$\begin{aligned}
x_* &= D^- u_* + (Q_0\bar{P}_1 + P_0\bar{Q}_1)w(u_*, \cdot) + Q_0\bar{Q}_1 D^-\big(u_* + D\bar{Q}_1 w(u_*, \cdot)\big)' \\
&= D^- u_* + z_{0*} + z_{1*}
\end{aligned} \tag{5.9a}$$

can be derived. Similar to the case of linear DAEs the abbreviations

$$\begin{aligned}
u_* &= D\bar{P}_1 x_* \\
z_{1*} &= P_0\bar{Q}_1 x_* = P_0\bar{Q}_1 w(u_*, \cdot) \\
z_{0*} &= Q_0 x_* \\
&= Q_0\bar{P}_1 w(u_*, \cdot) + Q_0\bar{Q}_1 D^-(D\bar{P}_1 D^-)' u_* + Q_0\bar{Q}_1 D^-(Dz_{1*})'
\end{aligned}$$

have been used. Observe that $u_* = D\bar{P}_1 x_*$ satisfies the ordinary differential equation

$$u' = (D\bar{P}_1 D^-)' u + D\bar{P}_1 w(u, \cdot) + (D\bar{P}_1 D^-)' D\bar{Q}_1 w(u, \cdot). \tag{5.9b}$$

This equation will be called inherent regular ODE. Once u_* is determined, the component z_{1*} is given in terms of u_* and the mapping w from Lemma 5.3. Finally, Dz_{1*} needs to be differentiated in order to calculate $z_{0*} = Q_0 x_*$.

Similar to Section 4.2 and [88] the inherent regular ODE (5.9b) can be studied without assuming the existence of a solution x_* for the DAE (5.5).

Theorem 5.4. *Let the assumptions of Lemma 5.3 be satisfied. Then*

(a) *im $D\bar{P}_1 D^-$ is a (time-varying) invariant subspace of the inherent ODE (5.9b), i.e. $u(t_0) \in \mathrm{im}\,\big(D\bar{P}_1 D^-\big)(t_0)$ implies $u(t) \in \mathrm{im}\,\big(D\bar{P}_1 D^-\big)(t)$ for every $t \in \mathcal{I}$.*

(b) *If the subspaces $\mathrm{im}\,D\bar{P}_1 D^-$ and $\mathrm{im}\,D\bar{Q}_1 D^-$ are constant, then (5.9b) simplifies to $u' = D\bar{P}_1 w(u, \cdot)$.*

Proof. The proof uses the same techniques as [88, Theorem 2.2], but R needs to be replaced by $D\bar{P}_1 D^-$. The above result is also a special case of Lemma 6.6 from Section 6.1. Thus a proof can be found there as well.   □

In contrast to index-1 equations, the space $\mathcal{M}_0(t)$ introduced on page 74 is no longer filled with solutions. It is clear from the decoupling (5.9) that for every $x^0 \in \mathbb{R}^m$ the component $u_0 = D(t_0)P_1(x^0, t_0)x^0$ uniquely determines a solution. This is made more precise in the following result.

Theorem 5.5. *Let the assumptions of Lemma 5.3 be satisfied and let u be the solution of the inherent regular ODE (5.9b) with initial value $u(t_0) = DP_1(x^0, t_0)x^0$. If the mapping $t \mapsto D(t)Q_1(t)(x^0, t)\mathsf{w}(u(t), t)$ belongs to the class of C^1-functions, then there is a unique solution $x \in C_D^1(\mathcal{I}, \mathbb{R}^m)$ of the initial value problem*

$$A[Dx]' + b(Ux, \cdot) + \mathfrak{B}Tx = 0, \qquad DP_1(x^0, t_0)\big(x^0 - x(t_0)\big) = 0. \qquad (5.10)$$

Proof. The mapping w from Lemma 5.3 defines the inherent regular ODE (5.9b). Let u be the solution satisfying $u(t_0) = DP_1(x^0, t_0)x^0$. Then Theorem 5.4 ensures that $u(t)$ belongs to $\operatorname{im} D(t)\bar{P}_1(t)D^-(t)$ for every t. Define

$$\begin{aligned}
x &= D^- u + z_0 + z_1 \\
&= D^- u + (Q_0\bar{P}_1 + P_0\bar{Q}_1)\mathsf{w}(u, \cdot) + Q_0\bar{Q}_1 D^-\big(u + D\bar{Q}_1\mathsf{w}(u, \cdot)\big)'.
\end{aligned}$$

This mapping is indeed a solution of (5.10) since

$$\begin{aligned}
A(Dx)' + b(Ux, \cdot) + \mathfrak{B}Tx &= A(Dx)' + b(Ux, \cdot) + \mathfrak{B}Tx - F\big(u, \mathsf{w}(u, \cdot), \cdot\big) \\
&= (AD + \mathfrak{B}T)\bar{Q}_1 D^-(Dx)' + AD\bar{P}_1 D^-(Dx)' - AD\bar{P}_1\mathsf{w}(u, \cdot) = 0
\end{aligned}$$

and $DP_1(x^0, t_0)x(t_0) = u(t_0) = DP_1(x^0, t_0)x^0$. The decoupling procedure described above shows that this solution is unique.   □

Example 5.6. Consider again the circuit from Figure 4.2 on page 73. In contrast to Example 4.8 the node potentials will now be denoted by e_i in order to avoid confusion with the variable u of the inherent regular ODE (5.9b).

For simplicity choose

$$i(i_V, t) = t \cdot i_V, \qquad \text{and} \qquad v(t) \equiv -1 \qquad (5.11)$$

for $t \geq 0$ such that $\psi = 2 + \frac{\partial i(i_V, t)}{\partial i_V} = 2 + t$ and $\phi = \frac{\partial \psi(i_V, t)}{\partial t}\big/\psi(i_V, t)^2 = \frac{1}{(2+t)^2}$. The matrix sequence was already calculated in Example 4.8. The particular choice of $i(i_V, t)$, $v(t)$ adopted here leads to

$$B = \begin{bmatrix} 2(2+t)e_1 & 0 & 1+t \\ 0 & 0 & -1 \\ 1 & -1 & 0 \end{bmatrix}, \qquad Q_0 = \begin{bmatrix} 0 & 0 & 0 \\ 0 & 0 & 0 \\ 0 & 0 & 1 \end{bmatrix}, \qquad Q_1 = \frac{1}{2+t}\begin{bmatrix} 1+t & -1-t & 0 \\ -1 & 1 & 0 \\ -1 & 1 & 0 \end{bmatrix}$$

and

$$G_2 = \frac{1}{(2+t)^2} \begin{bmatrix} (2+t)^2+1+2e_1(1+t)(2+t)^2 & -1-2e_1(1+t)(2+t)^2 & (1+t)(2+t)^2 \\ 1 & (2+t)^2-1 & -(2+t)^2 \\ (2+t)^2 & -(2+t)^2 & 0 \end{bmatrix}.$$

Observe that G_2 is nonsingular for $t \geq 0$ with $\det G_2 = -(2+t)$ such that the index is $\mu = 2$. Having Q_0 and Q_1 at our disposal we can determine

$$N_0 \cap S_0 = \operatorname{im} Q_0 Q_1 = \operatorname{im} \begin{bmatrix} 0 & 0 & 0 \\ 0 & 0 & 0 \\ \frac{-1}{2+t} & \frac{1}{2+t} & 0 \end{bmatrix}$$

and thus $T = Q_0$ with $Tx = \begin{bmatrix} 0 & 0 & i_V \end{bmatrix}^\top$. Looking at Figure 4.2 it turns out that the current source is controlled by the index-2 variable i_V – a situation that will hardly occur in real applications. However, choosing different functions for $i(i_V, t)$ it is possible to study the precise consequences of how the index-2 components enter the equations. The nonlinear case will be addressed later in Example 6.2 in the next section. Here the choice (5.11) ensures that Tx appears linearly such that this section's results apply.

The MNA equations can be rewritten in the form (5.5),

$$\begin{bmatrix} 1 & 0 \\ 0 & 1 \\ 0 & 0 \end{bmatrix} \left(\begin{bmatrix} 1 & 0 & 0 \\ 0 & 1 & 0 \end{bmatrix} \begin{bmatrix} e_1 \\ e_2 \\ i_V \end{bmatrix} \right)' + \begin{bmatrix} (2+t)e_1^2 \\ 0 \\ e_1 - e_2 + 1 \end{bmatrix} + \begin{bmatrix} 0 & 0 & 1+t \\ 0 & 0 & -1 \\ 0 & 0 & 0 \end{bmatrix} \begin{bmatrix} 0 \\ 0 \\ i_V \end{bmatrix} = 0,$$

with a time-dependent matrix $\mathfrak{B}$. An initialisation is given by the triple $(y^0, x^0, t_0) = \left(\begin{bmatrix} -\frac{1}{2} & 0 \end{bmatrix}^\top, \begin{bmatrix} \frac{1}{2} & \frac{3}{2} & 0 \end{bmatrix}^\top, 0 \right)$ since $Ay^0 + b(Ux^0, t_0) + \mathfrak{B}(t_0)Tx^0 = 0$. The mapping F from (5.8) reads

$$F(u, w, t) = \begin{bmatrix} w_1 + (1+t)w_3 + \frac{1}{2+t}((2+t)u_1 + (1+t)(w_1 - w_2))^2 \\ w_2 - w_3 \\ u_1 - u_2 + w_1 - w_2 + 1 \end{bmatrix}.$$

Due to the index-2 condition it is possible to solve the system $F(u, w, t) = 0$ for

$$w = \mathsf{w}(u, t) = \frac{(1+u_1-u_2)(t+2) - [1+u_1-u_2+(u_2-1)(t+2)]^2}{(2+t)^2} \begin{bmatrix} 1 \\ 1 \\ 1 \end{bmatrix} - (1+u_1-u_2) \begin{bmatrix} 1 \\ 0 \\ 0 \end{bmatrix}$$

and the inherent regular ODE (5.9b) reads

$$\begin{bmatrix} u_1 \\ u_2 \end{bmatrix}' = \frac{1}{2+t}(u_1 + u_2(1+t) - t)(2 - u_1 - u_2(1+t) + t) \begin{bmatrix} 1 \\ 1 \end{bmatrix}, \quad u(0) = DP_1(t_0)x^0 = \begin{bmatrix} 1 \\ 1 \end{bmatrix}.$$

The two components u_1 and u_2 necessarily coincide and $u_1' = \frac{2u_1(1+t)-t}{2+t} - u_1^2$ yields the unique solution $u_1 = u_2 = 1$. Using (5.9) we derive the solution

$$z_1 = \begin{bmatrix} -\frac{1+t}{2+t} \\ \frac{1}{2+t} \\ 0 \end{bmatrix}, \quad z_0 = \begin{bmatrix} 0 \\ 0 \\ -\frac{1}{(2+t)^2} \end{bmatrix}, \quad x = D^- u + z_0 + z_1 = \frac{1}{2+t} \begin{bmatrix} 1 \\ 3+t \\ -\frac{1}{2+t} \end{bmatrix}.$$

Notice that neither $u = DP_1 x = \frac{e_1 + e_2(1+t)}{2+t} \begin{bmatrix} 1 \\ 1 \end{bmatrix}$ nor $z_1 = P_0 Q_1 x = \frac{e_1 - e_2}{2+t} \begin{bmatrix} 1+t \\ -1 \\ 0 \end{bmatrix}$ depends on the index-2 component $Tx = \begin{bmatrix} 0 & 0 & i_V \end{bmatrix}^\top$ – a fact stressing the decoupled structure of (5.9).   □

It is stressed that the decoupling procedure is not meant to be used for solving DAEs as we did in the previous example. The decoupling procedure is a mathematical tool that will be used in Chapter 9 to study numerical methods for differential algebraic equations.

Obviously there is a close relationship between the decoupling procedure studied in this section and the one considered in Section 4.2 for index-1 equations. In particular, if (5.5) was an index-1 DAE, then $\bar{Q}_1 = 0$ and $\bar{P}_1 = I$ provide a re-interpretation of this section's results also for index-1 DAEs. In this case we find $T = 0$ and $U = I$ such that considering DAEs with the special structure (5.5) is no restriction at all for index-1 equations. Finally, since (5.6) reduces to $u = Dx$ and $w = D^-(Dx)' + Q_0 x$, we find that the decoupling procedure for index-2 equations (5.5), when applied to index-1 DAEs, coincides with the formalism of Section 4.2.

Unfortunately, (5.6) is not suited for decoupling index-2 DAEs of the more general structure (5.1). Observe that due to $x_* = D^- u_* + (P_0 \bar{Q}_1 + Q_0 \bar{P}_1) w_* + Q_0 \bar{Q}_1 D^-(Dx_*)'$ the component $Ux_* = D^- u_* + (P_0 \bar{Q}_1 + U Q_0) w_*$ can be written in terms of u_* and w_*. However, this is not possible for x_* itself. Hence writing the original DAE (5.1) in terms of the new variables u_* and w_* turns out to be rather difficult if not impossible.

But still, a more refined splitting using three instead of two components makes it possible to obtain similar results also for (5.1). In the next chapter we will investigate this situation in more detail.

6

Properly Stated Index-2 DAEs

The object of this chapter is to study nonlinear index-2 DAEs with a properly stated leading term exhibiting the general structure

$$A(t)\big[d(x(t),t)\big]' + b\big(x(t),t\big) = 0. \tag{6.1}$$

Up to now different simplified versions of (6.1) were addressed in this thesis. Linear DAEs

$$A(t)\big[D(t)x(t)\big]' + B(t)x(t) = q(t)$$

were treated in Chapter 3. There basic ideas such as the properly stated leading term or the decoupling procedure were introduced. In Chapter 4 the concept of the tractability index was introduced for nonlinear DAEs. We saw that instead of treating DAEs of the form (6.1) it is often advantageous to study the equivalent enlarged system

$$A(t)\big[R(t)y(t)\big]' + b\big(x(t),t\big) = 0,$$
$$y(t) - d(x(t),t) = 0,$$

such that is suffices to treat DAEs of the form

$$A(t)\big[D(t)x(t)\big]' + b\big(x(t),t\big) = 0. \tag{6.2}$$

Index-1 equations of this type were analysed in Section 4.2, but in Chapter 5 we needed to restrict attention to DAEs with the special structure

$$A(t)\big[D(t)x(t)\big]' + b\big(U(t)x(t),t\big) + \mathfrak{B}(t)T(t)x(t) = 0 \tag{6.3}$$

in order to address index-2 equations. Here the index-2 variables Tx enter the equations linearly. DAEs having this structure are of particlar importance for scientific computing in circuit simulation as the equations of the charge-oriented modified nodal analysis are of this form [61]. Having this application area in mind we found the important structural condition that

$$N_0(t) \cap S_0(x,t) \quad \textit{does not depend on } x \tag{6.4}$$

to be satisfied as well. Hence a wide range of differential algebraic equations and in particular many electrical circuits are already covered by the results presented so far. Nevertheless, the scope of our investigations is not yet wide enough.

Example 6.1. Consider a DAE in Hessenberg form

$$y' = f(y, z), \qquad 0 = g(y), \tag{6.5}$$

having index-2, i.e. $g_y f_z$ is assumed to be nonsingular. This system can be cast into the form (6.2) by introducing

$$x = \begin{bmatrix} y \\ z \end{bmatrix}, \qquad A = \begin{bmatrix} I \\ 0 \end{bmatrix}, \qquad D = \begin{bmatrix} I & 0 \end{bmatrix}, \qquad b(x, t) = \begin{bmatrix} -f(y, z) \\ g(y) \end{bmatrix}.$$

The matrix sequence (4.6) from Section 4.1 yields

$$G_0 = AD = \begin{bmatrix} I & 0 \\ 0 & 0 \end{bmatrix}, \qquad Q_0 = \begin{bmatrix} 0 & 0 \\ 0 & I \end{bmatrix}, \qquad B = \begin{bmatrix} -f_y & -f_z \\ g_y & 0 \end{bmatrix}, \qquad G_1 = \begin{bmatrix} I & -f_z \\ 0 & 0 \end{bmatrix}.$$

It turns out that $N_0 = \ker G_0 = \{(y, z) \,|\, y = 0\}$ and $S_0 = \{(y, z) \,|\, B \begin{bmatrix} y \\ z \end{bmatrix} \in \operatorname{im} G_0\}$ $= \{(y, z) \,|\, g_y y = 0\}$ such that $N_0 \cap S_0 = N_0$ is always constant. $T = Q_0$ is a projector onto $N_0 \cap S_0$ and the index-2 variables $Tx = \begin{bmatrix} 0 \\ z \end{bmatrix}$ may enter the equation (6.5) in a nonlinear way. $\qquad \square$

The above example shows that DAEs in Hessenberg form may not be covered by the structure (6.3) even though they can be cast into the form (6.2). Hence, choosing the model problem (6.2) for our investigations, both Hessenberg DAEs and MNA equations will be covered.

In the sequel properly stated DAEs (6.2) will be addressed where the index-2 components may enter nonlinearly. The special case of DAEs in Hessenberg form will be discussed later in Section 6.2.

Allowing the index-2 components to enter the equations in a nonlinear way complicates matters considerably. New qualitative features, that haven't been present before, will have a major impact on the analysis.

Example 6.2. As a motivation consider once more the circuit from Figure 4.2. In the Example 5.6 on page 83 we chose $i(i_V, t) = t \cdot i_V$ such that the index-2 variables $Tx = \begin{bmatrix} 0 & 0 & i_V \end{bmatrix}^\top$ enter the equations linearly.

For $i(i_V, t) = i_V^3$ this is no longer true and the matrix sequence changes to

$$D = \begin{bmatrix} 1 & 0 & 0 \\ 0 & 1 & 0 \end{bmatrix} \qquad Q_0 = T = \begin{bmatrix} 0 & 0 & 0 \\ 0 & 0 & 0 \\ 0 & 0 & 1 \end{bmatrix}, \qquad Q_1 = \frac{1}{2 + 3i_V^2} \begin{bmatrix} 1 + 3i_V^2 & -1 - 3i_V^2 & 0 \\ -1 & 1 & 0 \\ -1 & 1 & 0 \end{bmatrix}.$$

In contrast to the situation from Example 5.6 the index-2 variable i_V has a clear influences on

$$u = DP_1 x = \frac{e_1 + e_2(1 + 3i_V^2)}{2 + 3i_V^2} \begin{bmatrix} 1 \\ 1 \end{bmatrix}, \qquad z_1 = P_0 Q_1 x = \frac{e_1 - e_2}{2 + 3i_V^2} \begin{bmatrix} 1 + 3i_V^2 \\ -1 \\ 0 \end{bmatrix}.$$

Consequently, a decoupled system such as (5.9) and in particular an inherent ordinary differential equation (5.9b) cannot be expected for general nonlinear DAEs. □

The above example shows that errors in the components z_0, z_1 may influence the dynamical part u. This was already observed numerically in [152]. For nonlinear DAEs (6.1) a coupling between u and Dz_1 will become visible. This forms a stark contrast to the results obtained in the previous section for DAEs of the form (6.3).

The coupled structure makes an analysis more complicated. However, a careful investigation of the equation's properties and a more refined decoupling procedure will make a result quite similar to (5.9) possible. The price we have to pay is to change from the inherent regular ODE to an inherent implicit index-1 DAE.

6.1 A Decoupling Procedure for Index-2 DAEs

We consider differential algebraic equations (DAEs)

$$A(t)\big[D(t)x(t)\big]' + b\big(x(t),t\big) = 0 \tag{6.6}$$

with continuous coefficients. As in the previous sections let $\mathcal{D} \subset \mathbb{R}^m$ be a domain and $\mathcal{I} \subset \mathbb{R}$ an interval. $A(t) \in \mathbb{R}^{m \times n}$ and $D(t) \in \mathbb{R}^{n \times m}$ are rectangular matrices in general. We assume that A and D are continuous matrix functions and that the leading term is properly stated in the sense of Definition 3.2.

The nonlinear function $b : \mathcal{D} \times \mathcal{I} \to \in \mathbb{R}^m$ is defined for $(x,t) \in \mathcal{D} \times \mathcal{I}$. It is assumed that b and the partial derivative b_x are continuous.

The index of differential algebraic equations (6.6) with a properly stated leading term is defined using admissible sequences (4.6) from Section 4.1. For convenience and for later reference these definitions are repeated here. Recall that pointwise for $t \in \mathcal{I}$, $x \in \mathcal{D}$ we introduced

$$\left.\begin{aligned}
&G_0(t) = A(t)D(t), && B(x,t) = b_x(x,t) \\
&N_0(t) = \ker G_0(t), \\
&S_0(x,t) = \big\{\, z \in \mathbb{R}^m \,|\, B(x,t)z \in \operatorname{im} G_0(t) \,\big\} \\
&G_1(x,t) = G_0(t) + B(x,t)Q_0(t), \\
&N_1(x,t) = \ker G_1(x,t), \\
&S_1(x,t) = \big\{\, z \in \mathbb{R}^m \,|\, B(x,t)z \in \operatorname{im} G_1(x,t) \,\big\}.
\end{aligned}\right\} \tag{6.7a}$$

The continuous projector function $Q_0 \in C(\mathcal{I}, \mathbb{R}^{m \times m})$ is chosen in such a way that $\operatorname{im} Q_0 = N_0$. Similarly let $Q_1 \in C(\mathcal{D} \times \mathcal{I}, \mathbb{R}^{m \times m})$ be a projector function onto N_1. The complementary projectors

$$P_0(t) = I - Q_0(t), \qquad\qquad P_1(x,t) = I - Q_1(x,t)$$

will be used frequently. The index of (6.6) is defined to be $\mu = 1$, if N_0 and S_0 intersect transversally, i.e. $N_0 \cap S_0 = \{0\}$ or, equivalently, if G_1 remains nonsingular on $\mathcal{D} \times \mathcal{I}$. For index-2 equations the intersection $N_0 \cap S_0$ has a constant positive dimension, but $N_1 \cap S_1 = \{0\}$.

In Definition 4.5 the index was defined in terms of the matrices G_i. Therefore consider

$$
\left.
\begin{aligned}
C(x^1, x, t) &= \left(DP_1D^-\right)_x(x,t)x^1 + \left(DP_1D^-\right)_t(x,t), \\
B_1(x^1, x, t) &= B(x,t)P_0(t) - G_1(x,t)D^-(t)C(x^1,x,t)D(t), \\
G_2(x^1, x, t) &= G_1(x,t) + B_1(x^1,x,t)Q_1(x,t)
\end{aligned}
\right\}
\quad (6.7b)
$$

for $t \in \mathcal{I}$, $x \in \mathcal{D}$ and $x^1 \in \mathbb{R}^m$. The generalised reflexive inverse $D^-(t)$ is again uniquely defined by requiring

$$
DD^-D = D, \qquad D^-DD^- = D^-, \qquad D^-D = P_0, \qquad DD^- = R,
$$

where the projector function R originates from the definition of the properly stated leading term.

According to Definition 4.4 the sequence (6.7) is admissible, if $\operatorname{rank} G_i = r_i$ is constant on the interval $\mathcal{I}$ for $i = 0, 1, 2$, the subspace $N_0 = \ker G_0$ satisfies $N_0 \subset \ker Q_1$ and

$$
Q_0 \in C(\mathcal{I}, \mathbb{R}^{m \times m}), \quad Q_1 \in C(\mathcal{D} \times \mathcal{I}, \mathbb{R}^{m \times m}), \quad DP_1D^- \in C^1(\mathcal{D} \times \mathcal{I}, \mathbb{R}^{n \times n}).
$$

The DAE (6.6) is regular with index $\mu = 2$ if there is an admissible sequence (6.7), such that $r_0 \leq r_1 < r_2 = m$. Thus $G_2(x^1, x, t)$ remains nonsingular on $\mathbb{R}^m \times \mathcal{D} \times \mathcal{I}$ and we have

$$
N_1(x,t) \oplus S_1(x,t) = \mathbb{R}^m
$$

as is shown e.g. in [75]. Hence, without restriction we will always assume that Q_1 is the canonical projector onto N_1 along S_1. Of course we have to require that this particular choice of Q_1 leads to an admissible sequence.

As in Chapter 5 let $T \in C(\mathcal{D} \times \mathcal{I}, \mathbb{R}^{m \times m})$ be a projector function such that

$$
\operatorname{im} T(x,t) = N_0(t) \cap S_0(x,t) = \operatorname{im} Q_0(t)Q_1(x,t).
$$

It was remarked earlier that T can be chosen such that

$$
Q_0T = T = TQ_0, \qquad P_0U = P_0 = UP_0, \qquad Q_1T = 0 \qquad (6.8)
$$

(see Lemma 5.2 and equation (5.3) on page 80). As usual, $U = I - T$ denotes the complementary projector function corresponding to T.

It is well known that in order to prove the existence of solutions for differential algebraic equations, the DAE has to satisfy certain structural conditions. In

[99] the DAE is assumed to satisfy a hypothesis based on the derivative array. In contrast to that, [122] requires $Q_1 G_2^{-1}\big(b(x,t) - b(P_0 x, t)\big) = 0$. Unfortunately, for the application in circuit simulation the former approach is not feasible as it uses the derivative array. On the other hand the latter requirement is too restrictive in the sense that there are DAEs arising from the modified nodal analysis that do not satisfy this condition [153]. Hence in [61, 153] the generalised structural condition

(A1) $N_0(t) \cap S_0(x,t)$ *does not depend on x*

was introduced. This condition already played a crucial role in Chapter 5.

Notice that for equations arising from the modified nodal analysis the space $N_0(t) \cap S_0(x,t)$ is found to be constant [61]. Thus for applications in circuit simulation (A1) always holds automatically. For DAEs from other application backgrounds, (A1) can be checked using linear algebra tools in practice [107].

Even though properly stated index-2 DAEs of the general structure (6.6) are considered in this chapter, we still need to require the structural condition (A1). It will turn out that (A1) is sufficient for proving the local existence and uniqueness of solutions. Thus in circuit simulation there is no need to check complicated conditions that guarantee the existence of solutions.

It will be pointed out later where this condition becomes necessary. One immediate consequence is that the projector T can be chosen to be independent of x.

6.1.1 Splitting of DAE Solutions

For the moment let us assume that there is a solution $x_\star \in C_D^1(\mathcal{I}, \mathbb{R}^m)$ of the regular index-2 DAE (6.6). Then, by definition,

$$A(t_0) y^0 + b(x^0, t_0) = 0 \tag{6.9}$$

provides a consistent initialisation with $(y^0, x^0, t_0) = \big((Dx_\star)'(t_0),\, x_\star(t_0),\, t_0\big)$.

Some of the matrix functions defined in the sequence (6.7) depend not only on the time t but also on the arguments x and x^1. In order to obtain suitable evaluation points we introduce an arbitrary mapping $\bar{x} \in C^1(\mathcal{I}, \mathbb{R}^m)$ satisfying

$$\bar{x}(t_0) = x^0, \qquad\qquad \big(\bar{x}(t), t\big) \in \mathcal{D} \times \mathcal{I} \qquad \forall\ t \in \mathcal{I}, \tag{6.10}$$

and consider

$$\bar{Q}_1(t) = Q_1\big(\bar{x}(t), t\big), \quad \bar{P}_1 = P_1\big(\bar{x}(t), t\big), \tag{6.11}$$
$$\bar{G}_1(t) = G_1\big(\bar{x}(t), t\big), \quad \bar{B}(t) = B\big(\bar{x}(t), t\big), \quad \bar{G}_2(t) = G_2\big(\bar{x}'(t), \bar{x}(t), t\big)$$

defined for $t \in \mathcal{I}$. Observe that the above construction ensures

$$C(\bar{x}', \bar{x}, \cdot) = \big(DP_1 D^-\big)_x(\bar{x}, \cdot)\, \bar{x}' + (DP_1 D^-)_t(\bar{x}, \cdot) = \tfrac{\mathrm{d}}{\mathrm{d}t}\big(DP_1 D^-\big)(\bar{x}, \cdot)$$

for the matrix C from (6.7b). In order for this relation to hold, $\bar{x} \in C^1(\mathcal{I}, \mathbb{R}^m)$ was chosen to be sufficiently smooth.

If the solution itself satisfied $x_* \in C^1(\mathcal{I}, \mathbb{R}^m)$, then $\bar{x} = x_*$ would be an obvious choice. The constant function $\bar{x} \equiv x^0$ may be considered as well. This particular choice was adopted in Chapter 5.

We are now in a position to define functions

$$u(t) = D(t)\bar{P}_1(t)x_*(t), \qquad w(t) = T(t)x_*(t), \qquad z(t) = \bar{Z}(t)x_*(t) \qquad (6.12)$$

for $t \in \mathcal{I}$. The matrix $\bar{Z}(t)$ is given by the mapping

$$\bar{Z} : \mathcal{I} \to \mathbb{R}^{m \times m}, \quad t \mapsto \bar{Z}(t) = P_0(t)\bar{Q}_1(t) + U(t)Q_0(t).$$

Using the properties from (6.8) it becomes clear that $\bar{Z}$ is again a projector function, i.e. $\bar{Z}^2 = \bar{Z}$. Figure 6.1 shows how u, w and z are obtained from x_* by successive splitting.

In contrast to similar splittings previously considered, $z_{0*} = Q_0 x_*$ is divided into its U and T part, respectively. $U z_{0*}$ is then merged with the component $z_{1*} = P_0 \bar{Q}_1 x_*$ in order to obtain $z = z_{1*} + U z_{0*} = \bar{Z} x_*$. On the other hand the component $w = T Q_0 x_* = T z_{0*}$ is considered as a new variable. The solution x_* can be written as

$$x_*(t) = D^-(t)u(t) + z(t) + w(t). \qquad (6.13)$$

This refined splitting of z_{0*} provides a clear separation of algebraic and differential components. Looking at (3.15) for linear equations, $z = z_{1*} + U z_{0*}$ can be calculated by an algebraic relation but in order to obtain $w = T z_{0*}$ the component $Dz_{1*} = Dz$ needs to be differentiated[1]. A similar situation is valid for nonlinear DAEs (6.6) as will become clear in due course.

As the solution x_* belongs to $C_D^1(\mathcal{I}, \mathbb{R}^m) = \{x \in C(\mathcal{I}, \mathbb{R}^m) \mid Dx \in C^1(\mathcal{I}, \mathbb{R}^n)\}$, the algebraic part $z \in C_D^1(\mathcal{I}, \mathbb{R}^m)$ has the same property. Hence we consider $v(t) = D(t)z(t) = D(t)\bar{Q}_1(t)x_*(t)$, such that it is possible to write

$$\left(Dx_*\right)' = \left(D\bar{P}_1 x_* + D\bar{Q}_1 x_*\right)' = u' + v'.$$

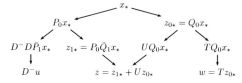

Figure 6.1: The relation between u, z, w and the solution x_*

---

[1]Recall that $DUQ_0 = -DTQ_0 = DP_0T = 0$ and thus $D\bar{Z} = D\bar{Q}_1$. Additionally $\operatorname{im} T = \operatorname{im} Q_0 \bar{Q}_1$ is satisfied such that $TQ_0\bar{Q}_1 = Q_0\bar{Q}_1$.

In order to decouple the DAE into its characteristic parts, we will rewrite (6.6) in terms of the new variables introduced above. Using the notation

$$f(y, x, t) = A(t)y + b(x, t)$$

the DAE (6.6) can be written as

$$
\begin{aligned}
0 &= f\Big((Dx_*)'(t),\, x_*(t),\, t \Big) \\
&= f\Big(u'(t) + v'(t),\, D^-(t)u(t) + z(t) + w(t),\, t \Big) \qquad\qquad (6.14\text{a}) \\
&= F\Big(u(t), w(t), z(t), u'(t), v'(t), t \Big), \qquad\qquad t \in \mathcal{I}.
\end{aligned}
$$

The function F is defined by

$$F(u, w, z, \eta, \zeta, t) = f\Big(\eta + \zeta,\, D^-(t)u + \bar{Z}(t)z + T(t)w,\, t \Big) \qquad (6.14\text{b})$$

where u, w, z, η and ζ are considered to be parameters. $\bar{Z}$ and T were introduced for convenience when defining F. Because of (6.12) they do not change (6.14a) at all but will be quite useful when calculating derivatives of F.

Lemma 6.3. *Let* (6.6) *be a regular DAE with index 2 on* $\mathcal{D} \times \mathcal{I}$. *Assume that* (6.9) *holds for a solution* $x_* \in C^1(\mathcal{I}, \mathbb{R}^m)$. *Choose* $\bar{x} \in C^1(\mathcal{I}, \mathbb{R}^m)$ *satisfying the condition* (6.10) *and define*

$$u_0 = u(t_0), \qquad w_0 = w(t_0), \qquad z_0 = z(t_0), \qquad \eta_0 = u'(t_0), \qquad \zeta_0 = v'(t_0)$$

for the functions u, w *and* z *from* (6.12). *If the structural condition* (A1) *holds, then locally around* $(u_0, w_0, z_0, \eta_0, \zeta_0, t_0)$ *equation* (6.14) *is equivalent to*

$$z(t) = \mathsf{z}\big(u(t), t\big), \quad u'(t) = \mathsf{f}\big(u(t), w(t), t\big), \quad w(t) = \mathsf{w}\big(u(t), v'(t), t\big) \quad (6.15)$$

with continuous functions z, f *and* w *being defined on neighbourhoods of* (u_0, t_0), (u_0, w_0, t_0) *and* (u_0, ζ_0, t_0), *respectively.*

In order to prove this result one splits the function F from (6.14b) using an approach similar to the one depicted in Figure 6.1. Two applications of the implicit function theorem yield the mappings z and w. The function

$$\mathsf{f}(u, w, t) = \big(D\bar{P}_1 D^- \big)'(t)\Big(u + \mathsf{v}(u, t) \Big) \qquad\qquad (6.16)$$
$$- \big(D\bar{P}_1 \bar{G}_2^{-1} \big)(t)\, b\Big(D^-(t)u + \mathsf{z}(u, t) + T(t)w, t \Big)$$

can be written in terms of the original data. The mapping $\mathsf{v}(u, t) = D(t)\mathsf{z}(u, t)$ is given once z is known. The detailed proof of Lemma 6.3 will be carried out in Section 6.3.

6.1.2 Local Existence and Uniqueness of DAE Solutions

From now on we drop the assumption that there is a solution of (6.6). However, we assume that the triple $(y^0, x^0, t_0) \in \operatorname{im} D(t_0) \times \mathcal{D} \times \mathcal{I}$ satisfies

(A2) $A(t_0)y^0 + b(x^0, t_0) = 0$.

This initialisation (y^0, x^0, t_0) doesn't need to be consistent, i.e. apriori we do not require that there is a solution passing through x^0. However, we will use (y^0, x^0, t_0) for the construction of a consistent initialisation (y_0, x_0, t_0). This process can be compared to the step-by-step construction of consistent initial values in [61].

Additionally assume that there is a mapping

(A3) $\bar{x} \in C^1(\mathcal{I}, \mathbb{R}^m), \qquad \bar{x}(t_0) = x^0, \qquad \big(\bar{x}(t), t\big) \in \mathcal{D} \times \mathcal{I} \quad \forall \ t \in \mathcal{I}$.

Then the matrix functions defined in (6.11) are used to introduce

$$u_0 = D(t_0)\bar{P}_1(t_0)x^0, \qquad w_0 = T(t_0)x^0, \qquad z_0 = \bar{Z}(t_0)x^0. \tag{6.17}$$

Notice that $x^0 = D^-(t_0)u_0 + z_0 + w_0$ holds by construction. We will study the equation

$$F(u, w, z, \eta, \zeta, t) = 0 \tag{6.18}$$

without assuming that there is a solution of the original DAE. Observe that the parameters η and ζ replace the derivatives u' and v', respectively.

The results obtained in this section are based on the proof of Lemma 6.3. Hence the results will only be quoted here. Full proofs are given in Section 6.3. The first statement provides a function $\mathbf{z}$ similar to the one obtained in Lemma 6.3.

Lemma 6.4. *Let (6.6) be a regular DAE with index 2 on $\mathcal{D} \times \mathcal{I}$. Assume that the structural condition (A1) as well as (A2) and (A3) hold. Then the function*

$$\bar{Z}(t)\bar{G}_2^{-1}(t)F(u, w, z, \eta, \zeta, t) + \big(I - \bar{Z}(t)\big)z =: \hat{F}_1(u, z, t)$$

is independent of w, η, ζ and there is $r_z > 0$ and a continuous function

$$\mathbf{z} : B_{r_z}(u_0, t_0) \to \mathbb{R}^m, \qquad \qquad with \qquad \qquad \mathbf{z}(u_0, t_0) = z_0,$$

such that $\hat{F}_1\big(u, \mathbf{z}(u, t), t\big) = 0$ for every $(u, t) \in B_{r_z}(u_0, t_0)$.

When proving this lemma, the assumption (A1) is crucial as it guarantees that $\hat{F}_1$ is indeed independent of w (see Section 6.3 for more details).

Using the function $\mathbf{z}$ from Lemma 6.4 we introduce $\mathbf{v}(u, t) = D(t)\mathbf{z}(u, t)$ and define $\mathbf{f}(u, w, t)$ as in (6.16). Recall that we had $u'(t) = \mathbf{f}\big(u(t), w(t), t\big)$ in Lemma 6.3. The function $\mathbf{f}$ obtained here will have the same significance. Thus in (6.18) we replace η by $\mathbf{f}$ and z by $\mathbf{z}$, respectively. The resulting equation is studied in the next lemma.

Lemma 6.5. *Let* (6.6) *be a regular DAE with index 2 on* $\mathcal{D} \times \mathcal{I}$. *Assume that the structural condition* (A1) *as well as* (A2) *and* (A3) *hold. Consider*

$$\hat{F}_2(u, w, \zeta, t) = T(t)\bar{G}_2^{-1}(t)F\big(u, w, \mathbf{z}(u,t), \mathbf{f}(u,w,t), \zeta, t\big) + \big(I - T(t)\big)w$$

where $\mathbf{z}$ *is the mapping from Lemma 6.4 and* $\mathbf{f}$ *is given by* (6.16). *If we define* $\zeta_0 = y^0 - \mathbf{f}(u_0, w_0, t_0)$, *then there is* $r_w > 0$ *and a continuous function*

$$\mathbf{w} : B_{r_w}(u_0, \zeta_0, t_0) \to \mathbb{R}^m, \qquad with \qquad \mathbf{w}(u_0, \zeta_0, t_0) = w_0,$$

such that $\hat{F}_2\big(u, \mathbf{w}(u, \zeta, t), \zeta, t\big) = 0$ *for every* $(u, \zeta, t) \in B_{r_w}(u_0, \zeta_0, t_0)$.

The mappings $\mathbf{z}$, $\mathbf{v}$, $\mathbf{f}$ and $\mathbf{w}$ introduced above allow the construction of a solution. To this end we need to consider the following system of differential algebraic equations

$$z = \mathbf{z}(u, t), \qquad\qquad v = \mathbf{v}(u, t) = D(t)\mathbf{z}(u, t),$$
$$u' = \mathbf{f}(u, w, t), \qquad\qquad w = \mathbf{w}(u, v', t).$$

Inserting $\mathbf{w}$ into $\mathbf{f}$, it turns out that we have to deal with the implicit DAE

$$u' = \mathbf{f}\big(u, \mathbf{w}(u, v', t), t\big) =: f(u, v', t) \tag{6.19a}$$
$$v = D(t)\mathbf{z}(u, t) \qquad =: g(u, t). \tag{6.19b}$$

Once u and v are known, the remaining components of the solution $x = D^- u + z + w$ are given by $z = \mathbf{z}(u, t)$, $w = \mathbf{w}(u, v', t)$.

Lemma 6.6. *Let* (6.6) *be a regular DAE with index 2 on* $\mathcal{D} \times \mathcal{I}$. *Assume that the structural condition* (A1) *as well as* (A2) *and* (A3) *hold. Let* $\mathbf{z}$, $\mathbf{f}$ *and* $\mathbf{w}$ *be the functions obtained from Lemma 6.4 and 6.5.*

(a) *The implicit DAE* (6.19) *has at most (differentiation) index 1.*

(b) *For every consistent initial condition* $\big(u(t_0), v(t_0)\big) = \big(u_0, g(u_0, t_0)\big)$ *there is a unique solution of* (6.19).

(c) *If* $u_0 \in \operatorname{im} D(t_0)\bar{P}_1(t_0)$, *then* $u(t) \in \operatorname{im} D(t)\bar{P}_1(t)$ *for every* t *where the solution exists.*

Proof. To see (a) it suffices to note that $I - \frac{\partial f}{\partial v'}\frac{\partial g}{\partial u} = I - f_{v'}g_u$ is nonsingular in a neighbourhood of (u_0, ζ_0, t_0) (see Remark 6.10 later on page 108). Of course, if $f_{v'}$ vanishes, then (6.19a) represents an ordinary differential equation.

In order to prove (b) differentiate (6.19b) and insert the result into (6.19a). This yields

$$u' = f\big(u, g_u(u, t)u' + g_t(u, t), t\big) = \hat{f}(u, u', t)$$

and due to (a) it is possible to solve for u'. Thus (6.19b) is equivalent to an ordinary differential equation $u' = \mathcal{F}(u, t)$. It remains to solve the initial value problem $u' = \mathcal{F}(u, t)$, $u(t_0) = u_0$ to see that $\big(u(t), g(u(t), t)\big)$ is the unique solution.

When proving (c), ideas can be employed that were originally used to study the case of linear DAEs [116]. Let (u, v) be a solution of the index-1 system (6.19) satisfying $u(t_0) \in \operatorname{im} D(t_0)\bar{P}_1(t_0)$. Multiplication of (6.19a) by the projector function $I - D\bar{P}_1 D^-$ yields

$$(I - D\bar{P}_1 D^-)(t)\, u'(t) = -\bigl(I - D\bar{P}_1 D^-\bigr)'(t)\,\bigl(D\bar{P}_1 D^-\bigr)(t)\, u(t)$$

since $D\bar{P}_1 D^- \mathsf{v}(u, \cdot) = D\bar{P}_1 P_0 \mathsf{z}(u, \cdot) = D\bar{P}_1 P_0 \bar{Z}\mathsf{z}(u, \cdot) = 0$. This means that $\hat{u} = (I - D\bar{P}_1 D^-)u$ satisfies the linear ODE

$$\hat{u}' = \bigl(I - D\bar{P}_1 D^-\bigr)'\hat{u}. \tag{6.20}$$

The initial condition $u(t_0) = u_0 \in \operatorname{im} D(t_0)\bar{P}_1(t_0)$ implies $\hat{u}(t_0) = 0$ and the solution of (6.20) is identically zero, i.e. $u(t) \in \operatorname{im} D(t)\bar{P}_1(t)$ for every t. $\square$

Starting from Lemma 6.6 it is straightforward to construct a solution of the original index-2 system (6.6). We collect the result in the following theorem.

Theorem 6.7. *Let* (6.6) *be a regular index-2 DAE on* $\mathcal{D}\times\mathcal{I}$ *with a properly stated leading term. Assume that the following conditions hold:*

(A1) $N_0(t) \cap S_0(x, t)$ *does not depend on* x

(A2) $\exists\, (y^0, x^0, t_0) \in \operatorname{im} D(t_0)\times\mathcal{D}\times\mathcal{I}$ *such that* $A(t_0)y^0 + b(x^0, t_0) = 0$,

(A3) $\exists\, \bar{x} \in C^1(\mathcal{I}, \mathbb{R}^m)$ *such that* $\bar{x}(t_0) = x^0$ *and* $\bigl(\bar{x}(t), t\bigr) \in \mathcal{D}\times\mathcal{I} \ \ \forall\, t \in \mathcal{I}$,

(A4) *the derivatives* $\frac{\partial b}{\partial x}$, $\frac{\mathrm{d}D}{\mathrm{d}t}$ *and* $\frac{\partial}{\partial t}\bigl(\bar{Z}\bar{G}_2^{-1}b\bigr)$ *exist and are continuous.*

Then there is a unique local solution of the initial value problem

$$A(t)\bigl[D(t)x(t)\bigr]' + b\bigl(x(t), t\bigr) = 0, \qquad D(t_0)P_1(x^0, t_0)\bigl(x(t_0) - x^0\bigr) = 0. \tag{6.21}$$

Proof. The mappings z and $\mathsf{v}(u, t) = D(t)\mathsf{z}(u, t)$ are obtained from Lemma 6.4. Similarly, let w be the mapping defined in Lemma 6.5. Then due to Lemma 6.6 there is a local solution of the implicit index-1 system

$$\begin{aligned} u' &= \mathbb{f}\bigl(u, \mathsf{w}(u, v', t), t\bigr), \qquad u(t_0) = u_0 := D(t_0)P_1(x^0, t_0)x^0, \\ v &= D(t)\mathsf{z}(u, t), \qquad\qquad\ v(t_0) = g(u_0, t_0) \end{aligned}$$

existing for $t \in \mathcal{I}_\varepsilon = (t_0 - \varepsilon, t_0 + \varepsilon) \cap \mathcal{I}$ with some $\varepsilon > 0$. Using this solution $\bigl(u(t), v(t)\bigr)$ we define

$$x_*(t) = D^-(t)u(t) + \mathsf{z}\bigl(u(t), t\bigr) + \mathsf{w}\bigl(u(t), v'(t), t\bigr). \tag{6.22}$$

It remains to check that (6.22) is indeed a solution.

Due to $u_0 \in \operatorname{im}(D\bar{P}_1)(t_0)$ and Lemma 6.6 we have $u(t) \in \operatorname{im}(D\bar{P}_1)(t)$ for every $t \in \mathcal{I}_\varepsilon$. Recall that $R(t) = (DD^-)(t)$ is the projector function related to the properly stated leading term and $Ru = u$ shows that

$$Dx_* = Ru + D\bar{Z}\mathsf{z}(u, \cdot) + DT\mathsf{w}(u, \cdot) = u + \mathsf{v}(u, \cdot).$$

Therefore (A4) ensures that Dx_* is a C^1 mapping[2]. In particular we have $D\bar{P}_1 x_* = u$, $\bar{Z}x_* = \mathsf{z}(u, \cdot)$ and $Tx_* = \mathsf{w}(u, v', \cdot)$ and, finally,

$$\bar{Z}\bar{G}_2^{-1}[A(Dx_*)' + b(x_*, \cdot)] = \hat{F}_1\big(u, \mathsf{z}(u, \cdot), \cdot\big) \qquad = 0,$$
$$T\bar{G}_2^{-1}[A(Dx_*)' + b(x_*, \cdot)] = \hat{F}_2\big(u, \mathsf{w}(u, v', \cdot), v', \cdot\big) \qquad = 0,$$
$$P_0\bar{P}_1\bar{G}_2^{-1}[A(Dx_*)' + b(x_*, \cdot)] = D^-\Big(u' - \mathsf{f}\big(u, \mathsf{w}(u, v', \cdot), \cdot\big)\Big) = 0$$

(see Remark 6.11 later on). Due to the splitting $I = \bar{Z} + T + P_0\bar{P}_1$ of the identity, we can conclude that $A(Dx_*)' + b(x_*, \cdot) = 0$ such that x_* is indeed a solution of (6.6). Since $D(t_0)P_1(x^0, t_0)x_*(t_0) = u(t_0) = u_0 = D(t_0)P_1(x^0, t_0)x^0$, this solution satisfies the initial value problem (6.21).

If there was another solution, say $\hat{x}_*$, then one could decouple x_* and $\hat{x}_*$ as described in Section 6.1.1. Therefore the corresponding $D\bar{P}_1$ parts u and $\hat{u}$ solve the same inherent index-1 system (6.19) and are therefore equal. Because of Lemma 6.3, z and $\hat{z}$ as well as w and $\hat{w}$ are also equal, respectively, such that x_* and $\hat{x}_*$ coincide.   □

The smoothness required in order to be able to construct the solution, is given when the function v is differentiable with respect to t. The condition (A4) on D and $\bar{Z}\bar{G}_2^{-1}b$ in Theorem 6.7 guarantees this fact but it may be unnecessary strong in general.

The theorem is given for index-2 DAEs. Using $Q_1 = 0$ and $P_1 = I$ it turns out that for index-1 equations the component v vanishes because of $D\bar{Z} = 0$. Thus (6.19a) reduces to the inherent regular ordinary differential equation (4.8b) and compared to Theorem 4.10 no additional smoothness is required.

Example 6.8. For illustration we want to employ the decoupling procedure described above for explicitly constructing a solution of the DAE

$$\begin{array}{r} x_1' + x_2^2 - x_3(1 + x_3) = g_{\varepsilon, t_*}(t) \\ x_2' \qquad\qquad - x_3 = g_{\varepsilon, t_*}(t) \\ x_1 - x_2 = 0 \end{array} \;\Leftrightarrow\; \begin{bmatrix} 1 & 0 \\ 0 & 1 \\ 0 & 0 \end{bmatrix}\left(\begin{bmatrix} 1 & 0 & 0 \\ 0 & 1 & 0 \end{bmatrix}\begin{bmatrix} x_1 \\ x_2 \\ x_3 \end{bmatrix}\right)' + \begin{bmatrix} x_2^2 - x_3(1 + x_3) - g_{\varepsilon, t_*}(t) \\ -x_3 - g_{\varepsilon, t_*}(t) \\ x_1 - x_2 \end{bmatrix} = 0.$$

The mapping g_{ε, t_*} appearing on the right-hand side is continuous but not differentiable,

$$g_{\varepsilon, t_*}(t) = \begin{cases} 0 & , \quad 0 \le t < t_* - \varepsilon, \\ \frac{t - t_*}{2\varepsilon} + \frac{1}{2} & , \quad t_* - \varepsilon \le t < t_* + \varepsilon, \\ 1 & , \quad t_* + \varepsilon \le t. \end{cases}$$

When simulating electrical circuits, piecewise linear function such as g_{ε, t_*} with small $\varepsilon > 0$ are often used to model independent sources that are switched on

---

[2]Due to the construction of z, the partial derivative v_u always exists. Since $\bar{Z}\bar{G}_2^{-1}b$ is smooth, $\phi_u(t) = \mathsf{z}(u, t)$ is a C^1-mapping for every fixed u. As $D \in C^1$ as well, we have $\phi_u \in C_D^1$. Thus the partial derivative v_t exists and is continuous.

at $t = t_*$. In general a treatment on the subintervals $[0, t_*)$ and $[t_*, \infty)$ is not feasible, as the switching point t_* may not be known in advance.

According to (A2) we need an initialisation to start with. Here we may use

$$(y^0, x^0, t_0) = \left(\begin{bmatrix} -\frac{5}{4} \\ -\frac{1}{2} \end{bmatrix}, \begin{bmatrix} 1 \\ 1 \\ -\frac{1}{2} \end{bmatrix}, 0 \right) \qquad \text{since} \qquad Ay^0 + b(x^0, t_0) = 0.$$

It is straightforward to calculate the matrix sequence (6.7) as

$$G_0 = \begin{bmatrix} 1 & 0 & 0 \\ 0 & 1 & 0 \\ 0 & 0 & 0 \end{bmatrix}, \quad B = \begin{bmatrix} 0 & 2x_2 & -2x_3-1 \\ 0 & 0 & -1 \\ 1 & -1 & 0 \end{bmatrix}, \quad N_0 = \text{span} \left\{ \begin{bmatrix} 0 \\ 0 \\ 1 \end{bmatrix} \right\}, \quad Q_0 = \begin{bmatrix} 0 & 0 & 0 \\ 0 & 0 & 0 \\ 0 & 0 & 1 \end{bmatrix}$$

$$G_1 = G_0 + BQ_0 = \begin{bmatrix} 1 & 0 & -2x_3-1 \\ 0 & 1 & -1 \\ 0 & 0 & 0 \end{bmatrix}, \qquad N_1 = \text{span} \left\{ \begin{bmatrix} 2x_3+1 \\ 1 \\ 1 \end{bmatrix} \right\}$$

$$S_0 = \text{span} \left\{ \begin{bmatrix} 1 \\ 1 \\ 0 \end{bmatrix}, \begin{bmatrix} 0 \\ 0 \\ 1 \end{bmatrix} \right\} = S_1$$

As $N_0 \cap S_0 = N_0$ is independent of x, the structural condition (A1) holds. A projector onto this subspace is given by $T = \begin{bmatrix} 0 & 0 & 0 \\ 0 & 0 & 0 \\ 0 & 0 & 1 \end{bmatrix}$.

Also, due to $N_1 \cap S_1 = \{0\}$ for $x_3 \neq 0$, the index is 2. Calculating the canonical projector $Q_1 = \frac{1}{2x_3} \begin{bmatrix} 2x_3+1 & -2x_3-1 & 0 \\ 1 & -1 & 0 \\ 1 & -1 & 0 \end{bmatrix}$ onto N_1 along S_1 we find that

$$G_2(x^1, x, t) = \frac{1}{2x_3} \begin{bmatrix} 2x_3(x_2+x_3)-x_3^1 & -2x_2x_3+x_3^1 & -2x_3(2x_3+1) \\ -x_3^1 & 2x_3+x_3^1 & -1 \\ 2x_3 & -2x_3 & 0 \end{bmatrix}.$$

This matrix depends on x, t and the auxiliary variable x^1. Choosing $\bar{x}(t) \equiv x^0$, such that (A3) is satisfied, we find

$$\bar{G}_2(t) = \begin{bmatrix} -1 & 2 & 0 \\ 0 & 1 & -1 \\ 1 & -1 & 0 \end{bmatrix}, \qquad \bar{Q}_1 = \begin{bmatrix} 0 & 0 & 0 \\ -1 & 1 & 0 \\ -1 & 1 & 0 \end{bmatrix}, \qquad \bar{Z} = \begin{bmatrix} 0 & 0 & 0 \\ -1 & 1 & 0 \\ 0 & 0 & 0 \end{bmatrix}.$$

Instead of the original DAE we turn to investigate (6.18) written in terms of the new variables u, w, z, η, ζ and t,

$$F(u, w, z, \eta, \zeta, t) = \begin{bmatrix} \eta_1 + \zeta_1 - w_3{}^2 - w_3 + (u_2 - z_1 + z_2)^2 - g_{\varepsilon, t_*}(t) \\ \eta_2 + \zeta_2 - w_3 - g_{\varepsilon, t_*}(t) \\ -u_2 + z_1 - z_2 + u_1 \end{bmatrix} = 0.$$

Using the matrix functions defined above we split this mapping into different parts in order to determine $\mathbf{z}$, $\mathbf{f}$ and $\mathbf{w}$. Observe that the function

$$\hat{F}_1(u, z, t) = \begin{bmatrix} z_1 \\ u_2 - u_1 + z_2 \\ z_3 \end{bmatrix} = 0$$

as defined in Lemma 6.4 takes a quite simple form here. In particular it allows the determination of $z = \mathbf{z}(u, t) = [0, \ u_1 - u_2, \ 0]^{\top}$ and therefore we get $\mathbf{v}(u, t) = D(t)\mathbf{z}(u, t) = [0, \ u_1 - u_2]^{\top}$. Notice that $\mathbf{v}(u, t)$ is continuously differentiable with respect to both arguments. The mapping $\mathbf{f}(u, w, t) = \begin{bmatrix} w_3{}^2 + w_3 - u_1{}^2 + g_{\varepsilon, t_*}(t) \\ w_3{}^2 + w_3 - u_1{}^2 + g_{\varepsilon, t_*}(t) \end{bmatrix}$ is defined according to (6.16) and

$$\hat{F}_2(u, w, \zeta, t) = \begin{bmatrix} w_1 \\ w_2 \\ \zeta_1 - \zeta_2 - w_3^2 + u_1^2 \end{bmatrix} = 0$$

from Lemma 6.5 fixes the component $w = \mathsf{w}(u, \zeta, t) = [0,\ 0,\ -\sqrt{\zeta_1 - \zeta_2 + u_1^2}]^\top$. Observe that $u_0 = [1\ 1]^\top$, $w_0 = [0\ 0\ -\tfrac{1}{2}]^\top$ and thus $\zeta_0 = y^0 - \mathsf{f}(u_0, w_0, t_0) = [0, \tfrac{3}{4}]^\top$. Hence we needed to choose the negative sign for the root in order to guarantee $w_0 = \mathsf{w}(u_0, \zeta_0, t_0)$.

Finally we arrive at the implicit index-1 DAE

$$u' = \mathsf{f}\big(u, \mathsf{w}(u, v', \cdot), \cdot\big) = \begin{bmatrix} 1 \\ 1 \end{bmatrix} \big[g_{\varepsilon, t_*} + v_1' - v_2' - \sqrt{v_1' - v_2' + u_1^2} \big], \qquad u(t_0) = u_0 = \begin{bmatrix} 1 \\ 1 \end{bmatrix},$$

$$v = D\mathsf{z}(u, \cdot) \qquad\quad = \begin{bmatrix} 0 \\ u_1 - u_2 \end{bmatrix}, \qquad\qquad\qquad\qquad v(t_0) = \begin{bmatrix} 0 \\ 0 \end{bmatrix}.$$

The t arguments were omitted for better readability. Obviously $v(t) \equiv \begin{bmatrix} 0 \\ 0 \end{bmatrix}$ is uniquely determined by the initial data and we have to consider the ordinary differential equation $u'(t) = \begin{bmatrix} 1 \\ 1 \end{bmatrix} \big[g_{\varepsilon, t_*}(t) - u_1(t) \big]$, $u(t_0) = u_0$. The unique solution is given by

$$u_1(t) = u_2(t) = \begin{cases} \alpha(t) & , & 0 \le t < t_* - \varepsilon \\ \alpha(t) + \beta(t) & , & t_* - \varepsilon \le t < t_* + \varepsilon \\ \alpha(t) + \beta(t) + \gamma(t) & , & t_* + \varepsilon \le t \le \infty \end{cases}$$

where the functions α, β and γ are defined by

$$\alpha(t) = e^{-t}, \quad \beta(t) = \tfrac{1}{2} + \tfrac{1}{2\varepsilon}\big[t - t_* - 1 + e^{t_* - \varepsilon - t} \big], \quad \gamma(t) = 1 + \tfrac{1}{2\varepsilon}\big[e^{t_* - \varepsilon - t} - e^{t_* + \varepsilon - t} \big].$$

Following (6.22) we find

$$x_*(t) = D^-(t)u(t) + \mathsf{z}\big(u(t), t\big) + \mathsf{w}\big(u(t), v'(t), t\big) = u_1(t) \begin{bmatrix} 1 \\ -1 \end{bmatrix}.$$

Direct computation shows that x_* is indeed a solution. Notice that there was no need for the initialisation to be consistent. The (inconsistent) initialisation $(y^0, x^0, t_0) = \big([-\tfrac{5}{4}, -\tfrac{1}{2}]^\top, [1,\ 1, -\tfrac{1}{2}]^\top, 0\big)$ was used to start with, but during the course of the above computations a consistent initialisation $(y_0, x_0, t_0) = \big([-1, -1]^\top, [1,\ 1, -1]^\top, 0\big)$ was obtained. However, x_* satisfies the initial condition $D(t_0)\bar{P}_1(t_0)\big(x_*(t_0) - x_0\big) = 0$ as stated in Theorem 6.7. The solution obtained for $t_* = 1$ and $\varepsilon = \tfrac{1}{4}$ is depicted in Figure 6.2.   □

Even though the decoupling procedure can be used to actually solving a DAE, it is not meant to be used for this purpose. The above example illustrates how the different functions interrelate and how the implicit function theorem is used

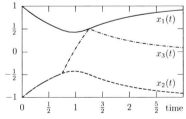

Figure 6.2: The components of the solution x for $t_* = 1$ and $\varepsilon = \tfrac{1}{4}$. Observe that x is only continuous, but $Dx = [x_1, x_2]^\top$ is smooth.

to solve for different components of the solution. For practical computations these manipulations are not recommended.

Nevertheless the decoupling procedure is an invaluable tool for proving the existence and uniqueness of solutions for nonlinear DAEs (6.6) making only mild smoothness assumptions. In Part III of this thesis we will see how the decoupling procedure can be successfully used to study numerical methods for index-2 DAEs as well. Of course there will be no explicit decoupling when doing serious computations. However, one has to ensure that a given method, when applied to (6.6), behaves as if it was integrating the index-1 system (6.19). This behaviour guarantees that qualitative properties of the solution are correctly reflected by the numerical approximation. In the context of linear DAEs an example of this situation was given in Section 3.3.

Before proving the obtained results in Section 6.3 and then moving on to studying numerical methods for nonlinear DAEs, some brief remarks about this new decoupling procedure are due.

6.1.3 Remarks about the Decoupling Procedure

In contrast to the case of linear DAEs [116, 120] or nonlinear index-1 equations [88] we did not obtain an inherent ordinary differential equation, but we derived the implicit DAE system (6.19) that governs the dynamical behaviour. Using the concept of the differentiation index it turned out that (6.19) has index 1. Of course, the transformation of higher index DAEs to equivalent ones having index 1 is a well-known theme in the theory of differential algebraic equations. But in contrast to classical approaches [10, 44, 99], we didn't use the derivative array at all. The concept of the tractability index made a more refined analysis of index-2 DAEs possible leading to lower smoothness requirements. This is of vital importance for practical applications.

The implicit system (6.19) is neither in Hessenberg form nor formulated with a properly stated leading term. It is easily seen that reformulations that fit into these classes of equations will have index 2 again. Thus one has to be careful when reformulating the implicit system and it turns out to be advantageous to consider (6.19) directly. A detailed study of general linear methods for (6.19) will be presented in Chapter 8.

The general approach taken here provides a unifying framework for the different classes of DAEs previously considered. Indeed, it turns out that the results given for

- linear DAEs having index 1 or 2,
- nonlinear index-1 equations and
- index-2 DAEs having the special structure (6.3)

can all be recovered as special cases.

Linear DAEs: If the DAE (6.6) is linear, then the sequence (6.7) depends neither on x nor on x^1. Hence the bar-notation introduced in (6.11) can be neglected. For linear DAEs the mapping $\mathfrak{f}$ defined in (6.16) therefore takes the form

$$\mathfrak{f}(u,w,\cdot) = (DP_1D^-)'u + (DP_1D^-)'\mathfrak{v}(u,\cdot) - DP_1G_2^{-1}BD^-u + DP_1G_2^{-1}q$$
$$- DP_1G_2^{-1}BZ\mathfrak{z}(u,\cdot) - DP_1G_2^{-1}BTw.$$

Using the properties (6.8) of U, T as well as the definition of B_1 from (6.7b) it turns out that

$$\begin{aligned}
DP_1G_2^{-1}BT &= DP_1G_2^{-1}BQ_0T = DP_1G_2^{-1}G_1Q_0T = DP_1G_2^{-1}G_2P_1Q_0T \\
&= DP_1T = DT - DQ_1T = 0, \\
DP_1G_2^{-1}BZ &= DP_1G_2^{-1}B(P_0Q_1 + UQ_0) \\
&= DP_1G_2^{-1}(BP_0)Q_1 + DP_1G_2^{-1}B(Q_0 - T) \\
&= DP_1G_2^{-1}(B_1 + G_2P_1D^-(DP_1D^-)'D)Q_1 + DP_1Q_0 \\
&= DP_1G_2^{-1}B_1Q_1 + DP_1D^-(DP_1D^-)'DQ_1 \\
&= (DP_1D^-)'DQ_1.
\end{aligned}$$

Hence the first equation (6.19a) of the implicit index-1 system reduces to

$$u' = (DP_1D^-)'u - DP_1G_2^{-1}BD^-u + DP_1G_2^{-1}q.$$

This is precisely the inherent regular ODE (3.11a'') for $\mu = 2$ from Section 3.2. It is important to notice, that $DP_1G_2^{-1}BT = 0$ ensures that the right-hand side is independent of w such that there is no coupling between the components u and v anymore.

Nonlinear index-1 equations: It was already remarked earlier, that in case of index-1 equations $Q_1 = 0$ and $P_1 = I$ imply that the component $\mathfrak{v}$ vanishes due to $D\bar{Z} = 0$. Therefore (6.19a) reduces to the inherent regular ODE (4.8b) as constructed in Section 4.2.

Nonlinear index-2 DAEs with the special structure (6.3): Given that the index-2 components $Tx = w$ enter the equations linearly, $\mathfrak{f}(u,w,\cdot)$ is again independent of w. This can be seen by computing

$$\begin{aligned}
\frac{\partial \mathfrak{f}}{\partial w}(u,w,\cdot) &= -D\bar{P}_1\bar{G}_2^{-1}\big(b_x(D^-u + \mathfrak{z}(u,\cdot),\cdot)U + \mathfrak{B}T\big)T \\
&= -D\bar{P}_1\bar{G}_2^{-1}\big(b_x(Ux_0,\cdot)U + \mathfrak{B}T\big)T = -D\bar{P}_1\bar{G}_2^{-1}\bar{B}T = 0.
\end{aligned}$$

Consequently, (6.19a) reduces to the ordinary differential equation

$$u' = (D\bar{P}_1D^-)'(u + \mathfrak{v}(u,t)) - D\bar{P}_1\bar{G}_2^{-1}b\big(D^-(t)u + \mathfrak{z}(u,t),\cdot\big). \tag{6.23}$$

In order to see that this equation coincides with the inherent regular ODE (5.9b), take $\bar{x} \equiv x^0$ and consider the solution x_* constructed in (6.22). Defining

$$\hat{w}(u, \cdot) = \bar{P}_1 D^- (Dx_*)' + (Q_0 + \bar{Q}_1)x_*$$
$$= \bar{P}_1 D^- \big(u + v(u, \cdot)\big)' + (Q_0 + \bar{Q}_1)z(u, \cdot) + Tw(u, v', \cdot)$$

we see that $\hat{w}$ is the unique solution of (5.8) (see Lemma 5.3 for more details). Using (6.23) one calculates

$$D\bar{P}_1\hat{w}(u, \cdot) = D\bar{P}_1 D^- \big(u + v(u, \cdot)\big)' + \underbrace{D\bar{P}_1 Q_0}_{=0} z(u, \cdot) + \underbrace{D\bar{P}_1 T}_{=0} w(u, v', \cdot)$$
$$= u' - (D\bar{P}_1 D^-)'(u + v(u, \cdot)) = -D\bar{P}_1\bar{G}_2^{-1} b\big(D^-(t)u + z(u, t), \cdot\big),$$
$$D\bar{Q}_1\hat{w}(u, \cdot) = D\bar{Q}_1\hat{z}(u, \cdot) = v(u, \cdot).$$

and comparing (5.9b) with (6.23) we see that the inherent regular ODE (5.9b) constructed in Section 5.1 is indeed just a special case of (6.19).

6.2 Application to Hessenberg DAEs

The important class of DAEs in Hessenberg form

$$\begin{aligned} y' &= f(y, z), \\ 0 &= g(y) \end{aligned} \qquad \Leftrightarrow \qquad \begin{bmatrix} I \\ 0 \end{bmatrix} \left(\begin{bmatrix} I & 0 \end{bmatrix} \begin{bmatrix} x_1 \\ x_2 \end{bmatrix} \right)' + \begin{bmatrix} -f(x_1, x_2) \\ g(x_1) \end{bmatrix} = 0 \qquad (6.24)$$

was already considered in Example 6.1. In this section we want to explore how the decoupling procedure introduced above applies to DAEs having this special Hessenberg structure. It is assumed that (6.24) has index 2 such that $g_y f_z$ remains nonsingular in the neighbourhood of a solution. Recall from Example 6.1 that the matrix sequence reads

$$G_0 = \begin{bmatrix} I & 0 \\ 0 & 0 \end{bmatrix}, \quad Q_0 = \begin{bmatrix} 0 & 0 \\ 0 & I \end{bmatrix}, \quad D^- = \begin{bmatrix} I \\ 0 \end{bmatrix}, \quad G_1 = \begin{bmatrix} I & -f_z \\ 0 & 0 \end{bmatrix}, \quad T = \begin{bmatrix} 0 & 0 \\ 0 & I \end{bmatrix} = Q_0.$$

A projector onto the kernel of G_1 is given by

$$Q_1 = \begin{bmatrix} f_z(g_y f_z)^{-1} g_y & 0 \\ (g_y f_z)^{-1} g_y & 0 \end{bmatrix}$$

and it turns out that[3]

$$G_2 = \begin{bmatrix} I - f_y f_z (g_y f_z)^{-1} g_y & -f_z \\ g_y & 0 \end{bmatrix}, \qquad Z = P_0 Q_1 + UQ_0 = \begin{bmatrix} f_z(g_y f_z)^{-1} g_y & 0 \\ 0 & 0 \end{bmatrix}.$$

Obviously the projector $\mathcal{Z} = f_z(g_y f_z)^{-1} g_y$ appears frequently, e.g.

$$G_2^{-1} = \begin{bmatrix} I - \mathcal{Z} & (I - \mathcal{Z})f_y \mathcal{Z} + f_z(g_y f_z)^{-1} \\ -(g_y f_z)^{-1} g_y & (g_y f_z)^{-1}(I - g_y f_y \mathcal{Z}) \end{bmatrix}.$$

---

[3]Instead of $G_2 = G_1 + B_1 Q_1$ with $B_1 = BP_0 - G_1 D^- CD$ (cf. (6.7)) the modified version $\tilde{G}_2 = G_1 + \tilde{B}_1 Q_1$, $\tilde{B}_1 = BP_0$ is used here in order to simplify expressions. From the proof of Lemma 5.3 we know that G_2 and $\tilde{G}_2$ have common rank.

It is straightforward to apply the decoupling procedure described in the previous section. For simplicity choose $\bar{x} \equiv x_0 = \left[\begin{smallmatrix} y_0 \\ z_0 \end{smallmatrix}\right]$ and recall from Lemma 6.4 that

$$\hat{F}_1(u,z) = \bar{Z}\bar{G}_2^{-1}b\big(D^- u + \bar{Z}z\big) + (I - \bar{Z})z = \left[\begin{smallmatrix} \bar{f}_z(\bar{g}_y\bar{f}_z)^{-1}g(u+\bar{Z}z_1)+(I-\bar{Z})z_1 \\ z_2 \end{smallmatrix}\right].$$

The bar notation indicates evaluation at $\bar{x} = x_0$ and u, $z = \left[\begin{smallmatrix} z_1 \\ z_2 \end{smallmatrix}\right]$ are newly introduced parameters. Solving $\hat{F}_1(u,z) = 0$ for z yields the map z such that $\hat{F}_1\big(u, \mathsf{z}(u)\big) = 0$. The function $\mathfrak{f}$ from (6.16) is given by

$$\mathfrak{f}(u,w) = (I - \bar{Z})\big[\bar{f}_y\bar{Z}g\big(u + \mathsf{z}_1(u)\big) - f\big(u + \mathsf{z}_1(u), w_2\big)\big].$$

Finally the mapping $\hat{F}_2$ from Lemma 6.5 is obtained by straightforward computation,

$$\hat{F}_2\big(u,w,\zeta\big) = \begin{bmatrix} I & 0 \\ 0 & (\bar{g}_y\bar{f}_z)^{-1} \end{bmatrix} \left[\begin{smallmatrix} w_1 \\ \bar{g}_y(f(u+\mathsf{z}(u),w_2)-\zeta_1)+(I-\bar{g}_y\bar{f}_y\bar{Z})g(u+\mathsf{z}(u)) \end{smallmatrix}\right].$$

Solving $\hat{F}_2(u,w,\zeta) = 0$ for $w = \mathsf{w}(u,\zeta)$ we arrive at the inherent index-1 system

$$u' = (I - \bar{Z})\big[\bar{f}_y\bar{Z}g\big(u + \mathsf{z}_1(u)\big) - f\big(u + \mathsf{z}_1(u), \mathsf{w}_2(u,v')\big)\big], \tag{6.25a}$$

$$v = D\mathsf{z}(u). \tag{6.25b}$$

It is clear from the above computations that the decoupling procedure can be applied to index-2 DAEs in Hessenberg form. Computing the partial derivative $\frac{\partial \hat{F}_1}{\partial z}(u_0, z_0) = \frac{\partial \hat{F}_2}{\partial w}(u_0, w_0, \zeta_0) = I$ shows that the existence of the functions z and w is guaranteed by the implicit function theorem. Hence the index-2 DAE (6.24) can be decoupled into the index-1 problem (6.25) and explicit representations for $z = \mathsf{z}(u, \cdot)$ and $w = \mathsf{w}(u, v')$. This decoupling is realised without differentiating the equations. Of course, in order to finally compute $w = \mathsf{w}(u, v')$ a differentiation is still necessary.

There is no doubt that (6.25) is far more complicated than the original formulation (6.24). When dealing with problems of Hessenberg type it is therefore advisable to stick to the original formulation (6.24) and to exploit this structure when studying numerical methods. This is done in e.g. in [84, 85, 140].

On the other hand, there are applications that do not give rise to equations in Hessenberg form. One prominent example are the equations of the charge-oriented modified nodal analysis.

In order to address these more general problems that are not of Hessenberg type, one has to invest considerable work such that the inherent structure can be revealed and then exploited for analysing numerical methods. The decoupling procedure is one attempt to go into this direction. We saw that this approach works for the simple DAE in Example 6.8 as well as for DAEs in Hessenberg form. The next section will give proofs for the general case.

6.3 Proofs of the Results

In this section proofs for Lemma 6.3, 6.4 and 6.5 will be given. The proofs presented here are rather technical, but the structure of our approach is clearly visible from Figure 6.3. The DAE (6.6) is rewritten in terms of new variables. The resulting equation (6.18), i.e. $F(u, w, z, \eta, \zeta, t) = 0$, is studied in detail. Notice that the derivatives u' and v' are replaced by parameters η and ζ, respectively. The mapping F is defined in (6.14b) on page 93.

Using the identity $I = \bar{Z} + P_0\bar{P}_1 + T$, this equation is split into three parts that can be dealt with one after the other.

① $\hat{F}_1(u, z, t) = 0$ together with Lemma 6.4 yields the mapping z. Here the structural condition (A1) is crucial as it allows the application of the implicit function theorem.

② Inserting $z = z(u, t)$ into the second part provides an explicit representation $u' = f(u, w, t)$.

③ Using the information about z and $\eta = u'$, the third part can be written as $\hat{F}_2(u, w, \zeta, t) = 0$ such that a further application of the implicit function theorem yields $w = w(u, \zeta, t)$. The details of this construction are covered by Lemma 6.5.

In order to derive the implicit DAE (6.19), we need to plug w back into f keeping in mind that ζ was representing v'.

The procedure sketched above will be carried out in detail. We start by giving a preliminary result.

Lemma 6.9. *Let* (6.6) *be a regular DAE with a properly stated leading term that has index 2 on* $\mathcal{D} \times \mathcal{I}$. *Let* (A1), (A2) *and* (A3) *hold. Then*

$$\bar{G}_2^{-1}(t)A(t)D(t) = \bar{P}_1(t)P_0(t), \qquad G_1(\xi, t)P_0(t) = \big(\bar{G}_2\bar{P}_1P_0\big)(t),$$

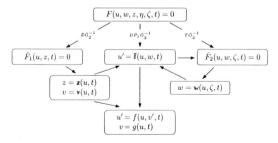

Figure 6.3: The roadmap to the proofs

hold for every $(\xi, t) \in \mathcal{D} \times \mathcal{I}$. *Furthermore we have*

$$\left(T\bar{G}_2^{-1}\right)(t_0)B(x^0, t_0)T(t_0) = T(t_0), \quad \left(\bar{Z}\bar{G}_2^{-1}\right)(t_0)B(x^0, t_0)\bar{Z}(t_0) = \bar{Z}(t_0).$$

If, in addition, the structural condition (A1) *is valid, then*

$$\operatorname{im} B(\xi, t)T(t) \subset \operatorname{im}\left(\bar{G}_2\bar{P}_1 P_0\right)(t) \quad \forall \ (\xi, t) \in \mathcal{D} \times \mathcal{I}.$$

Proof. Since $\bar{G}_2\bar{P}_1 P_0 = AD = G_1(\xi, \cdot)P_0$ for every ξ, the first two equations hold. The second two equations follow from

$$\bar{B}T = \bar{B}Q_0 T = \bar{G}_1 Q_0 T = \bar{G}_2\bar{P}_1 T,$$
$$\bar{Z}\bar{G}_2^{-1}\bar{B}\bar{Z} = P_0\bar{Q}_1\bar{G}_2^{-1}\bar{B}P_0\bar{Q}_1 + UQ_0\bar{G}_2^{-1}\bar{B}P_0\bar{Q}_1 + \bar{Z}\bar{G}_2^{-1}\bar{B}Q_0 UQ_0$$
$$= P_0\bar{Q}_1 + UQ_0 = \bar{Z},$$

respectively[4]. The argument t_0 was dropped for better readability. Notice that for $t \neq t_0$ we get $\bar{G}_2^{-1}(t)G_2(x^1, \xi, t) \neq I$ in general and the second two equations need not hold for $t \neq t_0$.

Given (A1), we have $\operatorname{im} T(t) \subset S_0(\xi, t)$ since $N_0(t) \cap S_0(x, t)$ is independent of x. We find $\operatorname{im} B(\xi, t)T(t) \subset \operatorname{im} G_0(t)$ which proves the last assertion of the lemma. $\quad\square$

Proof of Lemma 6.3. We assume that x_* is a solution of (6.6) and define the functions u, w, z according to (6.12),

$$u(t) = D(t)\bar{P}_1(t)x_*(t), \quad w(t) = T(t)x_*(t), \quad z(t) = \bar{Z}(t)x_*(t). \quad (6.12)$$

We know already that the DAE (6.6) can be equivalently written as

$$0 = f\Big(\big(Dx_*\big)'(t), x_*(t), t\Big) = F\Big(u(t), w(t), z(t), u'(t), v'(t), t\Big), \quad (6.14a)$$

where the mapping F is defined by

$$F(u, w, z, \eta, \zeta, t) = f\Big(\eta + \zeta, \ D^-(t)u + \bar{Z}(t)z + T(t)w, \ t\Big). \quad (6.14b)$$

Using the identity $I = P_0\bar{P}_1 + \bar{Z} + T = D^-D\bar{P}_1 + \bar{Z} + T$ as motivation we split (6.14b) into three parts

$$F_1(u, w, z, \eta, \zeta, t) = \quad \bar{Z}(t)\bar{G}_2^{-1}(t)F(u, w, z, \eta, \zeta, t) + \left(I - \bar{Z}(t)\right)z,$$
$$F_2(u, w, z, \eta, \zeta, t) = \quad T(t)\bar{G}_2^{-1}(t)F(u, w, z, \eta, \zeta, t) + \left(I - T(t)\right)w,$$
$$F_3(u, w, z, \eta, \zeta, t) = D(t)\bar{P}_1(t)\bar{G}_2^{-1}(t)F(u, w, z, \eta, \zeta, t).$$

---

[4]Recall $\bar{Q}_1 = \bar{Q}_1\bar{G}_2^{-1}\bar{B}P_0$, $\bar{B}P_0 = \bar{B}_1\bar{Q}_1 + \bar{G}_2\bar{P}_1 D^-(D\bar{P}_1 D^-)'D\bar{Q}_1$, $\bar{B}Q_0 = \bar{G}_2\bar{P}_1 Q_0$.

Observe that (6.12) and (6.14a) imply

$$F_i\big(u(t), w(t), z(t), u'(t), v'(t), t\big) = 0, \qquad i = 1, 2, 3, \qquad (6.27)$$

for $t \in \mathcal{I}$. We study these functions around $(u_0, w_0, z_0, \eta_0, \zeta_0, t_0)$ where

$$u_0 = u(t_0), \qquad w_0 = w(t_0), \qquad z_0 = z(t_0), \qquad \eta_0 = u'(t_0), \qquad \zeta_0 = v'(t_0).$$

As in (6.13) we have $x^0 = D^-(t_0)u_0 + z_0 + w_0$.

① Lemma 6.9 shows that (dropping the t argument)

$$F_1(u, w, z, \eta, \zeta, \cdot) = \bar{Z}\bar{G}_2^{-1} b\big(D^- u + \bar{Z}z + Tw, \cdot\big) + \big(I - \bar{Z}\big)z$$

does not depend on η nor ζ. Due to the structural condition (A1) F_1 is even independent of w as

$$F_{1,w}(u, w, z, \eta, \zeta, t) = \big(\bar{Z}\bar{G}_2^{-1}\big)(t)B\big(\xi(t), t\big)T(t) = 0 \qquad (6.28)$$

with $\xi(t) = D^-(t)u + \bar{Z}(t)z + T(t)w$ (see Lemma 6.9). We may redefine F_1 using the proper argument list:

$$\begin{aligned}
\hat{F}_1(u, z, \cdot) &= \bar{Z}\bar{G}_2^{-1} F(u, 0, z, 0, 0, \cdot) + \big(I - \bar{Z}\big)z \\
&= \bar{Z}\bar{G}_2^{-1} b\big(D^- u + \bar{Z}z, \cdot\big) + \big(I - \bar{Z}\big)z.
\end{aligned} \qquad (6.29)$$

Keep in mind that due to (6.28)

$$\hat{F}_1(u, z, \cdot) = \bar{Z}\bar{G}_2^{-1} b\big(D^- u + \bar{Z}z + w_0, \cdot\big) + \big(I - \bar{Z}\big)z \qquad (6.30)$$

is also valid. Using Lemma 6.9 again, we calculate

$$\hat{F}_{1,z}(u_0, z_0, t_0) = \bar{Z}(t_0)\bar{G}_2^{-1}(t_0)B(x^0, t_0)\bar{Z}(t_0) + \big(I - \bar{Z}\big) = I. \qquad (6.31)$$

(6.27) for $i = 1$ and (6.31) allow the application of the implicit function theorem and $z(t) = \mathbf{z}\big(u(t), t\big)$ is given as a function of $u(t)$ and t. The mapping $\mathbf{z}$ is defined locally around (u_0, t_0) and satisfies $\hat{F}_1\big(u, \mathbf{z}(u, t), t\big) = 0$ on a neighbourhood of (u_0, t_0). Thus

$$\mathbf{z}(u, t) = \bar{Z}(t)\mathbf{z}(u, t) \qquad (6.32)$$

is also valid since $0 = \big(I - \bar{Z}(t)\big)\hat{F}_1\big(u, \mathbf{z}(u, t), t\big) = \big(I - \bar{Z}(t)\big)\mathbf{z}(u, t)$. Due to $D(t)\mathbf{z}(u, t) = D(t)\bar{Z}\mathbf{z}(u, t) = D(t)\bar{Q}_1(t)\mathbf{z}(u, t)$ we arrive at

$$\begin{aligned}
v(t) &= D(t)\bar{Q}_1(t)z(t) = D(t)\bar{Q}_1(t)\bar{Z}(t)\,\mathbf{z}\big(u(t), t\big) \\
&= D(t)\mathbf{z}\big(u(t), t\big) = \mathbf{v}\big(u(t), t\big)
\end{aligned}$$

with the function $\mathbf{v}(u, t) = D(t)\mathbf{z}\big(u, t\big)$.

② Noting that

$$DP̄_1Ḡ_2^{-1}A(u+v)' = DP̄_1Ḡ_2^{-1}ADD^-(u+v)' = DP̄_1D^-(u+v)'$$
$$= u' - (DP̄_1D^-)'(u+v)$$

it turns out that F_3 provides an explicit representation of $u'(\cdot)$ in terms of $u(\cdot)$ and $w(\cdot)$. In particular, for $i = 3$ (6.27) is equivalent to

$$u'(t) = \mathbb{f}\big(u(t), w(t), t\big) \tag{6.33a}$$

with

$$\mathbb{f}(u,w,t) = (DP̄_1D^-)'(t)\Big(u + \mathbb{v}(u,t)\Big) \tag{6.33b}$$
$$- (DP̄_1Ḡ_2^{-1})(t)\, b\Big(D^-(t)u + \mathbb{z}(u,t) + T(t)w, t\Big).$$

③ Combining F_2 with the results obtained so far we get

$$\hat{F}_2\big(u(t), w(t), v'(t), t\big) = 0 \tag{6.34a}$$

where

$$\hat{F}_2\big(u, w, \zeta, t\big) = F_2\big(u, w, \mathbb{z}(u,t), \mathbb{f}(u,w,t), \zeta, t\big)$$
$$= (TḠ_2^{-1}A)(t)\big(\mathbb{f}(u,w,t) + \zeta\big) + (I - T(t))w \tag{6.34b}$$
$$+ (TḠ_2^{-1})(t)\, b\Big(D^-(t)u + \mathbb{z}(u,t) + T(t)w, t\Big)$$

is defined on a neighbourhood of (u_0, w_0, ζ_0, t_0). In order to apply the implicit function theorem once again, we need to show that $\hat{F}_{2,w}\big(u_0, w_0\zeta_0, , t_0\big)$ is nonsingular. After calculating

$$\mathbb{f}_w\big(u_0, w_0, t_0\big) = -(DP̄_1Ḡ_2^{-1})(t)B(x^0, t_0)T(t_0)$$

we obtain from Lemma 6.9

$$\hat{F}_{2,w}\big(u_0, w_0, \zeta_0, t_0\big)$$
$$= (TḠ_2^{-1}A)(t_0)\mathbb{f}_w(u_0, w_0, t_0) + I - T(t_0) + (TḠ_2^{-1})(t_0)B(x^0, t_0)T(t_0)$$
$$= -(TP̄_1P_0P̄_1Ḡ_2^{-1})(t)B(x^0, t_0)T(t_0) + I - T(t_0) + T(t_0).$$

This proves $\hat{F}_{2,w}\big(u_0, w_0, t_0\big) = I$, since $TP̄_1P_0P̄_1 = 0$. Thus the implicit function theorem shows that locally around (u_0, w_0, ζ_0, t_0) (6.34a) is equivalent to

$$w(t) = \mathbb{w}\big(u(t), v'(t), t\big)$$

with a continuous mapping $\mathbb{w}$. Similar to (6.32) $\mathbb{w}(u, \zeta, t) = T(t)\mathbb{w}(u, \zeta, t)$ holds.

The results of ①–③ show that (6.14) implies (6.15) from Lemma 6.3, i.e.

$$z(t) = \mathsf{z}\big(u(t),t\big), \quad u'(t) = \mathsf{f}\big(u(t),w(t),t\big), \quad w(t) = \mathsf{w}\big(u(t),v'(t),t\big).$$

To finish the proof we note that these mappings imply

$$\bar{G}_2^{-1} F(u,w,z,u',v',\cdot)$$
$$= \hat{F}_1(u,\mathsf{z}(u,\cdot),\cdot) + \hat{F}_2(u,\mathsf{w}(u,v',\cdot),v',\cdot) + D^-\big(u' - \mathsf{f}(u,\mathsf{w}(u,v',\cdot),\cdot)\big) = 0,$$

where t-arguments have once again been omitted.   □

Proof of Lemma 6.4. Again we drop the assumption that there is a solution. We require that (A1), (A2) and (A3) hold. Exactly as in (6.29) we define the mapping $\hat{F}_1 = \hat{F}_1(u,z,t)$ where u and z are considered to be parameters chosen in a neighbourhood of (u_0,z_0). Recall that u_0, w_0, z_0 are defined in (6.17) as

$$u_0 = D(t_0)\bar{P}_1(t_0)x^0, \qquad w_0 = T(t_0)x^0, \qquad z_0 = \bar{Z}(t_0)x^0.$$

We have

$$\hat{F}_1(u_0,z_0,t_0) = \big(\bar{Z}\bar{G}_2^{-1}\big)(t_0)\Big[A(t_0)y^0 + b(x^0,t_0)\Big] = 0$$

due to (6.30) and (A2). As in (6.31) we find $\hat{F}_{1,z}(u_0,z_0,t_0) = I$ and the implicit function theorem provides the function z satisfying $\hat{F}_1\big(u,\mathsf{z}(u,t),t\big) = 0$ on a neighbourhood of (u_0,t_0). Notice that z satisfies (6.32).   □

Proof of Lemma 6.5. Having z at our disposal we introduce $\mathsf{v}(u,t) = D(t)\mathsf{z}(u,t)$ and consider the mapping $\mathsf{f} = \mathsf{f}(u,w,t)$ defined in (6.33b). The function $\hat{F}_2 = \hat{F}_2(u,w,\zeta,t)$ can be defined as in (6.34b). Let $\zeta_0 = y^0 - \mathsf{f}(u_0,w_0,t_0)$ such that

$$\hat{F}_2(u_0,w_0,\zeta_0,t_0) = \big(T\bar{G}_2^{-1}\big)(t_0)\Big[A(t_0)y^0 + b(x^0,t_0)\Big] = 0.$$

We already calculated $\hat{F}_{2,w}\big(u_0,w_0,t_0\big) = I$ and therefore the implicit function theorem yields the mapping w with $\hat{F}_2\big(u,\mathsf{w}(u,\zeta,t),\zeta,t\big) = 0$ in a neighbourhood of (u_0,ζ_0,t_0). Again $\mathsf{w}(u,\zeta,t) = T(t)\mathsf{w}(u,\zeta,t)$ is satisfied.   □

Remark 6.10. The decoupling procedure discussed above leads to the implicit differential algebraic system

$$u' = \mathsf{f}\big(u,\mathsf{w}(u,v',t),t\big) =: f(u,v',t) \tag{6.35a}$$
$$v = D(t)\mathsf{z}(u,t) \qquad =: g(u,t). \tag{6.35b}$$

Differentiating (6.35b) and inserting into (6.35a) shows that

$$u' = f\big(u,g_u(u,t)u' + g_t(u,t),t\big). \tag{6.36}$$

In order to guarantee that (6.35) has index 1, as stated in Lemma 6.6, we have to ensure that (6.36) can be solved for u'. To this end consider the matrix $M(u, \zeta, t) = I - f_{v'}(u, \zeta, t)g_u(u, t)$ satisfying

$$M(u_0, \zeta_0, t_0) = I - \big(\mathsf{f}_w \mathsf{w}_\zeta \mathsf{v}_u\big)(u_0, \zeta_0, t_0)$$
$$= I + \big(D\bar{P}_1 \bar{G}_2^{-1} \bar{B}T\big)(t_0)\big(T\bar{G}_2^{-1}A\big)(t_0)\big(D\bar{Q}_1 D^-\big)(t_0) = I.$$

Thus M remains nonsingular also on a neighbourhood of (u_0, ζ_0, t_0) such that (6.35) has indeed index 1. $\qquad\square$

Remark 6.11. In Theorem 6.7 a solution x_* was constructed by first solving (6.35) and then defining $x_*(t) = D^-(t)u(t) + \mathsf{z}\big(u(t), t\big) + \mathsf{w}\big(u(t), v'(t), t\big)$. This function is a solution of (6.21) since

$$\bar{Z}\bar{G}_2^{-1}\left[A(Dx_*)' + b(x_*, \cdot)\right] = \hat{F}_1\big(u, \mathsf{z}(u, \cdot), \cdot\big) \qquad\quad = 0, \qquad (6.37a)$$
$$T\bar{G}_2^{-1}\left[A(Dx_*)' + b(x_*, \cdot)\right] = \hat{F}_2\big(u, \mathsf{w}(u, v', \cdot), v', \cdot\big) \qquad = 0, \qquad (6.37b)$$
$$P_0\bar{P}_1\bar{G}_2^{-1}\left[A(Dx_*)' + b(x_*, \cdot)\right] = D^-\Big(u' - \mathsf{f}\big(u, \mathsf{w}(u, v', \cdot), \cdot\big)\Big) = 0. \qquad (6.37c)$$

Here we want to remark that in order to see (6.37a), the relations $\bar{Z}\bar{G}_2^{-1}AD = 0$ and $\bar{Z}\bar{G}_2^{-1}b(\xi, \cdot) = \bar{Z}\bar{G}_2^{-1}b(U\xi, \cdot)$ have to be taken into account. The latter relation follows from the structural condition (A1) as was seen in (6.28). The property (6.32) was used as well. Similarly, $\mathsf{w}(u, \zeta, t) = T(t)\mathsf{w}(u, \zeta, t)$ and (6.34b) imply (6.37b). Finally, (6.37c) follows from (6.35a), $v(t) = \mathsf{v}\big(u(t), t\big)$ and the definition of f in (6.33b). $\qquad\square$

Part III

General Linear Methods for DAEs

7

General Linear Methods

The numerical solution of differential algebraic equations often poses difficulties for standard schemes. Hence there are many references studying numerical methods for DAEs. Linear multistep methods are addressed in [10, 71, 111] while [84, 104] are references for Runge-Kutta methods in the context of DAEs. In [75, 85, 101] both families of methods are investigated.

We saw in Section 3.3 that even for simple linear DAEs numerical methods might not work as expected. Recall from Example 3.10 on page 65 that the inherent dynamics of the system was integrated using the explicit Euler scheme even though the implicit method was applied to the original DAE.

In [89, 90] BDF methods and Runge-Kutta schemes were studied for linear DAEs. It was shown that properly stated leading terms allow a numerically qualified formulation ensuring that stability properties of these integration schemes carry over to the inherent regular ODE.

Even though these results are of key relevance for numerical computation, the approach of [89, 90] is confined to linear DAEs. Thus their scope is not wide enough to include the nonlinear DAEs arising in electrical circuit simulation using the modified nodal analysis.

In the introductory Chapter 2 we saw that both linear multistep and Runge-Kutta methods suffer from certain disadvantages when being used for integrated circuit design. In particular the damping behaviour of the BDF methods and artificial oscillations of numerical solutions obtained by the trapezoidal rule gave rise to serious difficulties (see Example 2.1 and 2.2). On the other hand, computational costs for fully-implicit Runge-Kutta methods are often prohibitive for the large scale DAE systems arising in circuit simulation (cf. Example 2.6). Although Runge-Kutta methods with diagonally implicit structure may be used, these methods suffer from a low stage order and are thus prone to the order reduction phenomenon (cf. Example 2.7).

The object of this work is to extend the results of [89, 90] in two directions. On the one hand general nonlinear DAEs of the form

$$A(t)\big[d\big(x(t),t\big)\big]' + b\big(x(t),t\big) = 0 \tag{7.1}$$

will be addressed such that DAEs in Hessenberg form and in particular the more general MNA equations are covered by the results. Additionally the wide class of general linear methods will be investigated in order to overcome the disadvantages associated with both linear multistep and Runge-Kutta methods.

In contrast to linear DAEs the nonlinear version (7.1) requires considerable effort in order to reveal the inherent structure necessary for an appropriate generalisation of the results from [89, 90]. The previous section showed that a decoupling procedure can be successfully employed (although technically quite complex). This approach lead to an inherent implicit index-1 system of the form

$$y' = f(y, z', t), \qquad z = g(y, t). \tag{7.2}$$

As a starting point, general linear methods are introduced for ordinary differential equations in this chapter. Implicit index-1 DAEs (7.2) are treated in Chapter 8. There order conditions and convergence results will be derived. Together with the decoupling procedure from the previous section, these investigations will lead to similar statements for properly stated nonlinear DAEs (7.1) in Chapter 9.

7.1 Consistency, Order and Convergence

General linear methods (GLMs) were introduced by Butcher [18] already in 1966. The original intent was to provide a unifying framework for linear multistep and Runge-Kutta methods. It turned out that many other schemes such as cyclic composition methods or predictor-corrector pairs can be cast into general linear form as well.

In Section 2.3 we saw that a general linear method $\mathcal{M} = [\mathcal{A}, \mathcal{U}, \mathcal{B}, \mathcal{V}]$ is characterised by four matrices $\mathcal{A} \in \mathbb{R}^{s \times s}$, $\mathcal{U} \in \mathbb{R}^{s \times r}$, $\mathcal{B} \in \mathbb{R}^{r \times s}$ and $\mathcal{V} \in \mathbb{R}^{r \times r}$. Given an ordinary differential equation

$$y'(t) = f(y(t), t), \qquad y(t_0) = y_0, \tag{7.3}$$

and r pieces of input information $y_1^{[n]}, \ldots, y_r^{[n]}$, at t_n, s internal stages

$$Y_i = h \sum_{j=1}^{s} a_{ij} f(Y_j, t_n + c_j h) + \sum_{j=1}^{r} u_{ij} y_j^{[n]}, \qquad i = 1, \ldots, s, \tag{7.4a}$$

are calculated. In order to proceed from t_n to $t_{n+1} = t_n + h$ using a stepsize h the input quantities are updated and another r quantities

$$y_i^{[n+1]} = h \sum_{j=1}^{s} b_{ij} f(Y_j, t_n + c_j h) + \sum_{j=1}^{r} v_{ij} y_j^{[n]}, \qquad i = 1, \ldots, r, \tag{7.4b}$$

are passed on to the next step. The numerical scheme (7.4) is often written in the more compact form

$$Y = h\mathcal{A}\,F(Y) + \mathcal{U}\,y^{[n]}, \qquad\qquad y^{[n+1]} = h\mathcal{B}\,F(Y) + \mathcal{V}\,y^{[n]} \qquad (7.5)$$

where the abbreviations

$$Y = \begin{bmatrix} Y_1 \\ \vdots \\ Y_s \end{bmatrix}, \qquad F(Y) = \begin{bmatrix} f(Y_1, t_n + c_1\,h) \\ \vdots \\ f(Y_s, t_n + c_s\,h) \end{bmatrix}, \qquad y^{[n]} = \begin{bmatrix} y_1^{[n]} \\ \vdots \\ y_r^{[n]} \end{bmatrix}$$

are used. This formulation of a GLM is due to Burrage and Butcher [13]. The internal stages $Y_i \approx y(t_n + c_i\,h)$ represent approximations to the exact solution at intermediate timepoints. Many different choices are possible for the external stages. We will assume that the components of $y^{[n]}$ are related to the exact solution by a weighted Taylor series, i.e.

$$y_i^{[n]} = \sum_{k=0}^{p} \alpha_{ik} h^k y^{(k)}(t_n) + \mathcal{O}(h^{p+1}). \qquad (7.6)$$

Recall that s and r denote the number of internal and external stages, respectively. In the introduction it was emphasised that the method's order p and its stage order q are further important parameters.

Definition 7.1. *A general linear method* $\mathcal{M} = [\mathcal{A}, \mathcal{U}, \mathcal{B}, \mathcal{V}]$ *has order p and stage order q if there are real parameters α_{ik} and c_j such that* (7.6) *implies*

$$Y_j = \sum_{k=0}^{p} \frac{(c_j h)^k}{k!} y^{(k)}(t_n) + \mathcal{O}(h^{q+1}), \qquad\qquad j = 1, \dots, s, \qquad (7.7a)$$

$$y_i^{[n+1]} = \sum_{k=0}^{p} \alpha_{ik} h^k y^{(k)}(t_{n+1}) + \mathcal{O}(h^{p+1}), \qquad\qquad i = 1, \dots, r. \qquad (7.7b)$$

In Definition 7.1 all components of the vector $y^{[n]}$ are required to be of order p. Although we will adopt this convention in the sequel, this restriction is not necessary. A good example are so-called Almost Runge-Kutta methods where $r = 3$, but $y_3^{[n]}$ is only a crude approximation to $h^2 y''(t_n)$. More details are given in [135].

Methods in Nordsieck form are of particular interest, as the formulation

$$y^{[n]} \approx \begin{bmatrix} y(t_n) \\ h\,y'(t_n) \\ \vdots \\ h^{r-1} y^{(r-1)}(t_n) \end{bmatrix}.$$

ensures that changing the stepsize for a GLM is nearly as easy as in case of Runge-Kutta methods. A simple rescaling of the Nordsieck vector ensures that the correct powers of the stepsize h are used at the beginning of a step. Observe that in contrast to Nordsieck's original formulation [127] factorials have been omitted for convenience.

For methods in Nordsieck form the parameters α_{ik} are given by

$$\alpha_{ik} = \begin{cases} 1 \,,\, i = k+1 \\ 0 \,,\, \text{otherwise} \end{cases}$$

such that, in particular, $\alpha_0 = e_1 = \begin{bmatrix} 1 & 0 & \cdots & 0 \end{bmatrix}^\top$ and $\alpha_1 = e_2 = \begin{bmatrix} 0 & 1 & \cdots & 0 \end{bmatrix}^\top$.

Example 7.2. Consider the 3 step BDF method. The coefficients are given in Table 2.1 on page 27 such that $y_{n+1} - \frac{18}{11}y_n + \frac{9}{11}y_{n-1} - \frac{2}{11}y_{n-2} = \frac{6}{11}hf(y_{n+1})$. Written in general linear form this scheme reads

$$\begin{bmatrix} Y \\ \hline y_{n+1} \\ y_n \\ y_{n-1} \\ y_{n-2} \end{bmatrix} = \begin{bmatrix} \frac{6}{11} & \frac{18}{11} & -\frac{9}{11} & \frac{2}{11} & 0 \\ \hline \frac{6}{11} & \frac{18}{11} & -\frac{9}{11} & \frac{2}{11} & 0 \\ 0 & 1 & 0 & 0 & 0 \\ 0 & 0 & 1 & 0 & 0 \\ 0 & 0 & 0 & 1 & 0 \end{bmatrix} \begin{bmatrix} hf(Y) \\ \hline y_n \\ y_{n-1} \\ y_{n-2} \\ y_{n-3} \end{bmatrix},$$

where the additional input quantity y_{n-3} is considered for convenience. Notice that y_{n-3} is not used for computing Y nor y_{n+1}. Since $y_n \approx y(t_n) + \mathcal{O}(h^4)$ and

$$y_{n-i} \approx y(t_n - ih) = \sum_{k=0}^{3} \frac{(-ih)^k}{k!} y^{(k)}(t_n) + \mathcal{O}(h^4), \qquad i = 1, \ldots, 3,$$

it turns out that the coefficients α_{ik} are given by

$$W = \begin{bmatrix} \alpha_0 & \alpha_1 & \alpha_2 & \alpha_3 \end{bmatrix} = \begin{bmatrix} 1 & 0 & 0 & 0 \\ 1 & -1 & \frac{1}{2} & -\frac{1}{6} \\ 1 & -2 & 2 & -\frac{4}{3} \\ 1 & -3 & \frac{9}{2} & -\frac{9}{2} \end{bmatrix}.$$

Using this matrix W it is possible to rewrite the BDF scheme in Nordsieck form. To this end the method $\mathcal{M} = [\mathcal{A}, \mathcal{U}, \mathcal{B}, \mathcal{V}]$ needs to be replaced by

$$\hat{\mathcal{M}} = \begin{bmatrix} \mathcal{A} & \mathcal{U}W \\ \hline W^{-1}\mathcal{B} & W^{-1}\mathcal{V}W \end{bmatrix} = \begin{bmatrix} \frac{6}{11} & 1 & \frac{5}{11} & -\frac{1}{22} & -\frac{7}{66} \\ \hline \frac{6}{11} & 1 & \frac{5}{11} & -\frac{1}{22} & -\frac{7}{66} \\ 1 & 0 & 0 & 0 & 0 \\ \frac{12}{11} & 0 & -\frac{12}{11} & -\frac{1}{11} & \frac{5}{11} \\ \frac{6}{11} & 0 & -\frac{6}{11} & -\frac{6}{11} & \frac{8}{11} \end{bmatrix}.$$

The method $\hat{\mathcal{M}}$ is in Nordsieck form with $\hat{W} = \begin{bmatrix} \hat{\alpha}_0 & \hat{\alpha}_1 & \hat{\alpha}_2 & \hat{\alpha}_3 \end{bmatrix} = I$. $\qquad \square$

The definition of order is visualised in Figure 7.1. In general, if $r > 1$, a starting procedure[1] has to be used in order to compute $y^{[0]}$ from the initial value $y_0 = y(t_0)$. The method $\mathcal{M}$ takes $y^{[0]}$ as input and calculates $y^{[1]}$ at the end of the first step. The local error is given by the difference $y^{[1]} - \hat{y}^{[1]}$, where $\hat{y}^{[1]}$ results from applying the starting procedure to the exact solution $y(t_1)$. Similarly a finishing procedure is used to recover the solution y_n from $y^{[n]}$ at the end of the integration. For methods in Nordsieck form the finishing procedure is trivial as the numerical result appears in the first component of $y^{[n]}$.

After several steps taken by the method the local errors accumulate to give the global error $y^{[n]} - \hat{y}^{[n]}$. If the method has order p then the global error is $n\,\mathcal{O}(h^{p+1}) = \mathcal{O}(h^p)$ provided that the method is stable.

Definition 7.3. *A general linear method $\mathcal{M} = [\mathcal{A},\mathcal{U},\mathcal{B},\mathcal{V}]$ is stable if the matrix $\mathcal{V}$ is power bounded, i.e. there is a constant C such that $\|\mathcal{V}^n\| \leq C$ for every n.*

The definition of stability is closely related to solving the trivial ODE $y' = 0$ where $Y = \mathcal{U}y^{[n]}$ and $y^{[n+1]} = \mathcal{V}y^{[n]}$. The requirements

$$y^{[n]} = \alpha_0\, y(t_n) + \mathcal{O}(h), \quad y^{[n+1]} = \alpha_0\, y(t_{n+1}) + \mathcal{O}(h), \quad Y = e\, y(t_n) + \mathcal{O}(h)$$

show that the method has to satisfy certain pre-consistency conditions.

Definition 7.4. *A method $\mathcal{M} = [\mathcal{A},\mathcal{U},\mathcal{B},\mathcal{V}]$ is pre-consistent if there is a vector α_0 called the pre-consistency vector, such that*

$$\mathcal{V}\alpha_0 = \alpha_0, \qquad\qquad \mathcal{U}\alpha_0 = e = \begin{bmatrix} 1 & \cdots & 1 \end{bmatrix}^{\top}.$$

Consider the one-dimensional differential equation $y' = 1$. The application of $\mathcal{M}$ yields $Y = h\mathcal{A}e + \mathcal{U}y^{[n]}$ and $y^{[n+1]} = h\mathcal{B}e + \mathcal{V}y^{[n]}$. The method has at least

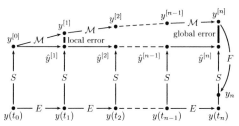

Figure 7.1: The local and global error for general linear methods. S and F denote the starting and finishing procedure, respectively. E is the exact solution operator that evaluates the exact solution at the next timepoint [26].

---

[1]Generalised Runge-Kutta methods are often used as starting methods [93]. In practice the vectors $y^{[n]}$ with $r > 1$ are built up gradually by using a variable order implementation (see Chapter 11).

order 1 for this problem, i.e.

$$y^{[n]} = \alpha_0 y(t_n) + \alpha_1 h y'(t_n) + \mathcal{O}(h^2), \quad y^{[n+1]} = \alpha_0 y(t_{n+1}) + \alpha_1 h y'(t_{n+1}) + \mathcal{O}(h^2),$$

provided that it satisfies the following consistency condition.

Definition 7.5. *A method* $\mathcal{M} = [\mathcal{A}, \mathcal{U}, \mathcal{B}, \mathcal{V}]$ *is called consistent if it is pre-consistent with pre-consistency vector* α_0 *and there is a vector* α_1 *such that*

$$\alpha_0 + \alpha_1 = \mathcal{B}e + \mathcal{V}\alpha_1.$$

The requirement of having stage order 1, i.e. $Y_i = y(t_n + c_i h) + \mathcal{O}(h^2) = y(t_n) + c_i h y'(t_n) + \mathcal{O}(h^2)$, leads to the additional condition

$$c = \mathcal{A}e + \mathcal{U}\alpha_1 \tag{7.8}$$

which fixes the vector c in terms of the method's coefficients and the consistency vector α_1.

A precise definition of convergence is rather complicated in this general setup. Let (7.3) be an initial value problem with $f : \mathbb{R}^m \times [t_0, t_{end}] \to \mathbb{R}^m$ being continuous and satisfying a Lipschitz condition for the first argument. A starting procedure $S : \mathbb{R}^m \times \mathbb{R} \to \mathbb{R}^{rm}$ is required and we assume that $S(y_0, h) \to \alpha_0 \otimes y_0$ as the stepsize h tends towards zero. Let $\eta(n)$ be the result computed by the method after n steps with stepsize $h = (t_{end} - t_0)/n$ starting from $S(y_0, h)$. The method is convergent if, for any such initial value problem and starting procedure, $\lim_{n \to \infty} \eta(n) = \alpha_0 \otimes y(t_{end})$.

It is proved in [25] that for ordinary differential equations convergence is equivalent to having stability and consistency.

The pre-consistency and consistency conditions as well as (7.8),

$$\mathcal{V}\alpha_0 = \alpha_0, \qquad \mathcal{U}\alpha_0 = e, \qquad \alpha_0 + \alpha_1 = \mathcal{B}e + \mathcal{V}\alpha_1, \qquad c = \mathcal{A}e + \mathcal{U}\alpha_1,$$

ensure that the general linear method $\mathcal{M} = [\mathcal{A}, \mathcal{U}, \mathcal{B}, \mathcal{V}]$ has order and stage order at least 1 for ordinary differential equations. In order to guarantee higher order p and stage order q the method's coefficients have to satisfy complicated expressions, the so-called order conditions. If the stage order q is required to be equal to the order p, these conditions simplify considerably [23, 26, 29].

Theorem 7.6. *A general linear method* $\mathcal{M} = [\mathcal{A}, \mathcal{U}, \mathcal{B}, \mathcal{V}]$ *has order p and stage order $q = p$ if and only if*

$$\mathcal{U}W = C - \mathcal{A}CK, \qquad\qquad \mathcal{V}W = WE - \mathcal{B}CK, \tag{7.9}$$

where

$$K = \begin{bmatrix} 0 & 1 & 0 & \cdots & 0 & 0 \\ 0 & 0 & 1 & \cdots & 0 & 0 \\ \vdots & \vdots & \vdots & \ddots & \vdots & \vdots \\ 0 & 0 & 0 & \cdots & 0 & 1 \\ 0 & 0 & 0 & \cdots & 0 & 0 \end{bmatrix}, \quad E = \begin{bmatrix} 1 & \frac{1}{1!} & \frac{1}{2!} & \cdots & \frac{1}{(p-1)!} & \frac{1}{p!} \\ 0 & 1 & \frac{1}{1!} & \cdots & \frac{1}{(p-2)!} & \frac{1}{(p-1)!} \\ \vdots & \vdots & \vdots & \ddots & \vdots & \vdots \\ 0 & 0 & 0 & \cdots & 1 & \frac{1}{1!} \\ 0 & 0 & 0 & \cdots & 0 & 1 \end{bmatrix} = \exp(K)$$

and $W = [\alpha_0 \cdots \alpha_p]$, $C = \begin{bmatrix} e & c & \frac{c^2}{2!} & \cdots & \frac{c^p}{p!} \end{bmatrix}$. *The exponentiation c^k is meant componentwise.*

Proof. Assume that the order and stage order is p, i.e. (7.6) implies (7.7) for $q = p$. Due to (7.7a), $Y_j = y(t_n + c_j h) + \mathcal{O}(h^{p+1})$, it follows that $h\, f(Y_j, t_n + c_j h) = h y'(t_n + c_j h) + \mathcal{O}(h^{p+2})$ such that a Taylor series expansion of (7.4a) yields

$$\sum_{k=0}^{p} \frac{c^k}{k!} h^k y^{(k)}(t_n) = \sum_{k=1}^{p} \frac{\mathcal{A}c^{k-1}}{(k-1)!} h^k y^{(k)}(t_n) + \sum_{k=0}^{p} \mathcal{U}\alpha_k h^k y^{(k)}(t_n) + \mathcal{O}(h^{p+1}).$$

Equating coefficients for like powers shows that

$$e = \mathcal{U}\alpha_0, \qquad \frac{1}{k!}c^k = \frac{1}{(k-1)!}\mathcal{A}c^{k-1} + \mathcal{U}\alpha_k, \qquad k = 1, \ldots, p. \qquad (7.10a)$$

Similarly, expanding (7.4b) leads to

$$\sum_{k=0}^{p} \alpha_k h^k y^{(k)}(t_{n+1}) = \sum_{k=0}^{p} \sum_{l=0}^{k} \frac{\alpha_l}{(k-l)!} h^k y^{(k)}(t_n) + \mathcal{O}(h^{p+1})$$

$$= \sum_{k=1}^{p} \frac{\mathcal{B}c^{k-1}}{(k-1)!} h^k y^{(k)}(t_n) + \sum_{k=0}^{p} \mathcal{V}\alpha_k h^k y^{(k)}(t_n) + \mathcal{O}(h^{p+1})$$

such that

$$\alpha_0 = \mathcal{V}\alpha_0, \qquad \sum_{l=0}^{k} \frac{\alpha_l}{(k-l)!} = \frac{1}{(k-1)!}\mathcal{B}c^{k-1} + \mathcal{V}\alpha_k, \qquad k = 1, \ldots, p. \qquad (7.10b)$$

The expressions (7.10) are equivalent to (7.9). On the other hand, if (7.9) holds, then the same Taylor series expansions show that the method has order and stage order p in the sense of Definition 7.1. $\qquad \square$

For methods in Nordsieck form with $s = r = p + 1$ Theorem 7.6 takes a particularly simple form. Due to $W = I$ it turns out that such a method has order p and stage order $q = p$ if the matrices $\mathcal{U}$ and $\mathcal{V}$ are chosen as

$$\mathcal{U} = C - \mathcal{A}CK, \qquad\qquad \mathcal{V} = E - \mathcal{B}CK. \qquad (7.11)$$

In particular, $\mathcal{A}$ may have a diagonally implicit structure. Thus, for any given order, there are diagonally implicit methods having stage order equal to the order. Of course for such a method to be of practical use, $\mathcal{V}$ needs to be stable. The construction of methods in Nordsieck form with $s = r = p + 1 = q + 1$ having excellent stability properties is carried out in [158].

7.2 Stability Analysis

Similar to the approach of Chapter 2 the linear stability behaviour of a general linear method $\mathcal{M} = [\mathcal{A}, \mathcal{U}, \mathcal{B}, \mathcal{V}]$ can be assessed by studying the linear scalar test equation

$$y'(t) = \lambda\, y(t) \tag{7.12}$$

where $\lambda \in \mathbb{C}$ is a complex number having negative real part. For (7.12) the numerical scheme (7.4) reads

$$\left.\begin{array}{rl} Y & = z\mathcal{A}Y + \mathcal{U}y^{[n]} \\ y^{[n+1]} & = z\mathcal{B}Y + \mathcal{V}y^{[n]} \end{array}\right\} \quad \Rightarrow \quad y^{[n+1]} = \big(\mathcal{V} + z\mathcal{B}(I - z\mathcal{A})^{-1}\mathcal{U}\big)y^{[n]}.$$

As usual the variable $z = \lambda h$ has been introduced.

Definition 7.7. *The matrix $M(z) = \mathcal{V} + z\mathcal{B}(I - z\mathcal{A})^{-1}\mathcal{U}$ is called stability matrix of the general linear method $\mathcal{M} = [\mathcal{A}, \mathcal{U}, \mathcal{B}, \mathcal{V}]$. The stability region is given by the set $\mathcal{G} = \{\, z \in \mathbb{C} \mid \sup_{n=1}^{\infty} \|M(z)^n\| < \infty \,\}$ and the stability function is defined as*

$$\Phi(w, z) = \det\big(w\,I - M(z)\big).$$

The close relationship between stability function and stability region becomes obvious by considering the sets

$$\mathcal{H} = \{\, z \in \mathbb{C} \mid \exists\, w \in \mathbb{C},\ |w| \geq 1,\ \Phi(w, z) = 0 \,\},$$
$$\mathcal{H}_0 = \{\, z \in \mathbb{C} \mid \exists\, w \in \mathbb{C},\ |w| > 1,\ \Phi(w, z) = 0 \,\}.$$

It is easy to see that the complement $\bar{\mathcal{G}} = \mathbb{C} \setminus \mathcal{G}$ of the stability region satisfies $\bar{\mathcal{G}} \subset \mathcal{H}$ and $\bar{\mathcal{G}} \supset \mathcal{H}_0$. Hence the boundary of $\mathcal{H}$ reveals the shape of the stability region $\mathcal{G}$. In order to trace out the boundary of $\mathcal{H}$ one has to solve $\Phi(\exp(\theta i), z) = 0$ for z where θ varies in the interval $[0, 2\pi]$.

Definition 7.8. *The general linear method $\mathcal{M} = [\mathcal{A}, \mathcal{U}, \mathcal{B}, \mathcal{V}]$ is called A-stable if the left half of the complex plane $\mathbb{C}^- = \{\, z \in \mathbb{C} \mid \mathrm{Re}(z) < 0 \,\}$ is included in the stability region, i.e. $\mathbb{C}^- \subset \mathcal{G}$.*

If the sector $\{\, r\,e^{(\pi-\theta)i} \in \mathbb{C} \mid r \in \mathbb{R},\ r \geq 0,\ \theta \in [-\alpha, \alpha] \,\}$ is included in the stability region $\mathcal{G}$, then $\mathcal{M}$ is called $A(\alpha)$-stable.

The method is called L-stable if it is A-stable and the stability matrix evaluated at infinity, $M_\infty = \lim_{z \to \infty} M(z)$, has zero spectral radius, i.e. the spectrum of M_∞ is given by $\sigma(M_\infty) = \{0\}$.

Since $M(z)$ is a matrix valued function, L-stability could equally well be defined using the stronger requirement $M_\infty = 0$. However, this leaves only little freedom for the method and we will confine ourselves with Definition 7.8.

Runge-Kutta methods are known to often have excellent stability properties. This was already discussed in Chapter 2. It is thus a desirable property that a general linear method has the same stability region as a Runge-Kutta method. This situation occurs when

$$\Phi(w, z) = \det\left(w\,I - M(z)\right) = w^{r-1}\left(w - R(z)\right) \tag{7.13}$$

such that $M(z)$ has only one nonzero eigenvalue. $R(z)$ has the same significance as the stability function of a Runge-Kutta method. Consequently, a general linear method is said to have Runge-Kutta stability if its stability function satisfies the relation (7.13).

It is a complicated task to determine conditions on the method that guarantee Runge-Kutta stability. Nevertheless Wright [158] succeeded in constructing practical methods by ensuring that the sufficient conditions of *inherent* Runge-Kutta stability are met.

Definition 7.9. *A general linear method* $\mathcal{M} = [\mathcal{A}, \mathcal{U}, \mathcal{B}, \mathcal{V}]$ *with* $\mathcal{V}e_1 = e_1$ *has inherent Runge-Kutta stability if*

$$\mathcal{B}\mathcal{A} = X\mathcal{B}, \qquad\qquad \mathcal{B}\mathcal{U} \equiv X\mathcal{V} - \mathcal{V}X$$

for some matrix X *and*

$$\Phi(w, 0) = \det(w\,I - \mathcal{V}) = w^{r-1}(w - 1).$$

The notation $A \equiv B$ indicates that the two matrices are equal with the possible exception of the first row.

The following theorem shows that inherent Runge-Kutta stability is indeed sufficient for Runge-Kutta stability. This result can be found in [25] or [158] as well.

Theorem 7.10. *Inherent Runge-Kutta stability implies Runge-Kutta stability.*

Proof. In order to check for Runge-Kutta stability we need to compute the characteristic polynomial of the stability matrix $M(z)$ or, equivalently, of a matrix that is related to $M(z)$ by a similarity transformation. It turns out that

$$
\begin{aligned}
(I - zX)M(z)(I - zX)^{-1} &= (I - zX)\left(\mathcal{V} + z\mathcal{B}(I - z\mathcal{A})^{-1}\mathcal{U}\right)(I - zX)^{-1}\\
&= \left(\mathcal{V} - zX\mathcal{V} + z[\mathcal{B} - zX\mathcal{B}](I - z\mathcal{A})^{-1}\mathcal{U}\right)(I - zX)^{-1}\\
&\equiv \left(\mathcal{V} - z(\mathcal{B}\mathcal{U} + \mathcal{V}X) + z[\mathcal{B} - z\mathcal{B}\mathcal{A}](I - z\mathcal{A})^{-1}\mathcal{U}\right)(I - zX)^{-1}\\
&= \left(\mathcal{V} - z(\mathcal{B}\mathcal{U} + \mathcal{V}X) + z\mathcal{B}\mathcal{U}\right)(I - zX)^{-1}\\
&= (\mathcal{V} - z\mathcal{V}X)(I - zX)^{-1} = \mathcal{V}.
\end{aligned}
$$

Due to inherent Runge-Kutta stability, $\mathcal{V}$ has only one nonzero eigenvalue and so does $M(z)$ since $(I - zX)M(z)(I - zX)^{-1}$ is identical to $\mathcal{V}$ except for the first row. $\qquad\square$

Methods with inherent Runge-Kutta stability seem very likely to be good candidates for practical computations. Methods with $s = r = p + 1 = q + 1$ can be constructed using only linear operations [158]. It is possible for $\mathcal{A}$ to have singly diagonally implicit structure such that these methods can be implemented efficiently. The stability for variable stepsize implementations is investigated in [38] and highly accurate error estimators are provided in [40]. Many implementation issues such as stage predictors and details of the Newton iteration are addressed by Huang in [93]. This work led to a fixed order implementation of general linear methods having inherent Runge-Kutta stability. The code was successfully used to solve stiff and nonstiff ordinary differential equations.

Instead of the linear scalar test problem (7.12) nonlinear ODEs

$$y'(t) = f\big(y(t), t\big), \qquad\qquad y(t_0) = y_0, \qquad (7.14a)$$

can be considered as well. Assume that f satisfies the one-sided Lipschitz condition

$$\langle u - v, f(u, t) - f(v, t)\rangle \leq \nu \|u - v\| \quad \forall\, u, v \in \mathbb{R}^m, \ t \in [t_0, t_{end}], \quad (7.14b)$$

where $\|\cdot\|$ is a norm in $\mathbb{R}^m$ and $\langle\cdot, \cdot\rangle$ the corresponding inner product. Provided that (7.14b) holds, any two solutions $y(t)$, $z(t)$ satisfy

$$\|y(t) - z(t)\| \leq \|y(t_0) - z(t_0)\| \exp\big(\nu(t - t_0)\big)$$

(see e.g. [85]). For every $\nu \leq 0$ the distance between these two solutions is thus a non-increasing function of t.

A Runge-Kutta method is called B-stable if it mimics this behaviour in the sense that

$$\|y_{n+1} - z_{n+1}\| \leq \|y_n - z_n\| \qquad (7.15)$$

for every problem (7.14a) satisfying (7.14b) with $\nu = 0$. Studying nonlinear stability for Runge-Kutta methods [12, 19] was motivated by the fundamental works of Dahlquist on nonlinear stability for one-leg methods in 1976 [50]. For any linear multistep method

$$\sum_{i=0}^{k} \alpha_i\, y_{n+1-i} = h \sum_{i=0}^{k} \beta_i\, f(y_{n+1-i}, t_{n+1-i}) \qquad (7.16a)$$

its one-leg counterpart is given by

$$\sum_{i=0}^{k} \alpha_i\, y_{n+1-i} = h\, f\Big(\sum_{i=0}^{k} \beta_i\, y_{n+1-i},\, t_n \Big). \qquad (7.16b)$$

In [50] Dahlquist introduced the concept of G stability, i.e. he derived conditions guaranteeing that numerical solutions obtained by a one-leg method behave in the sense of (7.15) for a special norm $\| \cdot \|_G$ constructed in the paper. Later Dahlquist was able to prove that a linear multistep method (7.16a) is A-stable if and only if the corresponding one-leg method (7.16b) is G-stable [51, 85].

A series of papers [13, 20, 21] lead to a criterion which generalises both G-stability for one-leg methods and B-stability for Runge-Kutta methods. This criterion is based on the matrix

$$
H = \begin{bmatrix} D\mathcal{A} + \mathcal{A}^\top D - \mathcal{B}^\top G\mathcal{B} & D\mathcal{U} - \mathcal{B}^\top G\mathcal{V} \\ \mathcal{U}^\top D - \mathcal{V}^\top G\mathcal{B} & G - \mathcal{V}^\top G\mathcal{V} \end{bmatrix}.
$$

Stable behaviour for a general linear method in the sense of (7.15) for problems (7.14a) satisfying (7.14b) with $\nu = 0$ is guaranteed if there is a matrix G and a diagonal matrix D both being positive definite such that H is positive semi-definite.

More information is given in the extensive reference [25].

8

Implicit Index-1 DAEs

In this chapter implicit differential algebraic equations (DAEs)

$$y' = f(y, z'), \qquad z = g(y), \tag{8.1}$$

are considered. It is assumed that the matrix $\mathsf{M} = I - f_{z'} g_y$ remains nonsingular in a neighbourhood of the solution. Thus the DAE (8.1) has index 1.

Equations of this type were obtained when decoupling properly stated index-2 DAEs in Chapter 6. Given a nonlinear DAE of the form

$$A(t)\big[D(t)x(t)\big]' + b\big(x(t), t\big) = 0 \tag{8.2}$$

we saw that the components $u = D\bar{P}_1 x$ and $v = D\bar{Q}_1 x$ of the solution need to satisfy the inherent index-1 equation

$$u' = f\big(u, \mathsf{w}(u, v', t), t\big), \qquad v = D(t)\mathsf{z}(u, t). \tag{8.3}$$

z and w are obtained by the implicit function theorem. Details are given in Section 6.1. (8.3) can be cast into the form (8.1) by considering

$$y = \begin{bmatrix} u \\ t \end{bmatrix}, \quad z = v, \quad f(y, z') = \begin{bmatrix} f\big(u, \mathsf{w}(u, v', t), t\big) \\ 1 \end{bmatrix}, \quad g(y) = D(t)\mathsf{z}(u, t).$$

Investigating implicit DAEs (8.1) can be regarded as the key step in studying numerical methods for properly stated index-2 DAEs (8.2). Once order conditions and convergence results are available for (8.1), the corresponding results for the more general equation (8.2) can be obtained using the decoupling procedure from Chapter 6. The details of this approach will be carried out in Chapter 9.

Nevertheless, implicit DAEs (8.1) constitute an interesting problem class in their own right as DAEs of this type arise frequently in applications.

Example 8.1. Consider the simple linear RCL circuit depicted in Figure 8.1. The modified nodal analysis leads to

$$Ce_1' - Ce_2' + \tfrac{1}{R}e_1 = 0,$$
$$-Ce_1' + Ce_2' + i_L = 0,$$
$$Li_L' - e_2 = 0.$$

Figure 8.1: A simple RCL circuit

This is an index-1 system, where e_i are node potentials and i_L represents the current through the inductive branch. Choosing new variables $y = \begin{pmatrix} e_2 \\ i_L \end{pmatrix}$ and $z = e_1$ these equations can be written equivalently as

$$y' = \begin{bmatrix} 0 & \frac{-1}{C} \\ \frac{1}{L} & 0 \end{bmatrix} y + \begin{bmatrix} 1 \\ 0 \end{bmatrix} z' = f(y, z'), \qquad z = \begin{bmatrix} 0 & R \end{bmatrix} y = g(y). \qquad \square$$

For more general circuits f and g may of course be nonlinear. Equations of the type (8.1) were also used as a model problem in [104] to study the local order conditions for Runge-Kutta methods. Notice that when reformulating (8.1) in Hessenberg form the index typically increases to 2.

Runge-Kutta methods for implicit DAEs and equations in Hessenberg form are studied thoroughly in the literature [84, 104, 140] but general linear methods have not yet been studied extensively in the context of DAEs. In this chapter classical results obtained for Runge-Kutta methods will be generalised to this wider class of integration schemes.

Section 8.1 is devoted to studying order conditions for general linear methods when applied to implicit index-1 equations (8.1). These order conditions are completely new but results for Runge-Kutta methods are still contained as special cases. Combining the order conditions with a stability analysis leads to a convergence result in Section 8.2. Here, the stability matrix – in contrast to a stability function in case of Runge-Kutta methods – plays a decisive role. Finally the accuracy of the stages and stage derivatives is investigated in Section 8.3. The results obtained there will be seminal when addressing DAEs of the form (8.2) in Chapter 9.

8.1 Order Conditions for the Local Error

General linear methods have not yet been studied for implicit index-1 systems (8.1). Given a method $\mathcal{M} = [\mathcal{A}, \mathcal{U}, \mathcal{B}, \mathcal{V}]$, the numerical scheme (7.5) can be modified for (8.1) such that

$$Y_i' = f(Y_i, Z_i'), \qquad i = 1, \dots, s, \qquad Z_i = g(Y_i), \qquad i = 1, \dots, s,$$
$$Y = h\,\mathcal{A}\,Y' + \mathcal{U}\,y^{[n-1]}, \qquad\qquad Z = h\,\mathcal{A}\,Z' + \mathcal{U}\,z^{[n-1]}$$
$$y^{[n]} = h\,\mathcal{B}\,Y' + \mathcal{V}\,y^{[n-1]}, \qquad\qquad z^{[n]} = h\,\mathcal{B}\,Z' + \mathcal{V}\,z^{[n-1]}.$$

In addition to the stages Y, Z stage derivatives Y', Z' are used to formulate the numerical scheme. Observe that $f : \mathbb{R}^{m_1} \times \mathbb{R}^{m_2} \to \mathbb{R}^{m_1}$ and $g : \mathbb{R}^{m_1} \to \mathbb{R}^{m_2}$ such that

$$Y = \begin{bmatrix} Y_1 \\ \vdots \\ Y_s \end{bmatrix}, \quad Y' = \begin{bmatrix} Y'_1 \\ \vdots \\ Y'_s \end{bmatrix} \in \mathbb{R}^{s \cdot m_1}, \quad Z = \begin{bmatrix} Z_1 \\ \vdots \\ Z_s \end{bmatrix}, \quad Z' = \begin{bmatrix} Z'_1 \\ \vdots \\ Z'_s \end{bmatrix} \in \mathbb{R}^{s \cdot m_2}.$$

s denotes the number of internal stages. The numerical scheme can be written equivalently in the simplified form

$$Y = h\,\mathcal{A}\,\tilde{f}(Y, Z') + \mathcal{U}\,y^{[n-1]}, \qquad \tilde{g}(Y) = h\,\mathcal{A}\,Z' + \mathcal{U}\,z^{[n-1]}, \qquad (8.4a)$$

$$y^{[n]} = h\,\mathcal{B}\,\tilde{f}(Y, Z') + \mathcal{V}\,y^{[n-1]}, \qquad z^{[n]} = h\,\mathcal{B}\,Z' + \mathcal{V}\,z^{[n-1]}, \qquad (8.4b)$$

where the abbreviations

$$\tilde{f}(Y, Z') = \begin{bmatrix} f(Y_1, Z'_1) \\ \vdots \\ f(Y_s, Z'_s) \end{bmatrix}, \qquad\qquad \tilde{g}(Y) = \begin{bmatrix} g(Y_1) \\ \vdots \\ g(Y_s) \end{bmatrix}$$

have been used.

In this section we derive necessary and sufficient conditions for the local error to be of a given order p. Compared to ordinary differential equations additional order conditions have to be satisfied since (8.1) is a DAE with index 1. The order conditions are obtained by expanding the exact and the numerical solution in a Taylor series, respectively, and comparing terms.

8.1.1 Taylor Expansion of the Exact Solution

The exact solution's Taylor expansion for a problem quite similar to (8.1) was derived in [104] by Kværnø. Hence this subsection rests, to a large extent, on the results already presented there. The Taylor series expansion will be written as a (generalised) B-series based on rooted trees. Due to the structure of the implicit index-1 equation (8.1) derivatives of y and z will be involved. Thus, trees with two types of vertices are required. The different types of vertices will be referred to as black nodes ($\bullet$) and white ones ($\circ$). Rooted trees consisting of these nodes form the basis for analysing the numerical solution and for deriving order conditions later on. Thus it is not possible to just quote the results of [104] but we need to summarise the key points necessary for the subsequent sections. Compared to [104] we will use a slightly different approach when introducing the rooted trees. It is aimed at making the theory easier accessible.

For convenience we start by defining sets of tall trees. Tall trees consist of only one type of vertex and have no ramifications,

$$TT_y = \left\{ \; \bullet, \; \text{\textbf{:}}, \; \text{\textbf{:}}, \; \text{\textbf{:}}, \; \dots \; \right\}, \qquad TT_z = \left\{ \; \circ, \; \text{\textdollar}, \; \text{\textdollar}, \; \text{\textdollar}, \; \dots \; \right\}.$$

The order $|\tau|$ of a tall tree τ is defined as the number of its nodes. Derivatives of z can be associated with trees in the following way:

$$z' = g_y y', \qquad\qquad z'' = g_{yy}(y', y') + g_y y'',$$

$$z''' = g_{yyy}(y', y', y') + 3\, g_{yy}(y', y'') + g_y y''', \qquad\qquad \ldots$$

All trees appearing here share a white root representing derivatives of g. The black tall trees connected to the root represent derivatives of y. These trees are collected in the following set of special trees,

$$ST_z = \{\emptyset\} \cup \left\{ [\tau_1, \ldots, \tau_k]_z = \overset{\tau_1\cdots\tau_k}{\curlyvee} \ \middle| \ k \geq 1, \ \tau_i \in TT_y \right\}.$$

The trees are 'special' in the sense that higher derivatives of y are still involved. The empty tree was included for convenience. Notice that we write $[\tau_1, \ldots, \tau_k]_{root}$ to indicate that the trees τ_i are connected to a white root ($root = z$) or a black root ($root = y$). The particular sequence of subtrees does not matter. Each permutation of the subtrees yields the same tree. Often pictorial representations $[\tau_1, \ldots, \tau_k]_z = \overset{\tau_1\cdots\tau_k}{\curlyvee}$ will be used. Here, the leafs indicate the black roots of the tall trees τ_i.

For a given tree $\sigma \in ST_z$ the order $|\sigma|$ is defined as the number of its black nodes. Thus, $z^{(p)}$ can be written in terms of trees with order p. This will be made more precise in the following result.

Lemma 8.2. *For each $\sigma \in ST_z$ define $G_S(\sigma)$ by*

$$G_S(\emptyset) = g, \qquad\qquad G_S\left(\overset{\tau_1\cdots\tau_k}{\curlyvee}\right) = g_{ky}(y^{(|\tau_1|)}, \ldots, y^{(|\tau_k|)})$$

with $g_{ky} = \frac{\partial^k g}{\partial y^k}$. Then the p-th derivative of the exact solution is given by

$$z^{(p)} = \sum_{\sigma \in ST_z, |\sigma| = p} \beta(\sigma)\, G_S(\sigma),$$

where $\beta(\sigma)$ is the number of times $G_S(\sigma)$ appears in the Taylor series expansion of the exact solution. □

$G_S(\sigma)$ is called an elementary differential. Strictly speaking, for every $\sigma \in ST_z$ the elementary differential $G_S(\sigma)$ is a mapping $G_S(\sigma) : \mathbb{R}^{m_1} \times \mathbb{R}^{m_2} \to \mathbb{R}^{m_2}$. Consistent initial values (y_0, z_0) determine the unique solution (y, z) such that for instance $G_S(\text{\textbf{8}})(y_0, z_0) = g_y(y_0)\, y'(t_0) = z'(t_0)$. For better readability the y- and z-arguments are generally omitted.

As a next step derivatives of the exact solution y are considered. Using the chain rule it turns out that

$$y' = f \qquad\qquad y'' = f_y y' + f_{z'} z''$$

$$y''' = f_{yy}(y', y') + 2 f_{yz'}(z'', y') + f_y y'' + f_{z'z'}(z'', z'') + f_{z'} z'''$$

It is straightforward to associate the different terms with trees as indicated above. The tall trees $\tau \in TT_y$ represent y-derivatives $y^{(|\tau|)}$. Similarly, $\sigma \in TT_z$ represents $z^{(|\sigma|+1)}$. Be careful not to mix up the different counting for y- and z-derivatives.

The highest derivative of z can be *replaced* by the expressions already calculated. This leads to

$$y'' = f_y y' + \left[\ f_{z'} g_{yy}(y', y') + f_{z'} g_y y''\ \right]$$

$$y''' = f_{yy}(y', y') + 2 f_{yz'}(z'', y') + f_y y'' + f_{z'z'}(z'', z'')$$

$$+ \left[\ f_{z'} g_{yyy}(y', y', y') + 3 f_{z'} g_{yy}(y', y'') + f_{z'} g_y y'''\ \right]$$

When replacing z-derivatives, the corresponding subtrees are treated accordingly. Take e.g. $\sigma =$ ⧅ representing $f_{z'} z'''$. Since z''' can be written in terms of ⧅, ⧅ and ⧅, we needed to replace σ's single subtree with these three trees one after the other. Thus we arrived at ⧅, ⧅ and ⧅, but the tree ⧅ itself is no longer present in the expansion of y'''.

Due to the index-1 condition the matrix $M = I - f_{z'} g_y$ is nonsingular. This allows, as the next step, to *solve* for the highest y derivative.

$$y'' = M^{-1} f_y y' + \left[\ M^{-1} f_{z'} g_{yy}(y', y')\ \right]$$

$$y''' = M^{-1} f_{yy}(y', y') + 2 M^{-1} f_{yz'}(z'', y') + M^{-1} f_y y'' + M^{-1} f_{z'z'}(z'', z'')$$

$$+ \left[\ M^{-1} f_{z'} g_{yyy}(y', y', y') + 3 M^{-1} f_{z'} g_{yy}(y', y'')\ \right]$$

Observe that due to the multiplication with M^{-1} the trees ⧅, ⧅ etc. are no longer present. Also notice that the trees representing $y^{(p)}$ fall into two distinct classes.

There are trees that were present from the very beginning when calculating $y^{(p)}$ from $y' = f(y, z')$. But there are also those trees that originate from representing the highest z-derivative by trees from ST_z. The former trees will be collected in the set ST_{yy} while the latter ones belong to ST_{yz}.

$$ST_{yy} = \{\emptyset\} \cup \Big\{ \; [\tau_1, \dots, \tau_k, \sigma_1, \dots, \sigma_l]_y = {}_{\tau_k} \overset{\tau_k \; \sigma_l}{\bullet \cdots} {}_{\circ}^{\sigma_l} \; \Big|$$
$$k \geq 0, \; l \geq 0, \; (k, l) \neq (0, 1), \; \tau_i \in TT_y, \; \sigma_i \in TT_z \; \Big\},$$

$$ST_{yz} = \Big\{ \; [\sigma]_y \; \Big| \; \sigma \in ST_z, \; \sigma \neq [\tau]_z \; \text{for} \; \tau \in TT_y \; \Big\}.$$

The tree with only one vertex, $\bullet \in ST_{yy}$, is represented by $[\;]_y$. The trees in ST_{yy} have a black root. All subtrees are tall trees. But trees of the form $[\sigma]_y$ with $\sigma \in TT_z$ had to be excluded since they where *replaced* by the corresponding trees from ST_{yz}. The trees in ST_{yz}, in turn, are nothing but the ST_z trees attached to a new black root. However, trees of the form $[[\tau]_z]_y$ with a tall tree $\tau \in TT_y$ are not included as we *solved* for the highest y-derivative represented by τ. Table 8.1 gives pictorial representations for special trees of order up to 4. For any given tree τ let $|\tau|_\bullet$ be the number of its black nodes. Similarly denote the number of white nodes by $|\tau|_\circ$. Using this notation we define

$$ST_y = ST_{yy} \cup ST_{yz}, \qquad |\tau| = \begin{cases} |\tau|_\bullet - |\tau|_\circ & , \tau \in ST_y \\ |\tau|_\bullet - |\tau|_\circ + 1 & , \tau \in ST_z \end{cases}$$

and arrive at the following straightforward result.

Lemma 8.3. *Given the elementary differentials*

$$F_S(\tau) = \begin{cases} y & , \tau = \emptyset \\ \mathsf{M}^{-1} f_{ky\,lz'}(y^{(|\tau_1|)}, \dots, y^{(|\tau_k|)}, z^{(|\sigma_1|+1)}, \dots, z^{(|\sigma_l|+1)}) \, , \, \tau = {}_{\tau_k}\overset{\tau_k \; \sigma_l}{\bullet \cdots}{}_\circ^{\sigma_l} \in ST_{yy} \\ \mathsf{M}^{-1} f_{z'} g_{ky}(y^{(|\tau_1|)}, \dots, y^{(|\tau_k|)}) & , \tau = \overset{\tau_1 \cdots \tau_k}{\mathsf{Y}} \; \in ST_{yz} \end{cases}$$

we have

$$y^{(p)} = \sum_{\tau \in ST_y, |\tau| = p} \beta(\tau) \, F_S(\tau).$$

Again $\beta(\tau)$ is the number of times $F_S(\tau)$ appears in the exact solution's Taylor series expansion. □

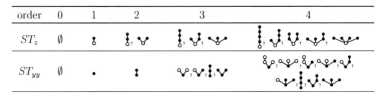

order	0	1	2	3	4
ST_z	$\emptyset$	⚇	⚇, ⌄	⚇, ⚈, ⌄	⚇, ⚈, ⚈, ⚈, ⌄
ST_{yy}	$\emptyset$	•	⚇	⌄, ⌄, ⚇, ⌄	⌄, ⌄, ⌄, ⌄, ⌄, ⌄, ⚇, ⌄, ⌄

Table 8.1: Special trees of order up to 4

Lemma 8.2 and 8.3 enable a representation of derivatives $y^{(p)}$, $z^{(p)}$ in terms of elementary differentials F_S and G_S, respectively. As these representations are based on special trees, the elementary differentials themselves are defined using $y^{(k)}$, $z^{(k)}$ for k up to $p-1$. A recursive insertion of expressions for $y^{(k)}$, $z^{(k)}$ into those of $y^{(p)}$, $z^{(p)}$ will make it possible to write derivatives of y and z in terms of partial derivatives of f and g only. In order to make this process more transparent we introduce the following notation: For any set T of trees, T^k is the subset of T which consists of all trees with order k.

Definition 8.4. *Introduce the following sets of rooted trees,*

$$T_z = \{\emptyset\} \cup \left\{ \, [\hat{\tau}_1, \ldots, \hat{\tau}_k]_z \, \mid \, \exists \, \text{\raisebox{-2pt}{\includegraphics[height=1em]{x}}}^{\tau_1 \cdots \tau_k} \in ST_z \, : \, \hat{\tau}_i \in T_y^{|\tau_i|} \right\},$$

$$T_{yy} = \{\emptyset\} \cup \left\{ \, [\hat{\tau}_1, \ldots, \hat{\tau}_k, \hat{\sigma}_1, \ldots, \hat{\sigma}_l]_y \, \mid \, \exists \, \text{\raisebox{-2pt}{\includegraphics[height=1em]{x}}}^{\tau_k \, \sigma_l} \in ST_{yy} \, : \, \begin{matrix} \hat{\tau}_i \in T_y^{|\tau_i|}, \\ \hat{\sigma}_j \in T_z^{|\sigma_j|+1} \end{matrix} \right\},$$

$$T_{yz} = \left\{ \, \big[[\hat{\tau}_1, \ldots, \hat{\tau}_k]_z \big]_y \, \mid \, \exists \, \text{\raisebox{-2pt}{\includegraphics[height=1em]{x}}}^{\tau_1 \cdots \tau_k} \in ST_{yz} \, : \, \hat{\tau}_i \in T_y^{|\tau_i|} \right\},$$

$$T_y = T_{yy} \cup T_{yz}.$$

The order of a given tree τ is defined as $|\tau| = \begin{cases} |\tau|_\bullet - |\tau|_\circ & , \, \tau \in T_y \\ |\tau|_\bullet - |\tau|_\circ + 1 & , \, \tau \in T_z \end{cases}$.

The trees of order up to 3 are depicted in Table 8.2. Even though Definition 8.4 seems quite technical, the idea is simple: Each tree $\tau \in T_y$, $\sigma \in T_z$ is obtained from a corresponding special tree $sp(\tau) \in ST_y$, $sp(\sigma) \in ST_z$, respectively, by recursively substituting subtrees. As an example consider the tree $\text{\raisebox{-2pt}{\includegraphics[height=1em]{x}}} \in ST_y$. The subtree $\bullet$ represents y' and therefore it is replaced by each element of T_y^1, respectively. As it happens, this doesn't change the tree since $T_y^1 = \{\bullet\}$. In contrast to that, the subtree $\circ$ represents z'' and needs to be replaced by every tree from T_z^2. Thus the special tree $\text{\raisebox{-2pt}{\includegraphics[height=1em]{x}}}$ gives rise to

$$\text{\raisebox{-2pt}{\includegraphics[height=1em]{x}}}, \quad \text{\raisebox{-2pt}{\includegraphics[height=1em]{x}}}, \quad \text{\raisebox{-2pt}{\includegraphics[height=1em]{x}}} \quad \text{with} \quad sp\left(\text{\raisebox{-2pt}{\includegraphics[height=1em]{x}}}\right) = sp\left(\text{\raisebox{-2pt}{\includegraphics[height=1em]{x}}}\right) = sp\left(\text{\raisebox{-2pt}{\includegraphics[height=1em]{x}}}\right) = \text{\raisebox{-2pt}{\includegraphics[height=1em]{x}}}.$$

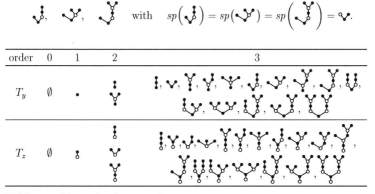

order	0	1	2	3
T_y	$\emptyset$	•	(tree)	(trees)
T_z	$\emptyset$	(tree)	(tree)	(trees)

Table 8.2: Trees of order up to 3

As indicated above, the unique special tree giving rise to $\tau \in T_y \cup T_z$ will be denoted as $sp(\tau)$.

It is obvious that the number of trees for a given order grows exponentially as can be seen in Table 8.3. However, the trees can be constructed efficiently using software tools such as MAPLE [9].

Not all of the trees in $T_y \cup T_z$ will lead to independent order conditions. To this end suitable simplifications will be considered in Section 8.1.2.

The same procedure of recursively substituting subtrees by lower order expressions can be applied to the elementary differentials as well.

Lemma 8.5. *Introduce the following elementary differentials*

$$
G(\sigma) = \begin{cases} z & , \sigma = \emptyset \\ g_{ky}\big(F(\tau_1),\ldots,F(\tau_k)\big) & , \sigma = \overset{\tau_1\cdots\tau_k}{\curlyvee} \in T_z \end{cases}
$$

$$
F(\tau) = \begin{cases} y & , \tau = \emptyset \\ f & , \tau = \bullet \\ \mathsf{M}^{-1} f_{kylz'}\big(F(\tau_1),\ldots,F(\tau_k),G(\sigma_1),\ldots,G(\sigma_l)\big) & , \tau = \overset{\tau_k\;\sigma_l}{\curlyvee}^{\sigma_1} \in T_y \end{cases}
$$

with $\mathsf{M} = I - f_{z'}g_y$. *Then*

$$
y^{(p)} = \sum_{\tau\,\in\,T_y,|\tau|\,=\,p} \alpha(\tau)F\big(\tau\big)(y_0,z_0), \qquad z^{(p)} = \sum_{\sigma\,\in\,T_z,|\sigma|\,=\,p} \alpha(\sigma)G\big(\sigma\big)(y_0,z_0). \tag{8.5}
$$

α *is the number of times that a given elementary differential appears in the exact solution's Taylor expansion.* α *can be calculated recursively as*

$$
\alpha(\tau) = \begin{cases} 1 & , \tau \in \{\emptyset,\; \bullet,\; \mathbf{\overset{\circ}{\bullet}} \} \\ \beta\big(sp(\tau)\big)\alpha(\tau_1)\cdots\alpha(\tau_k) & , \tau = \overset{\tau_1\cdots\tau_k}{\curlyvee} \in T_z,\; \tau = \overset{\tau_1\cdots\tau_k}{\curlyvee} \in T_{yz} \\ \beta\big(sp(\tau)\big)\prod_{i=1}^{k}\alpha(\tau_i)\prod_{j=1}^{l}\alpha(\sigma_j) & , \tau = \overset{\tau_k\;\sigma_l}{\curlyvee}^{\sigma_1} \in T_{yy} \end{cases}
$$

order p	1	2	3	4	5	6	$\sum_{p=1}^{6}$
ST_z	1	2	3	5	7	11	29
ST_{yy}	1	1	4	9	19	35	69
ST_{yz}	0	1	2	4	6	10	23
T_y	1	2	15	136	1458	17089	18701
T_z	1	3	18	157	1645	19132	20956

Table 8.3: The number of trees for a given order

Proof. From $y = F(\emptyset)$, $z = G(\emptyset)$, $y' = F(\bullet)$ and $z' = g_y \, y' = G(\begin{smallmatrix}\bullet\\\bullet\end{smallmatrix})$ we immediately get $\alpha(\tau) = 1$ for $\tau \in \{\emptyset, \ \bullet, \ \begin{smallmatrix}\bullet\\\bullet\end{smallmatrix}\}$. Let $p \geq 2$ be fixed and assume that (8.5) holds for $k \leq p$. From Lemma 8.3 it is known that

$$
\begin{aligned}
y^{(p)} &= \sum_{\tau_s \in ST_y^p} \beta(\tau_s) F_S(\tau_s) = \sum_{\tau_s \in ST_{yy}^p} \beta(\tau_s) F_S(\tau_s) + \sum_{\tau_s \in ST_{yz}^p} \beta(\tau_s) F_S(\tau_s) \\
&= \sum_{\substack{\tau_s \in ST_{yy}^p \\ \tau_s = [\tau_1, \dots, \tau_k, \sigma_1, \dots, \sigma_l]_y}} \beta(\tau_s) \mathsf{M}^{-1} f_{k y \, l z'}\big(y^{(|\tau_1|)}, \dots, y^{(|\tau_k|)}, z^{(|\sigma_1|+1)}, \dots, z^{(|\sigma_l|+1)}\big) \\
&\quad + \sum_{\substack{\tau_s \in ST_{yz}^p \\ \tau_s = [[\tau_1, \dots, \tau_k]_z]_y}} \beta(\tau_s) \mathsf{M}^{-1} f_{z'} g_{ky}\big(y^{(|\tau_1|)}, \dots, y^{(|\tau_k|)}\big).
\end{aligned} \tag{8.6}
$$

By construction we have $|\tau_i| < p$ and $|\sigma_j| < p - 1$. Thus, by the induction hypothesis, we get

$$
y^{(|\tau_i|)} = \sum_{\hat{\tau}_i \in T_y^{|\tau_i|}} \alpha(\hat{\tau}_i) F(\hat{\tau}_i), \qquad z^{(|\sigma_j|+1)} = \sum_{\hat{\sigma}_j \in T_z^{|\sigma_j|+1}} \alpha(\hat{\sigma}_j) G(\hat{\sigma}_j). \tag{8.7}
$$

Observe that τ_i and σ_j are tall trees. In order to distinguish them from $\hat{\tau}_i \in T_y$ and $\hat{\sigma}_j \in T_z$ we use the symbol $\hat{\ }$. Inserting (8.7) into (8.6) and using the multi-linearity of the derivatives, we obtain

$$
\begin{aligned}
y^{(p)} &= \sum_{\substack{\tau_s \in ST_{yy}^p \\ \tau_s = [\tau_1, \dots, \tau_k, \sigma_1, \dots, \sigma_l]_y}} \ \sum_{\hat{\tau}_1 \in T_y^{|\tau_1|}} \cdots \sum_{\hat{\sigma}_l \in T_z^{|\sigma_l|+1}} \beta(\tau_s) \alpha(\hat{\tau}_1) \cdots \alpha(\hat{\tau}_k) \alpha(\hat{\sigma}_1) \cdots \alpha(\hat{\sigma}_l) \cdot \\
&\qquad\qquad \cdot \mathsf{M}^{-1} f_{ky \, lz'}\big(F(\hat{\tau}_1), \dots, F(\hat{\tau}_k), G(\hat{\sigma}_1), \dots, G(\hat{\sigma}_l)\big) \\
&\quad + \sum_{\substack{\tau_s \in ST_{yz}^p \\ \tau_s = [[\tau_1, \dots, \tau_k]_z]_y}} \ \sum_{\hat{\tau}_1 \in T_y^{|\tau_1|}} \cdots \sum_{\hat{\tau}_k \in T_y^{|\tau_k|}} \beta(\tau_s) \alpha(\hat{\tau}_1) \cdots \alpha(\hat{\tau}_k) \cdot \\
&\qquad\qquad \cdot \mathsf{M}^{-1} f_{z'} g_{ky}\big(F(\hat{\tau}_1), \dots, F(\hat{\tau}_k)\big) \\
&= \sum_{\substack{\tau \in T_{yy}^p \\ \tau = [\hat{\tau}_1, \dots, \hat{\tau}_k, \hat{\sigma}_1, \dots, \hat{\sigma}_l]_y}} \beta(sp(\tau)) \alpha(\hat{\tau}_1) \cdots \alpha(\hat{\sigma}_l) F(\tau) + \sum_{\substack{\tau \in T_{yz}^p \\ \tau = [[\hat{\tau}_1, \dots, \hat{\tau}_k]_z]_y}} \beta(sp(\tau)) \alpha(\hat{\tau}_1) \cdots \alpha(\hat{\tau}_k) F(\tau)
\end{aligned}
$$

The formula for $\alpha(\tau)$, $\tau \in T_y$, follows by comparing terms with (8.5). For $\sigma \in T_z$ the corresponding expression is proved similarly. $\qquad\square$

Remark 8.6. In [104] it is claimed that if $\tau = [\tau_1, \dots, \tau_k, \sigma_1, \dots, \sigma_l]_y \in T_y$, then $\alpha(\tau) = \beta\big(sp(\tau)\big) \alpha(\tau_1) \cdots \alpha(\tau_k) \alpha(\sigma_1) \cdots \alpha(\sigma_l)$. The proof above shows that this is not correct for $\tau = [\sigma]_y \in T_{yz}$.

As an example consider $\tau = \begin{smallmatrix}\bullet\\\bullet\end{smallmatrix}$ with $sp(\tau) = \tau$. Recall from the computations on page 128 f. that $\beta\big(sp(\tau)\big) = \beta\big(sp(\begin{smallmatrix}\bullet\\\bullet\end{smallmatrix})\big) = 3$. According to [104] one calculates $\alpha(\tau) = 3\,\alpha(\begin{smallmatrix}\bullet\\\bullet\end{smallmatrix}) = 3 \cdot 3 \cdot \alpha(\bullet) \cdot \alpha(\begin{smallmatrix}\bullet\\\bullet\end{smallmatrix}) = 9$, which is wrong. Following Lemma 8.5, by contrast, we correctly find $\alpha(\tau) = 3 \cdot \alpha(\bullet) \cdot \alpha(\begin{smallmatrix}\bullet\\\bullet\end{smallmatrix}) = 3$. $\qquad\square$

Lemma 8.5 shows how to write the Taylor series expansion of the exact solution in terms of elementary differentials. In order to simplify notation we introduce (generalised) B-series in the following definition. These B-series will be the central tool in the subsequent analysis.

Definition 8.7. *Let T be a set of rooted trees. Denote by $F(\tau)$ corresponding elementary differentials. If $\boldsymbol{a} : T \to \mathbb{R}^n$ is a vector valued function, then the formal series*

$$B_T^F\big(\boldsymbol{a}; y, z\big) = \sum_{\tau \in T} \Big(\boldsymbol{a}(\tau) \otimes F\big(\tau\big)(y, z) \Big) \frac{\alpha(\tau)\, h^{|\tau|}}{|\tau|!}$$

is called B-series for the elementary weight function $\boldsymbol{a}$ at (y, z) with stepsize h.

Again, $\otimes$ denotes the Kronecker product which will often be omitted for convenience. If the type of the elementary differentials is clear from the context, say by specifying the set T, the superscript F will be dropped as well.

Using the result of Lemma 8.5 we arrive at the following B-series representations.

Corollary 8.8. *For $r \geq 1$ define the elementary weight functions*

$$\boldsymbol{S} : T_y \cup T_z \to \mathbb{R}^r, \quad \boldsymbol{S}_{i+1}(\tau) = \left\{ \begin{array}{ll} |\tau|! & , \ |\tau| = i \\ 0 & , \ otherwise, \end{array} \right. \quad i = 0, \dots, r-1,$$

$$\boldsymbol{E} : T_y \cup T_z \to \mathbb{R}^r, \quad \boldsymbol{E}_{i+1}(\tau) = \left\{ \begin{array}{ll} \frac{|\tau|!}{(|\tau|-i)!} & , \ |\tau| \geq i \\ 0 & , \ otherwise, \end{array} \right. \quad i = 0, \dots, r-1.$$

Let (y, z) be the exact solution of (8.1). Then, for $i = 0, \dots, r-1$, scaled derivatives of (y, z) are given by

$$h^i y^{(i)}(t_0) = B_{T_y}^F\big(\boldsymbol{S}_{i+1}; y(t_0), z(t_0)\big), \quad h^i y^{(i)}(t_0+h) = B_{T_y}^F\big(\boldsymbol{E}_{i+1}; y(t_0), z(t_0)\big),$$
$$h^i z^{(i)}(t_0) = B_{T_z}^G\big(\boldsymbol{S}_{i+1}; y(t_0), z(t_0)\big), \quad h^i z^{(i)}(t_0+h) = B_{T_z}^G\big(\boldsymbol{E}_{i+1}; y(t_0), z(t_0)\big).$$

Proof. With Lemma 8.5 and Definition 8.7 the representations of $h^i y^{(i)}(t_0)$ and $h^i z^{(i)}(t_0)$ are obvious. Recall that the exact solution at $t_0 + h$ is given by $y(t_0 + h) = \sum_{p=0}^{\infty} y^{(p)}(t_0) \frac{h^p}{p!}$. Thus we calculate

$$h^i y^{(i)}(t_0 + h) = \sum_{p=0}^{\infty} y^{(p+i)}(t_0) \frac{h^{p+i}}{p!} = \sum_{p=0}^{\infty} \sum_{\substack{\tau \in T \\ |\tau|=p+i}} \alpha(\tau)\, F(\tau) \frac{h^{p+i}}{p!}$$

$$= \sum_{\substack{\tau \in T \\ |\tau| \geq i}} \frac{|\tau|!}{(|\tau|-i)!} \alpha(\tau)\, F(\tau) \frac{h^{|\tau|}}{|\tau|!} = B_{T_y}^F\big(\mathbf{E}_{i+1}; y(t_0), z(t_0)\big). \qquad \square$$

With Corollary 8.8 a large step was taken towards deriving order conditions for general linear methods in the context of implicit index-1 DAEs – at least in case of methods in Nordsieck form. In this case, $\mathbf{S}$ represents the exact starting procedure with

$$\hat{y}^{[0]} = B_{T_y}\big(\mathbf{S}; y(t_0), z(t_0)\big), \qquad \hat{z}^{[0]} = B_{T_z}\big(\mathbf{S}; y(t_0), z(t_0)\big).$$

The result of applying $\mathbf{S}$ to the exact solution at $t_0 + h$ is given by

$$\hat{y}^{[1]} = B_{T_y}\big(\mathbf{E}; y(t_0), z(t_0)\big), \qquad \hat{z}^{[1]} = B_{T_z}\big(\mathbf{E}; y(t_0), z(t_0)\big).$$

It is clear from Figure 8.2 that in order to derive order conditions we need to find similar expressions for the numerical result $y^{[1]}$, $z^{[1]}$ after taking one step with the general linear method. The derivation of corresponding B-series representations is the aim of the next section.

8.1.2 Taylor Expansion of the Numerical Solution

Let $\mathcal{M} = [\mathcal{A},\mathcal{U},\mathcal{B},\mathcal{V}]$ be a general linear method having a nonsingular matrix $\mathcal{A}$. We assume that $\mathcal{M}$ is given in Nordsieck form and consider one step of the numerical scheme (8.4). Let the initial values $y(t_0) = y_0$, $z(t_0) = z_0$ be consistent. The input quantities are assumed to be exact, i.e. $y^{[0]} = B_{T_y}\big(\mathbf{S}; y(t_0), z(t_0)\big)$ as well as $z^{[0]} = B_{T_z}\big(\mathbf{S}; y(t_0), z(t_0)\big)$ are exact Nordsieck vectors at t_0. Our aim is to derive B-series representations for the quantities

$$Y = B_{T_y}\big(\mathbf{v}; y(t_0), z(t_0)\big), \qquad hZ' = B_{T_z}\big(\mathbf{k}; y(t_0), z(t_0)\big), \qquad (8.8a)$$
$$y^{[1]} = B_{T_y}\big(\mathbf{y}; y(t_0), z(t_0)\big), \qquad z^{[1]} = B_{T_z}\big(\mathbf{z}; y(t_0), z(t_0)\big) \qquad (8.8b)$$

involved in computing the numerical result. The definition of the elementary weight functions $\mathbf{v}$, $\mathbf{k}$, $\mathbf{y}$ and $\mathbf{z}$ will be given in due course. Before doing so a preliminary definition will prove useful.

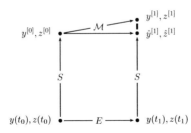

Figure 8.2: One step taken by a GLM $\mathcal{M}$ for the problem (8.1).

Definition 8.9. *For any given elementary weight function* $\boldsymbol{a} : T_y \to \mathbb{R}^s$ *with* $\boldsymbol{a}(\emptyset) = e = \begin{bmatrix} 1 & \cdots & 1 \end{bmatrix}^\top$ *define*

$$\boldsymbol{a}D : T_z \to \mathbb{R}^s, \qquad (\boldsymbol{a}D)(\sigma) = \begin{cases} e & , \sigma = \emptyset, \\ \boldsymbol{a}(\tau_1) \cdots \boldsymbol{a}(\tau_k) & , \sigma = \overset{\tau_1 \cdots \tau_k}{\vee}. \end{cases}$$

Observe that even though $\mathbf{a}$ is defined on T_y, $\mathbf{a}D$ operates on the set T_z. As $\mathbf{a}D$ produces results in $\mathbb{R}^s$, products $\mathbf{a}(\tau_1) \cdots \mathbf{a}(\tau_k)$ have to be understood component-wise. In particular $(\mathbf{a}D)(\bullet) = (\mathbf{a}D)([\,]_y) = \mathbf{a}(\emptyset) = e$.

Obviously, D is closely related to the derivative operator used in [25]. However, when dealing with implicit index-1 DAEs, D has a slightly different meaning. In the next lemma we show that D expresses the relationship $z = g(y)$ on the level of elementary weight functions.

Lemma 8.10. *Let* $\boldsymbol{v} : T_y \to \mathbb{R}^s$ *be an elementary weight function satisfying* $\boldsymbol{v}(\emptyset) = e = \begin{bmatrix} 1 & \cdots & 1 \end{bmatrix}^\top$. *If the initial values* (y_0, z_0) *are consistent, then the following relation holds*

$$\tilde{g}\Big(B_{T_y}(\boldsymbol{v}; y_0, z_0) \Big) = B_{T_z}(\boldsymbol{v}D; y_0, z_0).$$

Proof. Since $\mathbf{v}(\emptyset) = (\mathbf{v}D)(\emptyset) = e$ we have

$$\Big(\tilde{g}\Big(B_{T_y}(\mathbf{v}; y_0, z_0) \Big) \Big)_{h=0} = e \otimes g(y_0) = e \otimes z_0 = \Big(B_{T_z}(\mathbf{v}D; y_0, z_0) \Big)_{h=0}.$$

In order to prove the assertion we need to show that

$$\left(\frac{\mathrm{d}^p}{\mathrm{d}h^p} \tilde{g}\Big(B_{T_y}(\mathbf{v}; y_0, z_0) \Big) \right)_{h=0} = \left(\frac{\mathrm{d}^p}{\mathrm{d}h^p} B_{T_z}(\mathbf{v}D; y_0, z_0) \right)_{h=0}$$

holds for every $p \geq 1$. To this end it is convenient to introduce

$$\bar{Y}^{(k)} := \left(\frac{\mathrm{d}^k}{\mathrm{d}h^k} B_{T_y}(\mathbf{v}; y_0, z_0) \right)_{h=0} = \sum_{\hat{\tau} \in T_y^k} \alpha(\hat{\tau}) \mathbf{v}(\hat{\tau}) F(\hat{\tau})(y_0, z_0)$$

for $k \geq 1$. With this notation, Lemma 8.2, 8.5 and the multi-linearity of the derivative g_{ky} one obtains for $p \geq 1$

$$\left(\frac{\mathrm{d}^p}{\mathrm{d}h^p} \tilde{g}\Big(B_{T_y}(\mathbf{v}; y_0, z_0) \Big) \right)_{h=0} = \sum_{\sigma = [\tau_1, \ldots, \tau_k]_z \in ST_z^p} \beta(\sigma)\, \tilde{g}_{ky}\Big(\bar{Y}^{(|\tau_1|)}, \ldots, \bar{Y}^{(|\tau_k|)} \Big)$$

$$= \sum_{\substack{\sigma \in ST_z^p \\ \sigma = [\tau_1, \ldots, \tau_k]_z}} \sum_{\hat{\tau}_1 \in T_y^{|\tau_1|}} \cdots \sum_{\hat{\tau}_k \in T_y^{|\tau_k|}} \underbrace{\beta(\sigma)\alpha(\hat{\tau}_1) \cdots \alpha(\hat{\tau}_k)}_{=\alpha([\hat{\tau}_1, \ldots, \hat{\tau}_k]_z)} \cdot \underbrace{\mathbf{v}(\hat{\tau}_1) \cdots \mathbf{v}(\hat{\tau}_k)}_{=(\mathbf{v}D)([\hat{\tau}_1, \ldots, \hat{\tau}_k]_z)} \cdot$$

$$\cdot \underbrace{g_{ky}\Big(F(\hat{\tau}_1), \ldots, F(\hat{\tau}_k) \Big)}_{=G([\hat{\tau}_1, \ldots, \hat{\tau}_k]_z)}$$

$$= \sum_{\hat{\sigma} \in T_z^p} \alpha(\hat{\sigma})(\mathbf{v}D)(\hat{\sigma}) G(\hat{\sigma})(y_0, z_0)$$

which already proves the assertion. □

Lemma 8.10 establishes the relationship $z = g(y)$ on the level of elementary weight functions. As (8.1) is a coupled system of two equations, the equation $y' = f(y, z')$ has to be addressed as well. In order to do so we give the following definition.

Definition 8.11. *Given the general linear method* $\mathcal{M} = [\mathcal{A}, \mathcal{U}, \mathcal{B}, \mathcal{V}]$ *with non-singular* $\mathcal{A}$ *define the following elementary weight functions:*

$$\boldsymbol{v} : T_y \rightarrow \mathbb{R}^s, \qquad \boldsymbol{v}(\tau) = \mathcal{A}\, \boldsymbol{l}(\tau) + \mathcal{U}\, \boldsymbol{S}(\tau)$$

$$\boldsymbol{l} : T_y \rightarrow \mathbb{R}^s, \qquad \boldsymbol{l}(\tau) = \begin{cases} 0 & , \tau = \emptyset, \\ e & , \tau = \bullet \\ |\tau|\, \boldsymbol{v}(\tau_1) \cdots \boldsymbol{v}(\tau_k) \frac{k(\sigma_1)}{|\sigma_1|} \cdots \frac{k(\sigma_l)}{|\sigma_l|} & , \tau = \tau_1 \overset{\sigma_1}{\underset{\sigma_l}{\bigvee}} \end{cases}$$

$$\boldsymbol{k} : T_z \rightarrow \mathbb{R}^s, \quad (\boldsymbol{v}D)(\sigma) = \mathcal{A}\, \boldsymbol{k}(\sigma) + \mathcal{U}\, \boldsymbol{S}(\sigma).$$

$\boldsymbol{S}$ denotes the elementary weight function representing exact Nordsieck vectors $y^{[0]} = B_{T_y}(\boldsymbol{S}; y_0, z_0)$ and $z^{[0]} = B_{T_z}(\boldsymbol{S}; y_0, z_0)$. This map was already introduced in Corollary 8.8. Observe that $\boldsymbol{v}(\emptyset) = \mathcal{U}\,\boldsymbol{S}(\emptyset) = \mathcal{U}\,e_1 = e$ due to pre-consistency (cf. Definition 7.4). Thus $\boldsymbol{v}D$ can be constructed as in Definition 8.9. Indeed, $\boldsymbol{k}(\sigma)$ is defined in terms of $(\boldsymbol{v}D)(\sigma)$ and $\boldsymbol{S}(\sigma)$ since $\mathcal{A}$ is nonsingular by assumption.

Example 8.12. In order to get a feeling for how these elementary weight functions look like, we want to calculate $\boldsymbol{v}(\overset{\bullet}{\bigvee})$. For simplicity, assume that $r \geq 4$ such that $\begin{bmatrix} u_1 & u_2 & u_3 & u_4 & \cdots \end{bmatrix}$ denotes the columns of $\mathcal{U}$.

Descending into the recursion:

$$\mathbf{v}(\overset{\bullet}{\bigvee}) = \mathcal{A}\, \mathbf{l}(\overset{\bullet}{\bigvee}) + 6\, u_4$$
$$\downarrow$$
$$\mathbf{l}(\overset{\bullet}{\bigvee}) = \tfrac{3}{2}\, \mathbf{v}(\bullet)\, \mathbf{k}(\overset{\bullet}{\vee}) \qquad \longrightarrow \qquad \mathbf{k}(\overset{\bullet}{\vee}) = \mathcal{A}^{-1}\big(\mathbf{v}(\bullet)^2 - 2\, u_3\big)$$
$$\downarrow$$
$$\mathbf{v}(\bullet) = \mathcal{A}\, \mathbf{l}(\bullet) + u_2$$
$$\downarrow$$
$$\mathbf{l}(\bullet) = e$$

Ascending from the recursion:

$$\mathbf{v}(\overset{\bullet}{\bigvee}) = \mathcal{A}\left[\tfrac{3}{2}\, c\, \mathcal{A}^{-1}(c^2 - 2\, u_3)\right] + 6\, u_4$$
$$\uparrow$$
$$\mathbf{l}(\overset{\bullet}{\bigvee}) = \tfrac{3}{2}\, c\, \mathcal{A}^{-1}(c^2 - 2\, u_3) \qquad \longleftarrow \qquad \mathbf{k}(\overset{\bullet}{\vee}) = \mathcal{A}^{-1}(c^2 - 2\, u_3)$$
$$\uparrow$$
$$\mathbf{v}(\bullet) = \mathcal{A}\, e + u_2 =: c$$
$$\uparrow$$
$$\mathbf{l}(\bullet) = e$$

In the above calculations the vector $c = \mathbf{v}(\bullet) = \mathcal{A}e + u_2 = \mathcal{A}e + \mathcal{U}e_2$ was used as suggested by 7.8 in the previous section. Recall that $\alpha_1 = e_2 = \begin{bmatrix} 0 & 1 & 0 \cdots & 0 \end{bmatrix}^{\mathsf{T}}$ for methods in Nordsieck form. Vector-vector products such as c^2 have to be understood in a component-by-component sense.

Again, let us stress that these computations can be carried out efficiently using computer algebra tools such as MAPLE [9, 112]. □

Equipped with the elementary weight functions from Definition 8.11 we subject the equation $y' = f(y, z')$ to closer scrutiny.

Lemma 8.13. *Consider $\mathbf{v}$, $\mathbf{l}$ and $\mathbf{k}$ from Definition 8.11. Then the following relation holds*

$$h\, \tilde{f}\Big(B_{T_y}(\mathbf{v};\, y_0, z_0),\, \tfrac{1}{h} B_{T_z}(\mathbf{k};\, y_0, z_0) \Big) = B_{T_y}(\mathbf{l};\, y_0, z_0). \tag{8.9}$$

Before proving Lemma 8.13 some brief remarks are due. Lemma 8.13 together with Definition 8.11 reveals how B-series relate to the equation $y' = f(y, z')$. It comes as no surprise that the interaction of y, y' and z' complicates matters considerably. It is for this reason that Definition 8.11 is not that straightforward. However, Lemma 8.10 and 8.13 are the key results for proving that the quantities involved in the numerical scheme (8.4) are all B-series. As a guide through the following computations keep in mind that finally we want to prove (8.8), i.e. $\mathbf{v}$ represents the stages Y, but $\mathbf{l}$ and $\mathbf{k}$ represent the scaled stage derivatives $h\,Y'$ and $h\,Z'$, respectively.

Proof. The idea of the proof is similar to the one of Lemma 8.10. Notice that $\tfrac{1}{h} B_{T_z}(\mathbf{k};\, y_0, z_0)$ represents the formal series

$$\frac{1}{h} B_{T_z}(\mathbf{k};\, y_0, z_0) = \sum_{\sigma \in T_z, |\sigma| \geq 1} \mathbf{k}(\sigma) G(\sigma)(y_0, z_0) \frac{\alpha(\sigma)\, h^{|\sigma|-1}}{|\sigma|!}.$$

The zero-order term $\mathbf{k}(\emptyset) z_0 \tfrac{1}{h}$ is not present since $\mathbf{k}(\emptyset) = 0$. This follows immediately from the pre-consistency conditions in Definition 7.4. A similar remark applies to $\tfrac{1}{h} B_{T_y}(\mathbf{l};\, y_0, z_0)$ as well.

For $h = 0$ nothing is to show for (8.9) as $\mathbf{l}(\emptyset) = 0$. Calculating the first derivative we find

$$\left(\frac{\mathrm{d}}{\mathrm{d}h} \Big(h\, \tilde{f}\Big(B_{T_y}(\mathbf{v};\, y_0, z_0),\, \tfrac{1}{h} B_{T_z}(\mathbf{k};\, y_0, z_0) \Big) \Big) \right)_{h=0}$$

$$= \tilde{f}\big(e \otimes y_0, \mathbf{k}(\mathbf{\mathring{i}}) \otimes g_y(y_0) y'(t_0) \big)$$

$$= e \otimes f\big(y_0, z'(t_0) \big) = e \otimes y'(t_0) = \left(\frac{\mathrm{d}}{\mathrm{d}h}\, B_{T_y}(\mathbf{l};\, y_0, z_0) \right)_{h=0},$$

where $\mathbf{k}(\mathbf{\mathring{i}}) = e$ was used. This relation is a consequence of Definition 8.11. In order to prove a similar result for higher derivatives, we consider $\bar{Y}^{(k)}$ defined in the proof of Lemma 8.10 and introduce

$$\bar{Z}^{(k)} := \left(\frac{\mathrm{d}^{k-1}}{\mathrm{d}h^{k-1}} \frac{1}{h} B_{T_z}(\mathbf{k};\, y_0, z_0)\right)_{h=0} = \frac{1}{k} \sum_{\hat{\sigma} \in T_z^k} \alpha(\hat{\sigma}) \mathbf{k}(\hat{\sigma}) G(\hat{\sigma})(y_0, z_0)$$

for $k \geq 1$. We will calculate derivatives of the left hand side of (8.9). Using Lemma 8.5 and the multi-linearity of $f_{kylz'}$ we find

$$\left(\frac{\mathrm{d}^p}{\mathrm{d}h^p}\left(h\,\tilde{f}\Big(B_{T_y}(\mathbf{v};\, y_0, z_0),\, \tfrac{1}{h} B_{T_z}(\mathbf{k};\, y_0, z_0)\Big)\right)\right)_{h=0}$$

$$= p \sum_{\substack{\tau \in ST_{yy}^p \\ \tau = [\tau_1, \ldots, \tau_k, \sigma_1, \ldots, \sigma_l]_y}} \beta(\tau)\, f_{kylz'}\big(\bar{Y}^{(|\tau_1|)}, \ldots, \bar{Y}^{(|\tau_k|)}, \bar{Z}^{(|\sigma_1|+1)}, \ldots, \bar{Z}^{(|\sigma_l|+1)}\big) + p\, f_{z'} \bar{Z}^{(p)}$$

$$= p \sum_{\substack{\tau \in T_{yy}^p \\ \tau = [\hat{\tau}_1, \ldots, \hat{\tau}_k, \hat{\sigma}_1, \ldots, \hat{\sigma}_l]_y}} \alpha(\tau)\, \mathbf{v}(\hat{\tau}_1) \cdots \mathbf{v}(\hat{\tau}_k) \frac{\mathbf{k}(\hat{\sigma}_1)}{|\hat{\sigma}_1|} \cdots \frac{\mathbf{k}(\hat{\sigma}_l)}{|\hat{\sigma}_l|}\, \mathsf{M}\, F(\tau) + \sum_{\sigma \in T_z^p} \alpha(\sigma) \mathbf{k}(\sigma) f_{z'} G(\sigma)$$

$$= \sum_{\tau \in T_{yy}^p} \alpha(\tau)\, \mathbf{l}(\tau)\, \mathsf{M}\, F(\tau) + \sum_{\sigma \in T_z^p} \alpha(\sigma) \mathbf{k}(\sigma) f_{z'} G(\sigma). \tag{8.10}$$

The definition of $\mathbf{l}$ was already inserted here to simplify expressions. With (8.10) we are half way to the result, but we have to study the second term

$$\sum_{\sigma \in T_z^p} \alpha(\sigma) \mathbf{k}(\sigma) f_{z'} G(\sigma) = \sum_{\substack{\sigma = [\tau_1]_z \in T_z^p}} \alpha(\sigma) \mathbf{k}(\sigma) f_{z'} G(\sigma) + \sum_{\substack{\sigma \in T_z^p \\ \sigma = [\tau_1, \ldots, \tau_k]_z, k \geq 2}} \alpha(\sigma) \mathbf{k}(\sigma) f_{z'} G(\sigma) \tag{8.11}$$

in more detail. Recall that for $\sigma \in T_z$ we have $\tau = [\sigma]_y \in T_{yz}$ if and only if σ has at least two subtrees, i.e. $\sigma = [\tau_1, \ldots, \tau_k]_z$ with $k \geq 2$. In this situation we find $|\tau| = |\sigma|$, $\alpha(\tau) = \alpha(\sigma)$ and $\mathbf{l}(\tau) = \mathbf{k}(\sigma)$, such that

$$\sum_{\substack{\sigma \in T_z^p \\ \sigma = [\tau_1, \ldots, \tau_k]_z, k \geq 2}} \alpha(\sigma) \mathbf{k}(\sigma) f_{z'} G(\sigma) = \sum_{\tau \in T_{yz}^p} \alpha(\tau) \mathbf{l}(\tau)\, \mathsf{M}\, F(\tau). \tag{8.12}$$

On the other hand if $\sigma = [\tau]_z \in T_z$, then $\alpha(\sigma) = \alpha(\tau)$ (since $\beta(sp(\sigma)) = 1$), $\mathbf{S}(\sigma) = \mathbf{S}(\tau)$ and the definition of $\mathbf{v}$ yields $\mathbf{l}(\sigma) = \mathcal{A}^{-1}\big(\mathbf{v}(\tau) - \mathcal{U}\,\mathbf{S}(\tau)\big) = \mathbf{l}(\tau)$. Finally, $f_{z'} G(\sigma) = f_{z'} g_y F(\tau) = (I - \mathsf{M})\, F(\tau)$. Putting all this together yields

$$\sum_{\sigma = [\tau_1]_z \in T_z^p} \alpha(\sigma) \mathbf{k}(\sigma) f_{z'} G(\sigma) = \sum_{\tau \in T_y^p} \alpha(\tau)\, \mathbf{l}(\tau)\, (I - \mathsf{M})\, F(\tau). \tag{8.13}$$

Inserting (8.11), (8.12), (8.13) into (8.10) concludes the proof since

$$\left(\frac{\mathrm{d}^p}{\mathrm{d}h^p}\left(h\,\tilde{f}\Big(B_{T_y}(\mathbf{v};\, y_0, z_0),\, \tfrac{1}{h} B_{T_z}(\mathbf{k};\, y_0, z_0)\Big)\right)\right)_{h=0} = \sum_{\tau \in T_y^p} \alpha(\tau)\, \mathbf{l}(\tau)\, F(\tau). \quad \square$$

Taking a closer look at Lemma 8.10 and 8.13 the Taylor expansion of the numerical solution is now obvious. It remains to collect the result in the following theorem.

Theorem 8.14. *Let* $y^{[0]} = B_{T_y}(\boldsymbol{S}; y_0, z_0)$ *and* $z^{[0]} = B_{T_z}(\boldsymbol{S}; y_0, z_0)$ *be exact Nordsieck vectors. Consider one step taken by the general linear method* $\mathcal{M}$,

$$Y = h\mathcal{A}\,\tilde{f}(Y, Z', t_c) + \mathcal{U}\,y^{[0]}, \quad \tilde{g}(Y, t_c) = h\mathcal{A}\,Z' + \mathcal{U}\,z^{[0]}, \tag{8.14a}$$
$$y^{[1]} = h\mathcal{B}\,\tilde{f}(Y, Z', t_c) + \mathcal{V}\,y^{[0]}, \qquad z^{[1]} = h\mathcal{B}\,Z' + \mathcal{V}\,z^{[0]}. \tag{8.14b}$$

Then the stages Y, *the stage derivatives* hZ' *as well as the output quantities* $y^{[1]}$, $z^{[1]}$ *are B-series*

$$Y = B_{T_y}(\boldsymbol{v}; y_0, z_0), \qquad\qquad hZ' = B_{T_z}(\boldsymbol{k}; y_0, z_0), \tag{8.14c}$$
$$y^{[1]} = B_{T_y}(\boldsymbol{y}; y_0, z_0), \qquad\qquad z^{[1]} = B_{T_z}(\boldsymbol{z}; y_0, z_0), \tag{8.14d}$$

where

$$\boldsymbol{y}(\tau) = \mathcal{B}\,\boldsymbol{l}(\tau) + \mathcal{V}\,\boldsymbol{S}(\tau), \quad \tau \in T_y, \qquad \boldsymbol{z}(\sigma) = \mathcal{B}\,\boldsymbol{k}(\sigma) + \mathcal{V}\,\boldsymbol{S}(\sigma), \quad \sigma \in T_z.$$

The elementary weight functions $\boldsymbol{v}$, $\boldsymbol{l}$ *and* $\boldsymbol{k}$ *are given in Definition 8.11.*

Proof. Consider the elementary weight function $\mathbf{v} = \mathcal{A}\mathbf{1} + \mathcal{U}\mathbf{S}$ from Definition 8.11. Using B-series notation and applying Lemma 8.13 shows that

$$B_{T_y}(\mathbf{v}; y_0, z_0) = \mathcal{A}\,B_{T_y}(\mathbf{l}; y_0, z_0) + \mathcal{U}\,B_{T_y}(\mathbf{S}; y_0, z_0) \tag{8.15a}$$
$$= h\,f\Big(B_{T_y}(\mathbf{v}; y_0, z_0), \tfrac{1}{h}B_{T_z}(\mathbf{k}; y_0, z_0)\Big) + \mathcal{U}\,B_{T_y}(\mathbf{S}; y_0, z_0).$$

On the other hand, Lemma 8.10 and the definition of $\mathbf{k}$ yields

$$g\Big(B_{T_y}(\mathbf{v}; y_0, z_0)\Big) = B_{T_z}(\mathbf{v}D; y_0, z_0)$$
$$= \mathcal{A}\,B_{T_z}(\mathbf{k}; y_0, z_0) + \mathcal{U}\,B_{T_z}(\mathbf{S}; y_0, z_0). \tag{8.15b}$$

Thus, comparing (8.15) with (8.14a) reveals that the stages are B-series given by (8.14c). Due to linearity (8.14d) is an immediate consequence of (8.14b) and (8.14c). □

8.1.3 Order Conditions

The previous two sections were devoted to deriving B-series representations of both the analytical and the numerical solution. By comparing these Taylor series expansions it is possible to derive order conditions for general linear methods applied to implicit index-1 DAEs.

We assume that exact initial values $y_0 = y(t_0)$, $z_0 = z(t_0)$ are given. For general linear methods this information is not enough to start the integration, but we need to calculate initial input vectors $y^{[0]}$ and $z^{[0]}$. We assume that the process of calculating $y^{[0]}$, $z^{[0]}$ from $y_0 = y(t_0)$, $z_0 = z(t_0)$ can be described as B-series, i.e.

$$y^{[0]} = B_{T_y}^F\big(\mathbf{S}; y_0, z_0\big), \qquad\qquad z^{[0]} = B_{T_z}^G\big(\mathbf{S}; y_0, z_0\big)$$

where $\mathbf{S}$ is defined in Corollary 8.8. Using $y^{[0]}$ and $z^{[0]}$ as input quantities we employ a given general linear method to calculate output vectors $y^{[1]}$ and $z^{[1]}$. Recall from Definition 7.1 that the method has order p if

$$y^{[1]} = \hat{y}^{[1]} + \mathcal{O}(h^{p+1}), \qquad\qquad z^{[1]} = \hat{z}^{[1]} + \mathcal{O}(h^{p+1}).$$

Here, $\hat{y}^{[1]} = B_{T_y}\big(\mathbf{S}; y(t_0 + h), z(t_0 + h)\big)$ and $\hat{z}^{[1]} = B_{T_z}\big(\mathbf{S}; y(t_0 + h), z(t_0 + h)\big)$ are obtained by applying the starting procedure $\mathbf{S}$ to the exact solution at the next timepoint (see Figure 8.2 on page 135).

If we restrict attention to methods in Nordsieck form, Corollary 8.8 and Theorem 8.14 show that $y^{[1]}$, $z^{[1]}$ as well as $\hat{y}^{[1]}$, $\hat{z}^{[1]}$ can be written as B-series,

$$y^{[1]} = B_{T_y}^F\big(\mathbf{y}; y_0, z_0\big), \qquad\qquad \hat{y}^{[1]} = B_{T_y}^F\big(\mathbf{E}; y_0, z_0\big),$$
$$z^{[1]} = B_{T_z}^G\big(\mathbf{z}; y_0, z_0\big), \qquad\qquad \hat{z}^{[1]} = B_{T_z}^G\big(\mathbf{E}; y_0, z_0\big).$$

Thus order conditions can be derived by comparing elementary weight functions.

Theorem 8.15. Let $\mathcal{M} = [\mathcal{A}, \mathcal{U}, \mathcal{B}, \mathcal{V}]$ be a general linear method in Nordsieck form with a nonsingular matrix $\mathcal{A}$. $\mathcal{M}$ has order p for implicit index-1 DAEs (8.1) if and only if

$$\mathbf{y}(\tau) = \mathbf{E}(\tau) \qquad\qquad \forall\ \tau \in T_y\ \text{with}\ |\tau| \le p,$$
$$\mathbf{z}(\sigma) = \mathbf{E}(\sigma) \qquad\qquad \forall\ \sigma \in T_z\ \text{with}\ |\sigma| \le p. \qquad\qquad \square$$

If a general linear method is applied to implicit index-1 DAEs (8.1), it produces results with a local error of order $\mathcal{O}(h^{p+1})$ if and only if the conditions of Theorem 8.15 are satisfied. However, recall from Table 8.3 that the number of trees for a given order, and thus the number of order conditions, grows exponentially.

In the remainder of this section we focus on simplifying the result of Theorem 8.15. As a first step the close relationship between $\mathbf{l}$, $\mathbf{k}$ and $\mathbf{y}$, $\mathbf{z}$, respectively, is investigated. It will turn out that it suffices to consider trees in T_y only. The trees from T_z yield no additional order conditions.

Lemma 8.16. Let $\mathcal{M}$ be a general linear method in Nordsieck form with a nonsingular matrix $\mathcal{A}$. Then for every tree $\sigma \in T_z$ there is a tree $\tau \in T_y$ such that $|\sigma| = |\tau|$ and $\mathbf{k}(\sigma) = \mathbf{l}(\tau)$, $\mathbf{z}(\sigma) = \mathbf{y}(\tau)$.

Proof. Recall that $\mathbf{S}(\emptyset) = e_1 = \begin{bmatrix} 1 & 0 & \cdots & 0 \end{bmatrix}^\top$ is the first canonical unit vector for methods in Nordsieck form. The first column of $\mathcal{U}$ is given by $\mathcal{U} e_1 = e = \begin{bmatrix} 1 & \cdots & 1 \end{bmatrix}^\top$ due to pre-consistency (Definition 7.4).

Let σ be an arbitrary tree in T_z. For $\sigma = \emptyset$ nothing is to show since the relation $\mathbf{k}(\emptyset) = \mathcal{A}^{-1}\big((\mathbf{v}D)(\emptyset) - \mathcal{U} S(\emptyset)\big) = \mathcal{A}^{-1}\big(e - \mathcal{U} e_1\big) = 0 = \mathbf{l}(\emptyset)$.

For $|\sigma| \geq 1$ we distinguish two cases:

   (*a*) $\sigma = [\tau]_z$ for some $\tau \in T_y$.

        Observe that $|\sigma| = |\tau|$ and therefore $\mathbf{S}(\sigma) = \mathbf{S}(\tau)$. Calculating $\big(\mathbf{v}D\big)(\sigma) = \mathbf{v}(\tau) = \mathcal{A}\mathbf{l}(\tau) + \mathcal{U}\mathbf{S}(\tau)$ we find $\mathbf{k}(\sigma) = \mathcal{A}^{-1}\big((\mathbf{v}D)(\sigma) - \mathcal{U} S(\sigma)\big) = \mathbf{l}(\tau)$. This immediately implies $\mathbf{z}(\sigma) = \mathbf{y}(\tau)$ as well.

   (*b*) $\sigma = [\tau_1, \ldots, \tau_k]_z$ with $\tau_i \in T_y$ and $k \geq 2$.

        Let $\tau = [\sigma]_y$, i.e. we attach σ to a new black root. Then $\tau \in T_{yz}$ and, again, $|\tau| = |\sigma|$. Definition 8.11 yields $\mathbf{l}(\tau) = |\tau|\frac{\mathbf{k}(\sigma)}{|\sigma|} = \mathbf{k}(\sigma)$ and thus $\mathbf{z}(\sigma) = \mathbf{y}(\tau)$. $\qquad\qquad\qquad\Box$

If $\tau \in T_y$ is a tree corresponding to $\sigma \in T_z$ in the sense of Lemma 8.16, then τ and σ have the same order. Therefore $\mathbf{E}(\tau) = \mathbf{E}(\sigma)$ (cf. Corollary 8.8) and, due to $\mathbf{y}(\tau) = \mathbf{z}(\sigma)$, the elementary weight function $\mathbf{z}$ can be omitted from Theorem 8.15.

Another means of reducing the number of order conditions considerably is to study simplified trees. This was already done in [104].

Definition 8.17. *Every tree $\tau \in T_y$ can be associated with a simplified tree $\bar{\tau}$ as follows: If a white [black] node (except the root) has no ramifications and is followed by a black [white] node, then the tree can be simplified by removing these two nodes (see Figure 8.3). The simplified tree $\bar{\tau}$ corresponding to τ is the tree which is simplified as much as possible.*

$\bar{T}_y = \{\,\bar{\tau} \mid \tau \in T_y\,\}$ *is the set of simplified trees.*

Figure 8.3: Simplification of trees.

Pictorial representations of the simplified trees up to order 4 are given in Table 8.5 on page 144. Observe that simplifying a tree does not change its order.

The significance of the simplified trees is straightforward: The order conditions generated by $\bar{T}_y$ coincide with those generated by T_y.

Lemma 8.18. *Let $\tau \in T_y$. Then it holds that $\mathbf{y}(\tau) = \mathbf{y}(\bar{\tau})$.*

Proof. The proof is similar to the one carried out in [104]. Let $\tau \in T_y$ be a tree which contains $\eta = $ $= \big[[\tau_1, \ldots, \tau_k, \sigma_1, \ldots, \sigma_l]_y\big]_z$ as a (proper) subtree. The white root has no ramifications but is followed by a black node. Thus the first simplification rule applies. Notice that η is attached to one of τ's black nodes, i.e. $\xi = [\xi_1, \ldots, \xi_m, \eta, \eta_2, \ldots, \eta_n]_y$ is again a subtree of τ (where $\tau = \xi$ is possible). Since $\mathbf{y}(\tau)$ is calculated by evaluating $\mathbf{l}$ and $\mathbf{k}$ for τ's subtrees (depending on the subtrees' roots), in order to prove $\mathbf{y}(\tau) = \mathbf{y}(\bar{\tau})$ it suffices to show that $\mathbf{l}(\xi) = \mathbf{l}(\bar{\xi})$ where $\bar{\xi} = [\xi_1, \ldots, \xi_m, \tau_1, \ldots, \tau_k, \sigma_1, \ldots, \sigma_l, \eta_2, \ldots, \eta_n]_y$.

From Lemma 8.16 we know that

$$\mathbf{k}(\eta) = \mathbf{l}([\tau_1, \ldots, \tau_k, \sigma_1, \ldots, \sigma_l]_y) = |\eta| \prod_{i=1}^{k} \mathbf{v}(\tau_i) \prod_{j=1}^{l} \frac{\mathbf{k}(\sigma_j)}{|\sigma_j|}.$$

Therefore

$$
\begin{aligned}
\mathbf{l}(\xi) &= |\xi| \prod_{i=1}^{m} \mathbf{v}(\xi_i) \frac{\mathbf{k}(\eta)}{|\eta|} \prod_{j=2}^{n} \frac{\mathbf{k}(\eta_j)}{|\eta_j|} \\
&= |\xi| \prod_{i=1}^{m} \mathbf{v}(\xi_i) \prod_{i=1}^{k} \mathbf{v}(\tau_i) \prod_{j=1}^{l} \frac{\mathbf{k}(\sigma_j)}{|\sigma_j|} \prod_{j=2}^{n} \frac{\mathbf{k}(\eta_j)}{|\eta_j|} \\
&= \mathbf{l}\big([\xi_1, \ldots, \xi_m, \tau_1, \ldots, \tau_k, \sigma_1, \ldots, \sigma_l, \eta_2, \ldots, \eta_n]_y\big) = \mathbf{l}(\bar{\xi}).
\end{aligned}
$$

For the second simplification rule, assume that $\eta = $ $= \big[[\tau_1, \ldots, \tau_k]_z\big]_y \neq \tau$ is a subtree of τ. Two cases are possible: Either η is attached to a black node, such that $\xi = [\eta, \eta_2, \ldots, \eta_k, \sigma_1, \ldots, \sigma_l]_y$ is a subtree of τ, or η is connected to a white node and $\xi = [\eta, \eta_2, \ldots, \eta_k]_z$ is one of τ's subtrees.

In both cases $\mathbf{l}(\xi) = \mathbf{l}(\bar{\xi})$ and $\mathbf{k}(\xi) = \mathbf{k}(\bar{\xi})$ is shown along the lines of the reasoning above. The crucial point to notice is that $\mathbf{v}(\eta) = \mathbf{v}(\tau_1) \cdots \mathbf{v}(\tau_k)$.

Up to now only one simplification step was considered. However, it is clear that the arguments can be repeated recursively. □

With Lemma 8.18 we can state the final version of the order conditions.

Theorem 8.19. *Let $\mathcal{M}$ be a general linear method in Nordsieck form with a nonsingular matrix $\mathcal{A}$. $\mathcal{M}$ has order p for implicit index-1 DAEs (8.1) if and only if*

$$\boldsymbol{y}(\tau) = \boldsymbol{E}(\tau) \qquad \forall \ \tau \in \bar{T}_y \ \text{with} \ |\tau| \leq p.$$

Proof. We start from the order conditions of Theorem 8.15. It is, however, sufficient to require $\mathbf{y}(\tau) = \mathbf{E}(\tau)$ for all trees $\tau \in T_y$ with $|\tau| \leq p$ as Lemma 8.16 showed that the trees from T_z yield no additional order conditions.

Because of Lemma 8.18 we can restrict attention to trees from the smaller set $\bar{T}_y$ instead of T_y. □

The number of order conditions up to $p = 6$ is given in Table 8.4. Even though these numbers are still increasing exponentially, compared to Table 8.3 the number of order conditions is significantly reduced by considering simplified trees.

In order to write down the order conditions for a given general linear method $\mathcal{M} = [\mathcal{A}, \mathcal{U}, \mathcal{B}, \mathcal{V}]$ with s internal and r external stages, it is convenient to split

$$\mathcal{U} = \begin{bmatrix} u_1 & u_2 & \cdots & u_s \end{bmatrix} \qquad \text{and} \qquad \mathcal{V} = \begin{bmatrix} v_1 & v_2 & \cdots & v_r \end{bmatrix}$$

order	1	2	3	4	5	6	$\sum$
$\bar{T}_y$	1	2	6	21	81	336	447

Table 8.4: The number of order conditions

no.	order	τ	$\mathbf{y}(\tau)$	$=$	$\mathbf{E}(\tau)$
1	1		$\mathcal{B}e + v_2$	$=$	$\begin{bmatrix} 1 & 1 & 0 & 0 & 0 \end{bmatrix}^\top$
2	2		$2\mathcal{B}c + 2v_3$	$=$	$\begin{bmatrix} 1 & 2 & 2 & 0 & 0 \end{bmatrix}^\top$
3	2		$\mathcal{B}\mathcal{A}^{-1}(c^2 - 2u_3) + 2v_3$	$=$	$\begin{bmatrix} 1 & 2 & 2 & 0 & 0 \end{bmatrix}^\top$
4	3		$3\mathcal{B}(2\mathcal{A}c + 2u_3) + 6v_4$	$=$	$\begin{bmatrix} 1 & 3 & 6 & 6 & 0 \end{bmatrix}^\top$
5	3		$3\mathcal{B}c^2 + 6v_4$	$=$	$\begin{bmatrix} 1 & 3 & 6 & 6 & 0 \end{bmatrix}^\top$
6	3		$\mathcal{B}\mathcal{A}^{-1}(c(2\mathcal{A}c + 2u_3) - 6u_4) + 6v_4$	$=$	$\begin{bmatrix} 1 & 3 & 6 & 6 & 0 \end{bmatrix}^\top$
7	3		$\mathcal{B}\mathcal{A}^{-1}(c^3 - 6u_4) + 6v_4$	$=$	$\begin{bmatrix} 1 & 3 & 6 & 6 & 0 \end{bmatrix}^\top$
8	3		$\frac{3}{2}\mathcal{B}c\mathcal{A}^{-1}(c^2 - 2u_3) + 6v_4$	$=$	$\begin{bmatrix} 1 & 3 & 6 & 6 & 0 \end{bmatrix}^\top$
9	3		$\frac{3}{4}\mathcal{B}(\mathcal{A}^{-1}(c^2 - 2u_3))^2 + 6v_4$	$=$	$\begin{bmatrix} 1 & 3 & 6 & 6 & 0 \end{bmatrix}^\top$
10	4		$24\mathcal{B}(\mathcal{A}(\mathcal{A}c + u_3) + u_4) + 24v_5$	$=$	$\begin{bmatrix} 1 & 4 & 12 & 24 & 24 \end{bmatrix}^\top$
11	4		$12\mathcal{B}(\mathcal{A}c^2 + 2u_4) + 24v_5$	$=$	$\begin{bmatrix} 1 & 4 & 12 & 24 & 24 \end{bmatrix}^\top$
12	4		$8\mathcal{B}c(\mathcal{A}c + u_3) + 24v_5$	$=$	$\begin{bmatrix} 1 & 4 & 12 & 24 & 24 \end{bmatrix}^\top$
13	4		$4\mathcal{B}c^3 + 24v_5$	$=$	$\begin{bmatrix} 1 & 4 & 12 & 24 & 24 \end{bmatrix}^\top$
14	4		$6\mathcal{B}(\mathcal{A}c\mathcal{A}^{-1}(c^2 - 2u_3) + 4u_4) + 24v_5$	$=$	$\begin{bmatrix} 1 & 4 & 12 & 24 & 24 \end{bmatrix}^\top$
15	4		$6\mathcal{B}\mathcal{A}^{-1}(c(\mathcal{A}(\mathcal{A}c + u_3) + u_4) - 4u_5) + 24v_5$	$=$	$\begin{bmatrix} 1 & 4 & 12 & 24 & 24 \end{bmatrix}^\top$
16	4		$3\mathcal{B}\mathcal{A}^{-1}(c(\mathcal{A}c^2 + 2u_4) - 8u_5) + 24v_5$	$=$	$\begin{bmatrix} 1 & 4 & 12 & 24 & 24 \end{bmatrix}^\top$
17	4		$4\mathcal{B}\mathcal{A}^{-1}((\mathcal{A}c + u_3)^2 - 6u_5) + 24v_5$	$=$	$\begin{bmatrix} 1 & 4 & 12 & 24 & 24 \end{bmatrix}^\top$
18	4		$2\mathcal{B}\mathcal{A}^{-1}(c^2(\mathcal{A}c + u_3) - 12u_5) + 24v_5$	$=$	$\begin{bmatrix} 1 & 4 & 12 & 24 & 24 \end{bmatrix}^\top$
19	4		$\mathcal{B}\mathcal{A}^{-1}(c^4 - 24u_5) + 24v_5$	$=$	$\begin{bmatrix} 1 & 4 & 12 & 24 & 24 \end{bmatrix}^\top$
20	4		$\frac{4}{3}\mathcal{B}c\mathcal{A}^{-1}(c^3 - 6u_4) + 24v_5$	$=$	$\begin{bmatrix} 1 & 4 & 12 & 24 & 24 \end{bmatrix}^\top$
21	4		$\frac{8}{3}\mathcal{B}c\mathcal{A}^{-1}(c(\mathcal{A}c + u_3) - 3u_4) + 24v_5$	$=$	$\begin{bmatrix} 1 & 4 & 12 & 24 & 24 \end{bmatrix}^\top$
22	4		$4\mathcal{B}(\mathcal{A}c + u_3)\mathcal{A}^{-1}(c^2 - 2u_3) + 24v_5$	$=$	$\begin{bmatrix} 1 & 4 & 12 & 24 & 24 \end{bmatrix}^\top$
23	4		$2\mathcal{B}c^2\mathcal{A}^{-1}(c^2 - 2u_3) + 24v_5$	$=$	$\begin{bmatrix} 1 & 4 & 12 & 24 & 24 \end{bmatrix}^\top$
24	4		$3\mathcal{B}(\mathcal{A}((\mathcal{A}^{-1}(c^2 - 2u_3))^2) + 8u_4) + 24v_5$	$=$	$\begin{bmatrix} 1 & 4 & 12 & 24 & 24 \end{bmatrix}^\top$
25	4		$\frac{3}{2}\mathcal{B}\mathcal{A}^{-1}(c(\mathcal{A}c\mathcal{A}^{-1}(c^2 - 2u_3) + 4u_4) - 24u_5) + 24v_5$	$=$	$\begin{bmatrix} 1 & 4 & 12 & 24 & 24 \end{bmatrix}^\top$
26	4		$\frac{2}{3}\mathcal{B}\mathcal{A}^{-1}(c^2 - 2u_3)\mathcal{A}^{-1}(c^3 - 6u_4) + 24v_5$	$=$	$\begin{bmatrix} 1 & 4 & 12 & 24 & 24 \end{bmatrix}^\top$
27	4		$\frac{4}{3}\mathcal{B}\mathcal{A}^{-1}(c^2 - 2u_3)\mathcal{A}^{-1}(c(\mathcal{A}c + u_3) - 3u_4) + 24v_5$	$=$	$\begin{bmatrix} 1 & 4 & 12 & 24 & 24 \end{bmatrix}^\top$
28	4		$\mathcal{B}c(\mathcal{A}^{-1}(c^2 - 2u_3))^2 + 24v_5$	$=$	$\begin{bmatrix} 1 & 4 & 12 & 24 & 24 \end{bmatrix}^\top$
29	4		$\frac{3}{4}\mathcal{B}\mathcal{A}^{-1}(c(\mathcal{A}(\mathcal{A}^{-1}(c^2 - 2u_3))^2 + 8u_4) - 24u_5) + 24v_5$	$=$	$\begin{bmatrix} 1 & 4 & 12 & 24 & 24 \end{bmatrix}^\top$
30	4		$\frac{1}{2}\mathcal{B}(\mathcal{A}^{-1}(c^2 - 2u_3))^3 + 24v_5$	$=$	$\begin{bmatrix} 1 & 4 & 12 & 24 & 24 \end{bmatrix}^\top$

Table 8.5: Simplified trees of order up to 4 and corresponding order conditions.

columnwise. Definition 8.11 and Theorem 8.14 show how to calculate $\mathbf{y}$ for every tree $\tau \in \bar{T}_y$. The order condition corresponding to $\tau \in \bar{T}_y$ is obtained by equating $\mathbf{y}(\tau)$ and $\mathbf{E}(\tau)$. Since $\mathbf{E}(\tau)$ depends only on the order of τ, $\mathbf{E}$ needs to be calculated only once for every order.

Example 8.20. Let us investigate the order condition associated with the tree $\tau = $ ⋎. On page 137 we already calculated $\mathbf{l}(⋎) = \frac{3}{2} c \mathcal{A}^{-1}\big(c^2 - 2 u_3\big)$. Using the definition of $\mathbf{y}$ from Theorem 8.14 yields

$$\mathbf{y}(⋎) = \mathcal{B}\,\mathbf{l}(⋎) + \mathcal{V}\,\mathbf{S}(⋎) = \tfrac{3}{2}\,\mathcal{B}\,c\,\mathcal{A}^{-1}\big(c^2 - 2 u_3\big) + 6\,v_4.$$

Since $|\tau| = 3$, we find $\mathbf{E}(\tau) = e_1 + 3\,e_2 + 6\,(e_3 + e_4)$, where e_i denotes the i-th canonical unit vector. Thus the order condition corresponding to $\tau = $ ⋎ is given by

$$\tfrac{3}{2}\,\mathcal{B}\,c\,\mathcal{A}^{-1}\big(c^2 - 2 u_3\big) + 6\,v_4 = e_1 + 3\,e_2 + 6\,(e_3 + e_4). \qquad \square$$

The simplified trees of order up to 4 and their corresponding order conditions are given in Table 8.5. The conditions stated there were calculated for $s = r = 5$. However, they are valid for arbitrary s, r as well. In general one has to use the convention $u_k = 0$ for $k > s$ and $v_k = 0$ for $k > r$.

Unfortunately there is no "simple" procedure to read the order conditions directly from a given tree as is the case for Runge-Kutta methods. Due to the multivalue structure of the method it seems unavoidable to use a recursive approach for general linear methods.

Observe that each order condition is an equation in $\mathbb{R}^r$. Taking a closer look, it turns out that the order conditions derived for Runge-Kutta methods by Kværnø in [104] re-appear in the vector's first components. This is no surprise as Runge-Kutta methods are a special case of general linear methods with $r = 1$.

It is well known for Runge-Kutta methods that sufficiently high stage order guarantees that the order conditions for DAEs are satisfied [84]. A similar remark holds when general linear methods are applied to implicit index-1 DAEs. We will investigate this situation in more detail.

8.1.4 Order Conditions and Stage Order

We know from Theorem 7.6 that a general linear method $\mathcal{M} = [\mathcal{A}, \mathcal{U}, \mathcal{B}, \mathcal{V}]$ has order p and stage order $q = p$ for ordinary differential equations if and only if

$$\mathcal{U}W = C - \mathcal{A}CK, \qquad \mathcal{V}W = WE - \mathcal{B}CK.$$

The matrix $W = \begin{bmatrix} \alpha_0 \cdots \alpha_p \end{bmatrix}$ is closely related to Definition 7.1 where order and stage order are defined. Recall that for methods in Nordsieck form with

$s = r = p + 1$ this matrix is given by $W = I$. Thus general linear methods in Nordsieck form with $s = r$ have order and stage order $p = q = s - 1$ if and only if

$$\mathcal{U} = C - \mathcal{A}\,C\,K \qquad \mathcal{V} = E - \mathcal{B}\,C\,K. \tag{8.16}$$

The matrices C, E and K are defined in Theorem 7.1. In this section we will show that, given these definitions for $\mathcal{U}$ and $\mathcal{V}$, the method has order p for implicit index-1 DAEs (8.1) as well. We start by studying the elementary weight function $\mathbf{l}$.

If $\mathcal{U}$ and $\mathcal{V}$ are chosen in the special way (8.16), then $\mathbf{l}$ can be calculated simply as $\mathbf{l}(\tau) = C\,K\,\mathbf{S}(\tau)$. We prove this statement in the following result.

Lemma 8.21. *Let $\mathcal{M} = [\mathcal{A},\mathcal{U},\mathcal{B},\mathcal{V}]$ be a general linear method in Nordsieck form with $s = r = p + 1$ and $\mathcal{U} = C - \mathcal{A}\,C\,K$, $\mathcal{V} = E - \mathcal{B}\,C\,K$. Then*

$$\mathbf{l}(\tau) = C\,K\,\boldsymbol{S}(\tau) \qquad \forall\ \tau \in \bar{T}_y \ \text{ with } |\tau| \leq p. \tag{8.17}$$

Proof. Recall that $C = \begin{bmatrix} e & c & \frac{c^2}{2} & \cdots & \frac{c^p}{p!} \end{bmatrix}$ and $C\,K = \begin{bmatrix} 0 & e & c & \cdots & \frac{c^{p-1}}{(p-1)!} \end{bmatrix}$. Using the definition of the exact starting procedure $\mathbf{S}$ from Corollary 8.8 we find

$$C\,\mathbf{S}(\tau) = c^{|\tau|} \qquad \text{and} \qquad C\,K\,\mathbf{S}(\tau) = |\tau|\,c^{|\tau|-1} \qquad \forall\ |\tau| \leq p.$$

For $\tau = \emptyset$ and $\tau = \bullet$ it is straightforward to calculate $\mathbf{l}(\emptyset) = 0 = C\,K\,\mathbf{S}(\emptyset)$ and $\mathbf{l}(\bullet) = e = C\,K\,\mathbf{S}(\bullet)$, respectively. Assume that (8.17) holds for all trees τ with order $|\tau| \leq k$ where $1 \leq k < p$. Let τ be an arbitrary tree with order $|\tau| = k + 1$. Again we distinguish two cases:

(a) $\tau = \overset{\tau_i}{\underset{}{\bullet}}\!\!\overset{\sigma_i}{\underset{\sigma_i}{\mathbf{Y}}} \in T_{yy}$

The tree $|\tau|$ has order $k + 1$ by assumption and therefore $|\tau_i| \leq k$ and $|\sigma_j| \leq k - 1$. Evaluating $\mathbf{v}$ for the subtrees τ_i yields, due to the induction hypothesis,

$$\begin{aligned} \mathbf{v}(\tau_i) &= \mathcal{A}\,\mathbf{l}(\tau_i) + \mathcal{U}\,\mathbf{S}(\tau_i) \\ &= \mathcal{A}\,C\,K\mathbf{S}(\tau_i) + (C - \mathcal{A}\,C\,K)\,\mathbf{S}(\tau_i) = C\,\mathbf{S}(\tau_i). \end{aligned}$$

Since $|\sigma_j| = |[\sigma_j]_y|$ we have $\mathbf{k}(\sigma_j) = \mathbf{l}([\sigma_j]_y) = C\,K\,\mathbf{S}([\sigma_j]_y) = |\sigma_j|\,c^{|\sigma_j|-1}$. Combining these two results it turns out that

$$\begin{aligned} \mathbf{l}(\tau) &= |\tau| \prod_{i=1}^{k} \mathbf{v}(\tau_i) \prod_{j=1}^{l} \frac{\mathbf{k}(\sigma_j)}{|\sigma_j|} = |\tau| \prod_{i=1}^{k} c^{|\tau_i|} \prod_{j=1}^{l} c^{|\sigma_j|-1} \\ &= |\tau|\,c^{|\tau_1|+\cdots+|\tau_k|+(|\sigma_1|-1)+\cdots+(|\sigma_l|-1)}. \end{aligned}$$

Recall from Definition 8.4 that

$$\begin{aligned} |\tau| &= |\tau|_\bullet - |\tau|_\circ = 1 + \sum_{i=1}^{k} \left(|\tau_i|_\bullet - |\tau_i|_\circ \right) + \sum_{j=1}^{l} \left(|\sigma_i|_\bullet - |\sigma_i|_\circ \right) \\ &= 1 + \sum_{i=1}^{k} |\tau_i| + \sum_{j=1}^{l} (|\sigma_i| - 1) \end{aligned}$$

such that $\mathbf{l}(\tau) = |\tau|\,c^{|\tau|-1} = C\,K\,\mathbf{S}(\tau)$.

(b) $\tau = \overset{\tau_1 \cdots \tau_k}{\curlyvee} = [\sigma]_y \in T_{yz}$

In this case we have $|\tau| = |\sigma|$. Thus we can not apply the induction hypothesis directly but we have to take one step further into the recursion. Let $\sigma = \overset{\tau_1 \cdots \tau_k}{\curlyvee}$. Then $|\tau_i| \le k$ and

$$(\mathbf{v}D)(\sigma) = \prod_{i=1}^{k} \mathbf{v}(\tau_i) = \prod_{i=1}^{k} C\,\mathbf{S}(\tau_i) = c^{|\tau|} = C\,\mathbf{S}(\tau).$$

Observe that $\mathbf{l}(\tau) = \mathbf{k}(\sigma)$ and $\mathbf{S}(\tau) = \mathbf{S}(\sigma)$ such that

$$\begin{aligned}
\mathbf{l}(\tau) = \mathbf{k}(\sigma) &= \mathcal{A}^{-1}\big((\mathbf{v}D)(\sigma) - \mathcal{U}\,\mathbf{S}(\sigma)\big) \\
&= \mathcal{A}^{-1}\big(C\,\mathbf{S}(\tau) - (C - \mathcal{A}\,C\,K)\,\mathbf{S}(\tau)\big) = C\,K\,\mathbf{S}(\tau). \qquad \square
\end{aligned}$$

With Lemma 8.21 all the preparatory work is done and we can proceed to formulating the desired result that links the order conditions for implicit index-1 DAEs to the property of a general linear method to have high order and stage order for ordinary differential equations.

Theorem 8.22. *Let $\mathcal{M}$ be a general linear method in Nordsieck form satisfying $s = r > 1$. Assume that $\mathcal{M}$ has order and stage order equal to $p = q = s - 1$ for ordinary differential equations. Then $\mathcal{M}$ has order p for implicit index-1 DAEs (8.1).*

Proof. Since $\mathcal{M}$ has order and stage order $p = q = s-1$ for ordinary differential equations, we know from Theorem 7.1 that $\mathcal{U} = C - \mathcal{A}\,C\,K$ and $\mathcal{V} = E - \mathcal{B}\,C\,K$. Theorem 8.14 and Lemma 8.21 then yield

$$\mathbf{y}(\tau) = \mathcal{B}\,\mathbf{l}(\tau) + \mathcal{V}\,\mathbf{S}(\tau) = \mathcal{B}\,C\,K\,\mathbf{S}(\tau) + (E - \mathcal{B}\,C\,K)\,\mathbf{S}(\tau) = E\,\mathbf{S}(\tau)$$

for every tree with $|\tau| \le p$. In order to show that the order conditions for implicit index-1 DAEs are satisfied, it remains to show that $E\,\mathbf{S}(\tau) = \mathbf{E}(\tau)$. However, this is clear from the definition of $\mathbf{E}$ in Corollary 8.8 as the i-th component of the vector $E\,\mathbf{S}(\tau)$ reads

$$\big(E\,\mathbf{S}(\tau)\big)_i = \sum_{j=i}^{p+1} \frac{1}{(j-i)!}\delta_{|\tau|+1,j}|\tau|! = \frac{|\tau|!}{(|\tau|+1-i)!}$$

for $i - 1 \le |\tau| \le p$ and $\big(E\,\mathbf{S}(\tau)\big)_i = 0$ otherwise. $\qquad \square$

8.2 Convergence for Implicit Index-1 DAEs

The order conditions from Theorem 8.19 and 8.22 make it possible to derive general linear methods in Nordsieck form with a given *local* error when applied to implicit index-1 DAEs (8.1). However, when calculating the numerical solution the local error contributions accumulate over a series of steps taken by the method. Thus we have to study the *global* error

$$\epsilon_n^y = y^{[n]} - \hat{y}^{[n]} \qquad \text{and} \qquad \epsilon_n^z = z^{[n]} - \hat{z}^{[n]}.$$

Recall that $\hat{y}^{[n]}$ and $\hat{z}^{[n]}$ denote Nordsieck vectors obtained by applying the starting procedure **S** to the exact solution $y(t_n)$ and $z(t_n)$ at timepoint t_n, respectively (see Figure 8.4).

Only if the global error can be bounded in terms of the local error, the scheme is said to be stable and the numerical solution converges to the exact result if the stepsize h tends to zero.

The convergence results presented in this section follow ideas from [84, 104]. The existence of GLM solutions and the influence of perturbations is studied first. Then the main convergence result is proved in Theorem 8.27.

Hairer, Lubich and Roche [84] studied the application of Runge-Kutta methods to DAEs in Hessenberg form having index $\mu \in \{1, 2, 3\}$. Here, we investigate general linear methods for implicit index-1 systems (8.1) instead. Observe that a reformulation (8.1) in Hessenberg form typically increases the index to 2.

Fully implicit DAEs were studied by Kværnø in [104]. She considered Runge-Kutta methods for the numerical solution of index-1 DAEs $f(v, v') = 0$. The aim of this section is to generalise her results to the case of general linear methods.

Naturally, the mathematical techniques used in this section, in particular homotopy arguments and careful 'big-$\mathcal{O}$' calculations, are very similar to the ones used when analysing Runge-Kutta methods for fully implicit index-1 problems. Unfortunately even for Runge-Kutta methods the threefold result of existence of solutions, influence of perturbations and convergence is obtained only with considerable technical effort. Nevertheless, as these results are seminal for numerical computations it is essential to verify them for the class of general linear

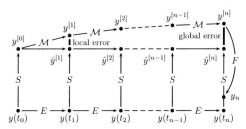

Figure 8.4: The global error for general linear methods. Only the y component is drawn for simplicity.

methods as well. It is a rather nice result that almost all ideas carry over from the Runge-Kutta to the general linear case.

However, there are also new aspects coming into the play. Namely the concept of consistency from Definition 7.4 and 7.5 is needed in Theorem 8.23 when proving the existence of solutions for the numerical scheme. Also, the requirement $|1 - b^\top \mathcal{A}^{-1} e| < 1$ for Runge-Kutta methods [10, 84, 104] has to be replaced by the power boundedness of the matrix $M_\infty = \mathcal{V} - \mathcal{B} \mathcal{A}^{-1} \mathcal{U}$. This matrix is the general linear method's stability matrix $M(z)$ evaluated at infinity (see Section 7.2 or [25]).

We consider the application of a general linear method $\mathcal{M} = [\mathcal{A}, \mathcal{U}, \mathcal{B}, \mathcal{V}]$ to the implicit index-1 equation (8.1). If $y^{[0]}$ and $z^{[0]}$ are some input vectors for the first step, we need to study the system

$$Y' = \tilde{f}(Y, Z'), \tag{8.18a}$$

$$Y = h\,\mathcal{A}\,Y' + \mathcal{U}\,y^{[0]}, \qquad\qquad \tilde{g}(Y) = h\,\mathcal{A}\,Z' + \mathcal{U}\,z^{[0]}, \tag{8.18b}$$

$$y^{[1]} = h\,\mathcal{B}\,\tilde{f}(Y, Z') + \mathcal{V}\,y^{[0]}, \qquad z^{[1]} = h\,\mathcal{B}\,Z' + \mathcal{V}\,z^{[0]}. \tag{8.18c}$$

$\tilde{f}(Y, Z') = \left[f(Y_1, Z_1')^\top, \ldots, f(Y_s, Z_s')^\top \right]^\top$ and $\tilde{g}(Y) = \left[g(Y_1)^\top, \ldots, g(Y_s)^\top \right]^\top$ again denote super vectors in $\mathbb{R}^{s\,m_1}$ and $\mathbb{R}^{s\,m_2}$, respectively. Similarly, recall that

$$Y = \begin{bmatrix} Y_1 \\ \vdots \\ Y_s \end{bmatrix} \in \mathbb{R}^{s\,m_1}, \qquad\qquad Z' = \begin{bmatrix} Z_1' \\ \vdots \\ Z_s' \end{bmatrix} \in \mathbb{R}^{s\,m_2},$$

As usual Kronecker products in (8.18) were left out to simplify notation. In contrast to (8.4) explicit representations of the stage derivatives $Y' = \tilde{f}(Y, Z')$ are used and Y' is not inserted into (8.18b). This will turn out to be useful when proving existence of solutions and convergence for the numerical scheme.

Theorem 8.23 (Existence of solutions). *Let $\mathcal{M}$ be a consistent general linear method with nonsingular matrix $\mathcal{A}$. Let the input vectors be given by*

$$y_i^{[0]} = \alpha_{0i}\nu + h\alpha_{1i}\mu + \mathcal{O}(h^2), \qquad z_i^{[0]} = \alpha_{0i}\eta + h\alpha_{1i}\zeta + \mathcal{O}(h^2)$$

for $i = 1, \ldots, r$ with $\nu, \mu, \eta, \zeta \in \mathbb{R}^n$. As in Definition 7.4, 7.5 the elements $\alpha_0, \alpha_1 \in \mathbb{R}^r$ denote the pre-consistency and consistency vector, respectively. Assume that

$$f(\nu, \zeta) = \mu + \mathcal{O}(h), \quad g(\nu) = \eta + \mathcal{O}(h^2), \quad g_y(\nu)\mu = \zeta + \mathcal{O}(h^2) \tag{8.19}$$

and that the matrices $(I - f_{z'}g_y)^{-1}$, f_y and $f_{z'}$ remain bounded in an h-independent neighbourhood of (ν, ζ). Then there exists a unique solution of (8.18a), (8.18b) satisfying

$$Y_i = \nu + \mathcal{O}(h), \qquad\qquad Z_i' = \zeta + \mathcal{O}(h).$$

Proof. First define[1] $\bar{y} = h(\mathcal{A}e)\,\mu + \mathcal{U}\,y^{[0]}$ and $\bar{z} = h(\mathcal{A}e)\,\zeta + \mathcal{U}\,z^{[0]}$ using the vector $e = [1,\ldots,1]^\top \in \mathbb{R}^s$. Consider the homotopy

$$Y' = \tilde{f}(Y,Z') + (\tau-1)\big(\tilde{f}(\bar{y}, e \otimes \zeta) - e \otimes \mu\big), \quad Y = h\mathcal{A}\,Y' + \mathcal{U}y^{[0]},$$
$$\tilde{g}(Y) = h\mathcal{A}\,Z' + \mathcal{U}z^{[0]} + (\tau-1)\big(\bar{z} - \tilde{g}(\bar{y})\big).$$

For $\tau = 0$ this system has the obvious solution $Y = \bar{y}$, $Y' = e \otimes \mu$ and $Z' = e \otimes \zeta$, but for $\tau = 1$ it is equivalent to (8.18a), (8.18b).

Considering Y, Y' and Z' as functions of τ we differentiate with respect to this parameter and obtain

$$\dot{Y}' = \{f_y\}\,\dot{Y} + \{f_{z'}\}\,\dot{Z}' + \tilde{f}(\bar{y}, e \otimes \zeta) - e \otimes \mu, \qquad \dot{Y} = h\mathcal{A}\,\dot{Y}',$$
$$\{g_y\}\,\dot{Y} = h\mathcal{A}\,\dot{Z}' + \bar{z} - \tilde{g}(\bar{y}),$$

where $\{f_y\} = \mathrm{blockdiag}\big(f_y(Y_1, Z_1'),\ldots,f_y(Y_s, Z_s')\big)$. The matrices $\{f_{z'}\}$ and $\{g_y\}$ are defined similarly. Due to $\dot{Y} = h\mathcal{A}\,\dot{Y}'$ we find that $\dot{Y}'$ and $\dot{Z}'$ satisfy

$$\dot{Y}' - h\{f_y\}\mathcal{A}\,\dot{Y}' = \{f_{z'}\}\,\dot{Z}' + \tilde{f}(\bar{y}, e \otimes \zeta) - e \otimes \mu,$$
$$\mathcal{A}^{-1}\{g_y\}\mathcal{A}\,\dot{Y}' = \dot{Z}' + \frac{1}{h}\mathcal{A}^{-1}\big(\bar{z} - \tilde{g}(\bar{y})\big).$$

Inserting the second equation into the first one, this system can be rewritten in matrix form

$$\begin{bmatrix} I - h\{f_y\}\mathcal{A} - \{f_{z'}\}\mathcal{A}^{-1}\{g_y\}\mathcal{A} & 0 \\ -\mathcal{A}^{-1}\{g_y\}\mathcal{A} & I \end{bmatrix} \begin{bmatrix} \dot{Y}' \\ \dot{Z}' \end{bmatrix} = \begin{bmatrix} \tilde{f}(\bar{y}, e \otimes \zeta) - e \otimes \mu - \frac{1}{h}\{f_{z'}\}\mathcal{A}^{-1}\big(\bar{z} - \tilde{g}(\bar{y})\big) \\ -\frac{1}{h}\mathcal{A}^{-1}\big(\bar{z} - g(\bar{y})\big) \end{bmatrix}.$$

Since $g_y(Y_i)a_{ij} = a_{ij}g_y(Y_j) + \mathcal{O}(d)$ provided that $\|Y_i - Y_j\| \le d$ with d independent of h, the block in the upper left corner can be written as

$$I - h\{f_y\}\mathcal{A} - \{f_{z'}\}\mathcal{A}^{-1}\{g_y\}\mathcal{A} = I - \{f_{z'}g_y\} - h\{f_y\}\mathcal{A} + \mathcal{O}(d). \quad (8.20)$$

By assumption $\mathsf{M} = I - f_{z'}g_y$ is nonsingular in a neighbourhood of (ν,ζ). Thus, (8.20) is also nonsingular for h and d sufficiently small.

Therefore the coefficient matrix of the above system has a bounded inverse. As (8.19) ensures that the right-hand side is $\mathcal{O}(h)$, standard arguments as in [84, theorem 4.1] show that $Y_i' = \mu + \mathcal{O}(h)$, $Z_i' = \zeta + \mathcal{O}(h)$. It remains to note that due to pre-consistency we have $\mathcal{U}\alpha_0 = e$ and therefore $\mathcal{U}y^{[0]} = e \otimes \nu + \mathcal{O}(h)$ which shows $Y = h\mathcal{A}Y' + \mathcal{U}y^{[0]} = e \otimes \nu + \mathcal{O}(h)$.

The uniqueness of the solution is obtained in the same way as in [84]. $\qquad\square$

---

[1]Using Kronecker products $\bar{y}$ reads $\bar{y} = h(\mathcal{A}e) \otimes \mu + (\mathcal{U} \otimes I_{m_1})y^{[0]}$ in more detail. A similar expression holds for $\bar{z}$.

When using the scheme (8.18) to solve (8.1) numerically, all the computations are performed on some computer architecture. Thus the computed results will not only suffer from inaccuracies due to the method (local and global error) but also from the fact that all calculations are performed in finite precision. The influence of these perturbations is studied in the next result.

Theorem 8.24 (Influence of perturbations). *Let the assumptions of theorem 8.23 be satisfied. Denote the solution of (8.18a), (8.18b) by Y', Y and Z'. Assume that $\bar{Y}'$, $\bar{Y}$, $\bar{Z}'$ satisfy the perturbed system*

$$\bar{Y}' = \tilde{f}(\bar{Y}, \bar{Z}') + \delta, \tag{8.21a}$$

$$\bar{Y} = h\mathcal{A}\,\bar{Y}' + \mathcal{U}\,\bar{y}^{[0]}, \qquad \tilde{g}(\bar{Y}) = h\mathcal{A}\,\bar{Z}' + \mathcal{U}\,\bar{z}^{[0]} + \theta, \tag{8.21b}$$

where the input vectors satisfy

$$\bar{y}^{[0]} - y^{[0]} = \mathcal{O}(h^2), \qquad \bar{z}^{[0]} - z^{[0]} = \mathcal{O}(h^2). \tag{8.22}$$

If the perturbations are of order $\delta = \mathcal{O}(h)$ and $\theta = \mathcal{O}(h^2)$ then there is a constant $C > 0$ such that

$$\left\| \begin{bmatrix} \bar{Y}'-Y' \\ \bar{Z}'-Z' \end{bmatrix} \right\| \leq C \left(\|\bar{y}^{[0]} - y^{[0]}\| + \|\delta\| + \tfrac{1}{h}\|\bar{z}^{[0]} - z^{[0]}\| \right. \tag{8.23}$$
$$\left. + \tfrac{1}{h}\|g_y(\eta,\zeta)\,(\bar{y}^{[0]} - y^{[0]})\| + \tfrac{1}{h}\|\theta\| \right)$$

provided that h is sufficiently small.

The norm used in (8.23) for super vectors is given by

$$\left\| \begin{bmatrix} \bar{Y}'-Y' \\ \bar{Z}'-Z' \end{bmatrix} \right\| = \max\left(\|\bar{Y}' - Y'\|, \|\bar{Z}' - Z'\| \right) = \max_{i=1}^{s} \left(\|\bar{Y}_i' - Y_i'\|, \|\bar{Z}_i' - Z_i'\| \right)$$

where $\| \cdot \|$ denotes some vector norm in $\mathbb{R}^{m_1}$ or $\mathbb{R}^{m_2}$, respectively.

Proof. The result of Theorem 8.24 is obtained very similar to the proof of Theorem 8.23, but a different homotopy is used:

$$Y' = \tilde{f}(Y, Z') + (1 - \tau)\,\delta,$$
$$\tilde{g}(Y) = h\mathcal{A}\,Z' + \mathcal{U}\,z^{[0]} + (1 - \tau)\big(\theta + \mathcal{U}\,(\bar{z}^{[0]} - z^{[0]})\big),$$
$$Y = h\mathcal{A}\,Y' + \mathcal{U}\,y^{[0]} + (1 - \tau)\,\mathcal{U}(\bar{y}^{[0]} - y^{[0]}).$$

For $\tau = 1$ this system is (8.18a), (8.18b), but for $\tau = 0$ it coincides with (8.21). See [84, 104] for more details. $\qquad\square$

Before we continue to study the propagation of errors over several consecutive steps we remark that (8.22) and (8.23) imply

$$\left\| \begin{bmatrix} \bar{Y}'-Y' \\ \bar{Z}'-Z' \end{bmatrix} \right\| \leq C_1 h \tag{8.24a}$$

for $\delta = 0$, $\theta = 0$. But if $\bar{y}^{[0]} - y^{[0]} = \mathcal{O}(h)$ and $\bar{z}^{[0]} - z^{[0]} = \mathcal{O}(h)$, we only have

$$\left\| \begin{bmatrix} \bar{Y}'-Y' \\ \bar{Z}'-Z' \end{bmatrix} \right\| \leq C_2. \tag{8.24b}$$

Lemma 8.25 (Error propagation). *Let the assumptions of Theorem 8.23 be satisfied. Denote exact input vectors[2] by $\hat{y}^{[n]}$, $\hat{z}^{[n]}$ and assume that*

$$y^{[n]} = \hat{y}^{[n]} + \epsilon_n^y, \quad z^{[n]} = \hat{z}^{[n]} + \epsilon_n^z \quad \text{with} \quad \|\epsilon_n^y\| = \mathcal{O}(h^p), \quad \|\epsilon_n^z\| = \mathcal{O}(h^p).$$

Let $(y^{[n+1]}, z^{[n+1]})$ be the numerical solution obtained after n steps. Then the global error satisfies the recursion

$$\epsilon_{n+1} = M_\infty \, \mathcal{S}_n \, \epsilon_n + \mathcal{V} \, \mathcal{P}_n \, \epsilon_n - d_{n+1} + \eta \tag{8.25}$$

where $M_\infty = \mathcal{V} - \mathcal{B}\mathcal{A}^{-1}\mathcal{U}$ is the method's stability matrix evaluated at infinity, d_{n+1} is the local discretisation error and

$$\eta = \begin{cases} \mathcal{O}(h^{p+1}), & \text{for } p \geq 2, \text{ or } p = 1 \text{ and } f \text{ linear in } z', \\ \mathcal{O}(h), & \text{for } p = 1 \text{ and } f \text{ nonlinear in } z'. \end{cases}$$

$\mathcal{S}_n$ and $\mathcal{P}_n$ are projector functions given by

$$\mathcal{S}_n = \begin{bmatrix} I - \mathsf{M}^{-1} & \mathsf{M}^{-1} f_{z'} \\ -g_y \mathsf{M}^{-1} & I + g_y \mathsf{M}^{-1} f_{z'} \end{bmatrix}, \qquad \mathcal{P}_n = I - \mathcal{S}_n,$$

where $\mathsf{M}^{-1} = (I - f_{z'} g_y)^{-1}$, $f_{z'}$ and g_y are evaluated at $\big(y(t_n), z'(t_n)\big)$.

Remark 8.26. In order to simplify matters, (8.25) is a slight abuse of notation. We have $\epsilon_n = \begin{bmatrix} \epsilon_n^y \\ \epsilon_n^z \end{bmatrix}$, $d_{n+1} = \begin{bmatrix} d_{n+1}^y \\ d_{n+1}^z \end{bmatrix}$ and similarly $\eta = \begin{bmatrix} \eta_f \\ \eta_g \end{bmatrix}$ such that $\epsilon_n, d_{n+1}, \eta \in \mathbb{R}^{r(m_1+m_2)}$. On the other hand

$$\mathcal{S}_n = \begin{bmatrix} \mathcal{S}_n^1 & \mathcal{S}_n^2 \\ \mathcal{S}_n^3 & \mathcal{S}_n^4 \end{bmatrix} \in \mathbb{R}^{(m_1+m_2)\times(m_1+m_2)}, \qquad \mathcal{P}_n = \begin{bmatrix} \mathcal{P}_n^1 & \mathcal{P}_n^2 \\ \mathcal{P}_n^3 & \mathcal{P}_n^4 \end{bmatrix} \in \mathbb{R}^{(m_1+m_2)\times(m_1+m_2)}$$

and $M_\infty, \mathcal{V} \in \mathbb{R}^{r \times r}$. Thus (8.25) has to be understood as the system

$$\begin{bmatrix} \epsilon_{n+1}^y \\ \epsilon_{n+1}^z \end{bmatrix} = \left(\begin{bmatrix} M_\infty \otimes \mathcal{S}_n^1 & M_\infty \otimes \mathcal{S}_n^2 \\ M_\infty \otimes \mathcal{S}_n^3 & M_\infty \otimes \mathcal{S}_n^4 \end{bmatrix} + \begin{bmatrix} \mathcal{V} \otimes \mathcal{P}_n^1 & \mathcal{V} \otimes \mathcal{P}_n^2 \\ \mathcal{V} \otimes \mathcal{P}_n^3 & \mathcal{V} \otimes \mathcal{P}_n^4 \end{bmatrix} \right) \begin{bmatrix} \epsilon_n^y \\ \epsilon_n^z \end{bmatrix} + \begin{bmatrix} \eta_f - d_{n+1}^y \\ \eta_g - d_{n+1}^z \end{bmatrix}.$$

Proof. Using exact input vectors we find

$$\hat{Y}' = \tilde{f}(\hat{Y}, \hat{Z}'), \tag{8.26a}$$

$$\hat{Y} = h\mathcal{A}\,\hat{Y}' + \mathcal{U}\,\hat{y}^{[n]}, \qquad \tilde{g}(\hat{Y}) = h\mathcal{A}\,\hat{Z}' + \mathcal{U}\,\hat{z}^{[n]} \tag{8.26b}$$

$$\hat{y}^{[n+1]} = h\mathcal{B}\,\hat{Y}' + \mathcal{V}\,\hat{y}^{[n]} + d_{n+1}^y, \qquad \hat{z}^{[n+1]} = h\mathcal{B}\,\hat{Z}' + \mathcal{V}\,\hat{z}^{[n]} + d_{n+1}^z \tag{8.26c}$$

---

[2]Recall that $\hat{y}_{i+1}^{[n]} = h^i y^{(i)}(t_n)$ and $\hat{z}_{i+1}^{[n]} = h^i z^{(i)}(t_n)$, $i = 1, \ldots, r$, for methods in Nordsieck form.

with local error d_{n+1}^y, d_{n+1}^z. On the other hand

$$Y' = \tilde{f}(Y, Z'), \tag{8.27a}$$

$$Y = h\mathcal{A}Y' + \mathcal{U}y^{[n]}, \qquad\qquad \tilde{g}(Y) = h\mathcal{A}Z' + \mathcal{U}z^{[n]} \tag{8.27b}$$

$$\hat{y}^{[n+1]} = h\mathcal{B}Y' + \mathcal{V}y^{[n]} + \epsilon_{n+1}^y, \qquad \hat{z}^{[n+1]} = h\mathcal{B}Z' + \mathcal{V}z^{[n]} + \epsilon_{n+1}^z \tag{8.27c}$$

with global error ϵ_{n+1}^y, ϵ_{n+1}^z. Introduce $\Delta Y' = Y' - \hat{Y}'$, $\Delta Z' = Z' - \hat{Z}'$ and assume that

$$\|\epsilon_n^y\| \leq \varepsilon, \qquad \|\epsilon_n^z\| \leq \varepsilon, \qquad \|\Delta Y'\| \leq \delta, \qquad \|\Delta Z'\| \leq \delta.$$

(8.24) implies $\delta = \mathcal{O}(h)$ for $p \geq 2$ and $\delta = \mathcal{O}(1)$ for $p = 1$. By assumption we have $\varepsilon = \mathcal{O}(h^p)$. Writing $Z' = \hat{Z}' + \Delta Z'$ and $Y = \hat{Y} + h\mathcal{A}\,\Delta Y' + \mathcal{U}\epsilon_n^y$, we may linearise f in $(\hat{Y}, \hat{Z}')$ such that

$$\tilde{f}(Y, Z') = \tilde{f}(\hat{Y}, \hat{Z}') + \{f_{z'}\}\Delta Z' + \eta_f.$$

$\{f_{z'}\} = \text{blockdiag}\big(f_{z'}(\hat{Y}_1, \hat{Z}_1'), \ldots, f_{z'}(\hat{Y}_s, \hat{Z}_s')\big)$ and so on are notations similar to the ones already introduced in the proof of Theorem 8.23. The term η_f contains higher order terms of the form

$$\{f_y\}\big(h\mathcal{A}\,\Delta Y' + \mathcal{U}\epsilon_n^y\big), \qquad\qquad \{f_{yy}\}\big(h\mathcal{A}\,\Delta Y' + \mathcal{U}\epsilon_n^y, h\mathcal{A}\,\Delta Y' + \mathcal{U}\epsilon_n^y\big),$$

$$\{f_{yz'}\}\big(h\mathcal{A}\,\Delta Y' + \mathcal{U}\epsilon_n^y, \Delta Z'\big), \quad \{f_{z'z'}\}\big(\Delta Z', \Delta Z'\big), \quad \ldots$$

Due to Theorem 8.23 the relations $\hat{Y}_i = y(t_n) + \mathcal{O}(h)$ and $\hat{Z}_i' = z'(t_n) + \mathcal{O}(h)$ imply $f_y(\hat{Y}_i, \hat{Z}_i') = f_y\big(y(t_n), z'(t_n)\big) + \mathcal{O}(h)$. Similar estimates hold for the other derivatives. It turns out that

$$\tilde{f}(Y, Z') = \tilde{f}(\hat{Y}, \hat{Z}') + f_{z'}\,\Delta Z' + \mathcal{O}\big(h\delta + \varepsilon + \delta^2 + h\delta^2\big). \tag{8.28}$$

Writing f_y, $f_{z'}$, g_y and so on has to be understood as evaluating the derivatives in $\big(y(t_n), z'(t_n)\big)$. Similar calculations show that

$$\tilde{g}(Y) = \tilde{g}\big(\hat{Y} + h\mathcal{A}\,\Delta Y' + \mathcal{U}\epsilon_n^y\big) = \tilde{g}(\hat{Y}) + \{g_y\}\big(h\mathcal{A}\,\Delta Y' + \mathcal{U}\epsilon_n^y\big) + \eta_g,$$

where η_g contains higher order terms of the form $\{g_{yy}\}\big(h\mathcal{A}\,\Delta Y' + \mathcal{U}\epsilon_n^y\big)^2$ and so on. With $\hat{Y}_i = y(t_n) + \mathcal{O}(h)$ we get $\eta_g = \mathcal{O}\big(h^2\delta^2 + h\delta\varepsilon + \varepsilon^2\big)$ and finally

$$\tilde{g}(Y) = \tilde{g}(\hat{Y}) + g_y \cdot \big(h\mathcal{A}\,\Delta Y' + \mathcal{U}\epsilon_n^y\big) + \mathcal{O}\big(h^2\delta^2 + h^2\delta + h\varepsilon\big). \tag{8.29}$$

Using (8.28), (8.29) together with (8.26), (8.27) we find

$$\left.\begin{aligned}
\Delta Y' &= \tilde{f}(Y, Z') - \tilde{f}(\hat{Y}, \hat{Z}') = f_{z'}\Delta Z' + \mathcal{O}\big(h\delta + \varepsilon + \delta^2 + h\delta^2\big), \\[4pt]
\Delta Z' &= \frac{1}{h}\mathcal{A}^{-1}\Big(\tilde{g}(Y) - \tilde{g}(\hat{Y}) - \mathcal{U}\big(z^{[n]} - \hat{z}^{[n]}\big)\Big) \\[4pt]
&= \frac{1}{h}\mathcal{A}^{-1}\Big(g_y \cdot \big(h\mathcal{A}\,\Delta Y' + \mathcal{U}\epsilon_n^y\big) - \mathcal{U}\epsilon_n^z\Big) + \mathcal{O}\big(h\delta^2 + h\delta + \varepsilon\big) \\[4pt]
&= g_y\,\Delta Y' + \frac{1}{h}g_y\,\mathcal{A}^{-1}\mathcal{U}\epsilon_n^y - \frac{1}{h}\mathcal{A}^{-1}\mathcal{U}\epsilon_n^z + \mathcal{O}\big(h\delta^2 + h\delta + \varepsilon\big)
\end{aligned}\right\} \tag{8.30}$$

The calculations performed here involve several Kronecker products [91]. For more clarity observe that

$$\mathcal{A}^{-1} \cdot g_y \cdot \mathcal{A}\,\Delta Y' = (\mathcal{A}^{-1} \otimes I_{m_2})(I_s \otimes g_y)(\mathcal{A} \otimes I_{m_1})\Delta Y'$$
$$= (\mathcal{A}^{-1}\mathcal{A} \otimes g_y)\Delta Y' = (I_s \otimes g_y)\Delta Y' \in \mathbb{R}^{s\,m_2}$$

and similarly

$$\mathcal{A}^{-1} \cdot g_y \cdot \mathcal{U}\,\epsilon_n^y = (\mathcal{A}^{-1} \otimes I_{m_2})(I_s \otimes g_y)(\mathcal{U} \otimes I_{m_1})\epsilon_n^y = (\mathcal{A}^{-1}\mathcal{U} \otimes g_y)\epsilon_n^y$$
$$= (I_s \otimes g_y)(\mathcal{A}^{-1}\mathcal{U} \otimes I_{m_1})\epsilon_n^y \in \mathbb{R}^{s\,m_2}.$$

In the equations on the previous page we wrote $g_y\,\Delta Y' = (I_s \otimes g_y)\Delta Y'$ and $g_y\,\mathcal{A}^{-1}\mathcal{U}\,\epsilon_n^y = (I_s \otimes g_y)(\mathcal{A}^{-1}\mathcal{U} \otimes I_n)\epsilon_n^y$ in order to shorten notations.

Writing the system (8.30) in matrix form yields

$$\begin{bmatrix} I & -f_{z'} \\ -g_y & I \end{bmatrix}\begin{bmatrix} \Delta Y' \\ \Delta Z' \end{bmatrix} = \begin{bmatrix} 0 & 0 \\ g_y & -I \end{bmatrix}\begin{bmatrix} \frac{1}{h}\mathcal{A}^{-1}\mathcal{U} & 0 \\ 0 & \frac{1}{h}\mathcal{A}^{-1}\mathcal{U} \end{bmatrix}\begin{bmatrix} \epsilon_n^y \\ \epsilon_n^z \end{bmatrix} + \mathcal{O}\big(h\delta + \varepsilon + \delta^2 + h\delta^2\big).$$

Using the notation $\mathsf{M} = I - f_{z'}g_y$, the inverse of the coefficient matrix reads

$$\begin{bmatrix} I & -f_{z'} \\ -g_y & I \end{bmatrix}^{-1} = \begin{bmatrix} \mathsf{M}^{-1} & \mathsf{M}^{-1}f_{z'} \\ g_y\mathsf{M}^{-1} & I + g_y\mathsf{M}^{-1}f_{z'} \end{bmatrix}$$

and we get an explicit representation for $\Delta Y'$, $\Delta Z'$. Namely,

$$\begin{bmatrix} \Delta Y' \\ \Delta Z' \end{bmatrix} = -\mathcal{S}_n \begin{bmatrix} \frac{1}{h}\mathcal{A}^{-1}\mathcal{U} & 0 \\ 0 & \frac{1}{h}\mathcal{A}^{-1}\mathcal{U} \end{bmatrix}\begin{bmatrix} \epsilon_n^y \\ \epsilon_n^z \end{bmatrix} + \mathcal{O}\big(h\delta + \varepsilon + \delta^2 + h\delta^2\big). \quad (8.31)$$

Notice that

$$\mathcal{S}_n = \mathcal{S}\big(y(t_n), z'(t_n)\big) = \begin{bmatrix} I - \mathsf{M}^{-1} & \mathsf{M}^{-1}f_{z'} \\ -g_y\mathsf{M}^{-1} & I + g_y\mathsf{M}^{-1}f_{z'} \end{bmatrix} \quad (8.32)$$

is a projector function. All matrix functions occurring in the definition of $\mathcal{S}_n$ have to be evaluated at $\big(y(t_n), z'(t_n)\big)$.

Since $\|\epsilon_n^y\| \leq \varepsilon$, $\|\epsilon_n^z\| \leq \varepsilon$ we infer from the inequality (8.31) that

$$\left\|\begin{bmatrix} \Delta Y' \\ \Delta Z' \end{bmatrix}\right\| \leq K\left(h\delta + \varepsilon + \delta^2 + h\delta^2 + \tfrac{\varepsilon}{h}\right)$$

with some constant K. As in [104, 111] it is possible to obtain an estimate for δ by considering the functional iteration $\delta = G(\delta) = K\left(h\delta + \varepsilon + \delta^2 + h\delta^2 + \tfrac{\varepsilon}{h}\right)$ using the initial value $\delta^{(0)} = C\,h^{p-1}$. It turns out that $\delta = \mathcal{O}(h^{p-1})$ for $p \geq 1$. Inserting this result into (8.31) yields

$$\begin{bmatrix} \Delta Y' \\ \Delta Z' \end{bmatrix} = -\mathcal{S}_n \begin{bmatrix} \frac{1}{h}\mathcal{A}^{-1}\mathcal{U} & 0 \\ 0 & \frac{1}{h}\mathcal{A}^{-1}\mathcal{U} \end{bmatrix}\begin{bmatrix} \epsilon_n^y \\ \epsilon_n^z \end{bmatrix} + \begin{cases} \mathcal{O}\big(h^p\big), & \text{for } p \geq 2, \\ \mathcal{O}(1), & \text{for } p = 1. \end{cases}$$

Observe that the term $\mathcal{O}(1)$ originates from $\mathcal{O}(\delta^2)$ in (8.31). Recall that the term $\mathcal{O}(\delta^2)$ comes from higher order terms of the type $\{f_{z'z'}\}(\Delta Z', \Delta Z')$. If f is linear in z', these terms do not appear and we can conclude

$$\begin{bmatrix} \Delta Y' \\ \Delta Z' \end{bmatrix} = -\mathcal{S}_n \begin{bmatrix} \frac{1}{h}\mathcal{A}^{-1}\mathcal{U} & 0 \\ 0 & \frac{1}{h}\mathcal{A}^{-1}\mathcal{U} \end{bmatrix} \begin{bmatrix} \epsilon_n^y \\ \epsilon_n^z \end{bmatrix} + \tilde{\eta} \tag{8.33}$$

with

$$\tilde{\eta} = \begin{cases} \mathcal{O}(h^p), & \text{for } p \geq 2, \text{ or } p = 1 \text{ and } f \text{ linear in } z', \\ \mathcal{O}(1), & \text{for } p = 1 \text{ and } f \text{ nonlinear in } z'. \end{cases}$$

It remains to insert (8.33) into (8.26c), (8.27c) to obtain

$$\begin{aligned}
\begin{bmatrix} \epsilon_{n+1}^y \\ \epsilon_{n+1}^z \end{bmatrix} &= \begin{bmatrix} h\mathcal{B} & 0 \\ 0 & h\mathcal{B} \end{bmatrix} \begin{bmatrix} \Delta Y' \\ \Delta Z' \end{bmatrix} + \begin{bmatrix} \mathcal{V} & 0 \\ 0 & \mathcal{V} \end{bmatrix} \begin{bmatrix} \epsilon_n^y \\ \epsilon_n^z \end{bmatrix} - \begin{bmatrix} d_{n+1}^y \\ d_{n+1}^z \end{bmatrix} \\
&= \mathcal{S}_n \begin{bmatrix} -\mathcal{B}\mathcal{A}^{-1}\mathcal{U} & 0 \\ 0 & -\mathcal{B}\mathcal{A}^{-1}\mathcal{U} \end{bmatrix} \begin{bmatrix} \epsilon_n^y \\ \epsilon_n^z \end{bmatrix} + \begin{bmatrix} \mathcal{V} & 0 \\ 0 & \mathcal{V} \end{bmatrix} \begin{bmatrix} \epsilon_n^y \\ \epsilon_n^z \end{bmatrix} - d_{n+1} + \eta
\end{aligned}$$

with

$$\eta = \begin{cases} \mathcal{O}(h^{p+1}), & \text{for } p \geq 2, \text{ or } p = 1 \text{ and } f \text{ linear in } z', \\ \mathcal{O}(h), & \text{for } p = 1 \text{ and } f \text{ nonlinear in } z'. \end{cases}$$

Introducing $\mathcal{P}_n = I - \mathcal{S}_n$ we get the desired result:

$$\begin{aligned}
\epsilon_{n+1} &= -\mathcal{B}\mathcal{A}^{-1}\mathcal{U}\mathcal{S}_n\epsilon_n + \mathcal{V}(\mathcal{S}_n\epsilon_n + \mathcal{P}_n\epsilon_n) - d_{n+1} + \eta \\
&= (\mathcal{V} - \mathcal{B}\mathcal{A}^{-1}\mathcal{U})\mathcal{S}_n\epsilon_n + \mathcal{V}\mathcal{P}_n\epsilon_n - d_{n+1} + \eta. \qquad \square
\end{aligned}$$

Our aim is to prove convergence of general linear methods for implicit index-1 DAEs (8.1). With Lemma 8.25 most of the work is already done. All that is left to do is to use Lemma 8.25 recursively to find a representation of the global error in terms of the local one and the initial data.

Theorem 8.27 (Convergence). *Let* $\mathcal{M} = [\mathcal{A}, \mathcal{U}, \mathcal{B}, \mathcal{V}]$ *be a general linear method with nonsingular* $\mathcal{A}$. *Assume that*

(a) $M_\infty = \mathcal{V} - \mathcal{B}\mathcal{A}^{-1}\mathcal{U}$ *and* $\mathcal{V}$ *are power bounded.*

(b) f, g *are sufficiently differentiable.*

(c) *The initial input vectors* $y^{[0]}$, $z^{[0]}$ *satisfy*

$$y^{[0]} = \hat{y}^{[0]} + \mathcal{O}(h^p), \qquad z^{[0]} = \hat{z}^{[0]} + \mathcal{O}(h^p) \tag{8.34}$$

with $p \geq 2$. $\hat{y}^{[n]}$, $\hat{z}^{[n]}$ *denote exact input vectors.*

(d) *The local truncation error satisfies* $d_{n+1} = \mathcal{O}(h^{p+1})$ *with* $p \geq 2$.

Then the method is convergent with order p.

Proof. We start from the error recursion

$$\epsilon_{n+1} = M_\infty \, \mathcal{S}_n \, \epsilon_n + \mathcal{V} \, \mathcal{P}_n \, \epsilon_n + \tilde{d}_{n+1} \tag{8.35}$$

derived in Lemma 8.25. For convenience $\tilde{d}_{n+1} = -d_{n+1} + \eta_{n+1}$ is used as an abbreviation. Recall that $\eta_{n+1} = \mathcal{O}(h^{p+1})$ and thus also $\tilde{d}_{n+1} = \mathcal{O}(h^{p+1})$. The components of the projectors $\mathcal{P}$ and $\mathcal{S}$ defined in (8.32) are smooth functions. Therefore we have

$$\mathcal{S}_{n+1}\mathcal{S}_n = \mathcal{S}_n + \mathcal{O}(h), \qquad\qquad \mathcal{P}_{n+1}\mathcal{P}_n = \mathcal{P}_n + \mathcal{O}(h),$$
$$\mathcal{S}_{n+1}\mathcal{P}_n = \mathcal{O}(h), \qquad\qquad \mathcal{P}_{n+1}\mathcal{S}_n = \mathcal{O}(h).$$

Multiplication of (8.35) by $\mathcal{P}_{n+1}$ and $\mathcal{S}_{n+1}$, respectively, yields

$$\begin{bmatrix} \mathcal{P}_{n+1}\epsilon_{n+1} \\ \mathcal{S}_{n+1}\epsilon_{n+1} \end{bmatrix} = \begin{bmatrix} \mathcal{V}\big(1 + \mathcal{O}(h)\big) & \mathcal{O}(h) \\ \mathcal{O}(h) & M_\infty\big(1 + \mathcal{O}(h)\big) \end{bmatrix} \begin{bmatrix} \mathcal{P}_n\epsilon_n \\ \mathcal{S}_n\epsilon_n \end{bmatrix} + \begin{bmatrix} \mathcal{P}_{n+1}\tilde{d}_{n+1} \\ \mathcal{S}_{n+1}\tilde{d}_{n+1} \end{bmatrix}$$

$$= A(h)^{n+1} \begin{bmatrix} \mathcal{P}_0 e_0 \\ \mathcal{S}_0 e_0 \end{bmatrix} + \sum_{i=0}^{n} A(h)^i \begin{bmatrix} \mathcal{P}_{n+1-i}\,\tilde{d}_{n+1-i} \\ \mathcal{S}_{n+1-i}\,\tilde{d}_{n+1-i} \end{bmatrix}$$

For sufficiently small h the matrix $A(h) = \begin{bmatrix} \mathcal{V}(1+\mathcal{O}(h)) & \mathcal{O}(h) \\ \mathcal{O}(h) & M_\infty(1+\mathcal{O}(h)) \end{bmatrix}$ is power bounded, given that $\mathcal{V}$ and $M_\infty = \mathcal{V} - \mathcal{B}\mathcal{A}^{-1}\mathcal{U}$ are power bounded themselves. In particular, standard arguments [104] show that

$$\|\epsilon_n\| \leq C \left(\|\mathcal{P}_0 e_0\| + \|\mathcal{S}_0 e_0\| + \frac{1}{h} \max_{i=1,\dots,n} \|\mathcal{P}_i \tilde{d}_i\| + \frac{1}{h} \max_{i=1,\dots,n} \|\mathcal{S}_i \tilde{d}_i\| \right).$$

The constant C is independent of h. Due to the assumptions (c) and (d) this stability inequality proves convergence of order (at least) p. $\square$

After proving this convergence result, let us briefly comment on the assumptions made when formulating Theorem 8.27.

Assumption (c) guarantees the existence of a solution for the first step of the numerical scheme (8.18) as (8.34) implies

$$y^{[0]} = \alpha_0 \, y(t_0) + h\alpha_1 \, y'(t_0) + \mathcal{O}(h^2), \quad z^{[0]} = \alpha_0 \, z(t_0) + h\alpha_1 \, z'(t_0) + \mathcal{O}(h^2)$$

with pre-consistency vector α_0 and consistency vector α_1 (see Theorem 8.23). The error recursion of Lemma 8.25 and the power boundedness of $\mathcal{V}$ and M_∞ ensure that $y^{[n]}$, $z^{[n]}$ are calculated with the same accuracy for $n \geq 1$. Thus the numerical scheme (8.18) is solvable also for $n \geq 1$. In order to guarantee (d) the order conditions from Theorem 8.19 have to be satisfied.

Comparing with the corresponding Theorem 3.3. in [104], which deals exclusively with Runge-Kutta methods, the assumption on $\mathcal{V}$ and M_∞ to be power bounded is the most obvious deviation. However, if $\mathcal{M}$ was itself a Runge-Kutta method, then $\mathcal{V} = 1$ and $M_\infty = 1 - b^\top \mathcal{A}^{-1} e = R(\infty)$. Thus for Runge-Kutta methods assumption (a) means that $|R(\infty)| \leq 1$.

Also note that we restricted attention to the case $p \geq 2$. This restriction is not necessary. Similar results as in [104] can be derived also for $p = 1$ using mere technical considerations.

8.3   The Accuracy of Stages and Stage Derivatives

Theorem 8.27 guarantees that the global error

$$\epsilon_n^y = y^{[n]} - \hat{y}^{[n]} = \mathcal{O}(h^p), \qquad\qquad \epsilon_n^z = z^{[n]} - \hat{z}^{[n]} = \mathcal{O}(h^p)$$

is of order p both for the y and the z component. It is interesting to investigate whether similar estimates hold for the global error

$$\Delta Y_{ni} = Y_{ni} - y(t_{n-1} + c_i h), \qquad\qquad \Delta Z'_{ni} = Z'_{ni} - z'(t_{n-1} + c_i h)$$

of the stages and stage derivatives as well. Y_n and Z'_n are calculated via

$$Y_n = h\mathcal{A}\tilde{f}(Y_n, Z'_n) + \mathcal{U}y^{[n-1]}, \qquad\qquad \tilde{g}(Y_n) = h\mathcal{A}Z'_n + \mathcal{U}z^{[n-1]},$$

where the input vectors $y^{[n-1]}$, $z^{[n-1]}$ are obtained by taking $n-1$ consecutive steps with the general linear method $\mathcal{M}$ starting from $y^{[0]}$, $z^{[0]}$ at $t = t_0$. In order to keep track of the current step number, Y_n and Z'_n are equipped with the subscript n.

From [84] it is well known that for Runge-Kutta methods applied to index-2 DAEs in Hessenberg form the global error of the z component is essentially given by the local error, i.e. if the local error is $\mathcal{O}(h^{p+1})$ then the global error is $\mathcal{O}(h^{p+1})$ as well. In this situation one immediately finds $\Delta Z'_n = \mathcal{O}(h^{\min(p,q)})$ from a perturbation result similar to Theorem 8.24.

Unfortunately, this standard approach for estimating the order of the derivative approximations is not feasible for implicit index-1 systems. In general the z component is calculated with global error of order $\mathcal{O}(h^p)$ only and convergence of order $\mathcal{O}(h^{p+1})$ can't be expected. This is illustrated in the following example.

Example 8.28. For $L \in \mathbb{R}$ consider the implicit DAE

$$\begin{aligned} y' &= f(y, z') = L y^2 - z', &\qquad t \in [0, 0.5], &\qquad y(0) = z(0) = 1, \\ z &= g(y) = y^3. \end{aligned}$$

Due to $I - f_{z'} g_y = 1 + 3y^2 > 0$ the index is indeed 1 and the exact solution is given by

$$y(t) = \frac{1}{6}\left(t L + 2 + \sqrt{(t L + 2)^2 + 12}\right), \qquad\qquad z(t) = y(t)^3.$$

This DAE can be solved numerically using the general linear method

$$
M = \left[
\begin{array}{ccc|ccc}
\frac{1}{4} & 0 & 0 & 1 & 0 & -\frac{1}{32} \\
\frac{1}{6} & \frac{1}{4} & 0 & 1 & \frac{1}{12} & -\frac{1}{24} \\
\frac{1}{6} & \frac{1}{2} & \frac{1}{4} & 1 & \frac{1}{12} & -\frac{1}{24} \\
\hline
\frac{1}{6} & \frac{1}{2} & \frac{1}{4} & 1 & \frac{1}{12} & -\frac{1}{24} \\
0 & 0 & 1 & 0 & 0 & 0 \\
0 & -2 & 2 & 0 & 0 & 0
\end{array}
\right], \qquad
c = \left[
\begin{array}{c}
\frac{1}{4} \\
\frac{1}{2} \\
1
\end{array}
\right]
$$

constructed in [158]. The method has order $p = 2$ and stage order $q = 2$ for ordinary differential equations. To verify this, calculate

$$
C = \left[\begin{array}{ccc} e & c & \frac{c^2}{2} \end{array}\right] = \left[
\begin{array}{ccc}
1 & \frac{1}{4} & \frac{1}{32} \\
1 & \frac{1}{2} & \frac{1}{16} \\
1 & 1 & \frac{1}{2}
\end{array}
\right], \qquad
E = \left[
\begin{array}{ccc}
1 & 1 & \frac{1}{2} \\
0 & 1 & 1 \\
0 & 0 & 1
\end{array}
\right]
$$

and use Theorem 7.6 to find that $\mathcal{U} = C - \mathcal{A}CK$ and $\mathcal{V} = E - \mathcal{B}CK$. Due to Theorem 8.22 the DAE order conditions are satisfied for order up to 2 as well.

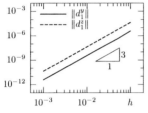

Figure 8.5 (a): Norm of the local error $d_1^y = y^{[1]} - \hat{y}^{[1]}$, $d_1^z = z^{[1]} - \hat{z}^{[1]}$.

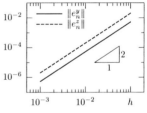

Figure 8.5 (b): Norm of the global error $\epsilon_n^y = y^{[n]} - \hat{y}^{[n]}$, $\epsilon_n^z = z^{[n]} - \hat{z}^{[n]}$ at $t_n = 0.5$

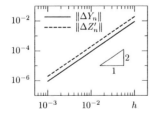

Figure 8.5 (c): Norm of the global error ΔY_n, $\Delta Z_n'$ at $t_n = 0.5$

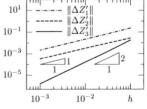

Figure 8.5 (d): Norm of the global error $\Delta Z_n'$ for $n = 1, 2, 3$

Figure 8.5: Numerical solution of Example 8.28 for $L = 1$ on $[0, 0.5]$ using fixed stepsizes. For Figure 8.5 (a) exact input vectors were used. All the other results were calculated from input vectors $y^{[0]} = \hat{y}^{[0]} + \mathcal{O}(h^p)$, $z^{[0]} = \hat{y}^{[0]} + \mathcal{O}(h^p)$ where $p = 2$ is the method's order.

Consequently, both the y and the z component are calculated with local error $d_n = \mathcal{O}(h^3)$. This is confirmed in Figure 8.5 (a). Since the matrices

$$
\mathcal{V} = \begin{bmatrix} 1 & \frac{1}{12} & -\frac{1}{24} \\ 1 & \frac{1}{12} & -\frac{1}{24} \\ 1 & \frac{1}{12} & -\frac{1}{24} \end{bmatrix}, \qquad M_\infty = \mathcal{V} - \mathcal{B}\mathcal{A}^{-1}\mathcal{U} = \begin{bmatrix} 0 & 0 & 0 \\ \frac{4}{3} & \frac{1}{3} & -\frac{1}{12} \\ \frac{16}{3} & \frac{4}{3} & -\frac{1}{3} \end{bmatrix}
$$

have eigenvalues $\sigma(\mathcal{V}) = \{1, 0, 0\}$, $\sigma(M_\infty) = \{0, 0, 0\}$, and are therefore power bounded, we expect convergence of order 2. Again, the numerical results from Figure 8.5 (b) confirm this result. Observe that the global error of the z component is indeed $\mathcal{O}(h^2)$ even though the local error is $\mathcal{O}(h^3)$.

However, this does not mean that the stage derivatives are calculated with a lower order. In fact, Figure 8.5 (c) shows that the stage derivatives Z'_n are calculated with order 2 as well. As seen from Figure 8.5 (d) this order of convergence is not exhibited after the first and second step but from the third step onwards. We will see that the reason for this behaviour is the fact that M_∞ is nilpotent with $M_\infty^2 = 0$. □

From Figure 8.5 (d) it is clear that in general, and in particular for the first few steps, one can only expect $\Delta Z'_n = \mathcal{O}(h^{\min(p-1,q)})$. This will be proved in the next lemma.

Given the additional assumption $M_\infty^{k_0} = 0$, it is nevertheless possible to show $\Delta Z'_n = \mathcal{O}(h^{\min(p,q)})$. This situation will be dealt with later in Lemma 8.30.

For simplicity, this section deals exclusively with methods in Nordsieck form.

Lemma 8.29. *Let the assumptions of Theorem 8.27 be satisfied. Given that the general linear method has stage order q for ordinary differential equations, the global error satisfies $\Delta Y_n = \mathcal{O}(h^{\min(p,q+1)})$ and $\Delta Z'_n = \mathcal{O}(h^{\min(p-1,q)})$ for the stages and stage derivatives, respectively.*

Even though Lemma 8.29 is a straightforward result, it is quite technical to prove. Again, B-series and elementary weight functions can be used. Recall that the method has s internal and r external stages. Thus matters become more complicated due to the fact that either $r < q + 1$ or $r \geq q + 1$ can hold. In order to treat both cases consistently, we introduce $\hat{r} = \min(r, q+1)$. As in Lemma 8.21 the matrices $C = \begin{bmatrix} e & c & \frac{c^2}{2} & \cdots & \frac{c^q}{q!} \end{bmatrix}$ and $K = \begin{bmatrix} 0 & I_q \\ 0 & 0 \end{bmatrix}$ will prove to be useful. In particular it is possible to write the stage order conditions in matrix form.

By assumption the method has stage order q for ordinary differential equations. Thus we know from Theorem 7.6 that the first $\hat{r}$ columns of $\mathcal{U}$ are fixed by the stage order conditions. Using MATLAB notation [123] to extract submatrices we obtain

$$
\mathcal{U}(\,:\,, 1:\hat{r}) = \big(C - \mathcal{A}\,C\,K\big)(\,:\,, 1:\hat{r}). \tag{8.36a}
$$

If $r \geq q+1$ then (8.36a) reads $\mathcal{U}(:, 1{:}q{+}1) = C - \mathcal{A}CK$ as in Theorem 7.6. However, in case of $r < q+1$ additional conditions $\mathcal{A}\, c^{k-1} = \frac{1}{k}\, c^k$ are satisfied for $r \leq k \leq q$ due to high stage order. These are exactly the $C(k)$-conditions used for analysing Runge-Kutta methods (see page 34 or [25]). In MATLAB notation we can write

$$\big(\mathcal{A}\, C\, K\big)(:, r+1 : q+1) = C(:, r+1 : q+1). \tag{8.36b}$$

Using the relations (8.36) we proceed to proving Lemma 8.29.

Proof. Let $\hat{Y}_n$ and $\hat{Z}'_n$ be the stages and stage derivatives calculated from exact input vectors, i.e.

$$\hat{Y}_n = h\mathcal{A}f(\hat{Y}_n, \hat{Z}'_n) + \mathcal{U}\hat{y}^{[n-1]}, \qquad g(\hat{Y}_n) = h\mathcal{A}\hat{Z}'_n + \mathcal{U}\hat{z}^{[n-1]},$$

The local error $\hat{Y}_{ni} - y(t_{n-1}+c_i h)$, $\hat{Z}'_{ni} - z'(t_{n-1}+c_i h)$ is studied first. The global error can then be bounded by finding estimates for $Y_{ni} - \hat{Y}_{ni}$ and $Z'_{ni} - \hat{Z}'_{ni}$.

(a) Exact input vectors are given as B-series $\hat{y}^{[n-1]} = B_{T_y}\big(\mathbf{S}; y(t_{n_1}), z(t_{n-1})\big)$ and $\hat{z}^{[n-1]} = B_{T_z}\big(\mathbf{S}; y(t_{n-1}), z(t_{n-1})\big)$ with the elementary weight function $\mathbf{S}$ from Corollary 8.8. Theorem 8.14 implies that

$$\hat{Y}_n = B_{T_y}\big(\hat{\mathbf{v}}; y(t_0), z(t_0)\big), \qquad h\hat{Z}'_n = B_{T_z}\big(\hat{\mathbf{k}}; y(t_0), z(t_0)\big),$$

where $\hat{\mathbf{v}}$ and $\hat{\mathbf{k}}$ are the weight functions given in Definition 8.11. If $\tau \in T_y$ with $|\tau| < \hat{r}$, then $\hat{\mathbf{v}}(\tau) = C\,\mathbf{S}(\tau)$ is obtained similarly to Lemma 8.21 using (8.36a). On the other hand, if $\hat{r} \leq |\tau| \leq q$, then the term $\mathcal{U}\mathbf{S}(\tau)$ vanishes such that

$$\hat{\mathbf{v}}(\tau) = \mathcal{A}\hat{\mathbf{l}}(\tau) + \mathcal{U}\mathbf{S}(\tau) = \mathcal{A}\hat{\mathbf{l}}(\tau) = \mathcal{A}CK\mathbf{S}(\tau) = C\,\mathbf{S}(\tau)$$

due to (8.36b). We have shown that $\hat{\mathbf{v}}(\tau) = C\,\mathbf{S}(\tau)$ for all trees with $|\tau| \leq q$ such that $\hat{Y}_i = y(t_{n-1}+c_i h) + \mathcal{O}(h^{q+1})$ follows. The corresponding result for $h\,\hat{Z}'_i$ is obtained similarly by keeping in mind that $\hat{\mathbf{l}}([\sigma]_y) = \hat{\mathbf{k}}(\sigma)$ for $[\sigma]_y \in T_{yz}$ and $\hat{\mathbf{l}}(\tau) = \hat{\mathbf{k}}([\tau]_z)$ for $[\tau]_z \in T_z$ (Lemma 8.16).

(b) In order to derive bounds for $Y_{ni} - \hat{Y}_{ni}$ and $Z'_{ni} - \hat{Z}'_{ni}$ the stages Y_n as well as the stage derivatives $h\,Z'_n$ computed after n steps are written as

$$Y_n = B_{T_y}\big(\mathbf{v}; y(t_0), z(t_0)\big), \qquad hZ'_n = B_{T_z}\big(\mathbf{k}; y(t_0), z(t_0)\big).$$

Recall that $\mathbf{v}$ and $\mathbf{k}$ were introduced in Definition 8.11,

$$\mathbf{v} : T_y \to \mathbb{R}^s, \qquad \mathbf{v}(\tau) = \mathcal{A}\mathbf{l}(\tau) + \mathcal{U}\,\mathbf{S}^{n-1}(\tau)$$

$$\mathbf{l} : T_y \to \mathbb{R}^s, \qquad \mathbf{l}(\tau) = \begin{cases} 0 & , \tau = \emptyset, \\ e & , \tau = \bullet \\ |\tau|\mathbf{v}(\tau_1)\cdots\mathbf{v}(\tau_k)\dfrac{\mathbf{k}(\sigma_1)}{|\sigma_1|}\cdots\dfrac{\mathbf{k}(\sigma_l)}{|\sigma_l|} & , \tau = \end{cases}$$

$$\mathbf{k} : T_z \to \mathbb{R}^s, \qquad \big(\mathbf{v}D\big)(\sigma) = \mathcal{A}\mathbf{k}(\sigma) + \mathcal{U}\,\mathbf{S}^{n-1}(\sigma),$$

where the starting procedure $\mathbf{S}^{n-1}$ represents both $y^{[n-1]}$ and $z^{[n-1]}$. Using $\big(y(t_0), z(t_0)\big)$ as the new reference point for the B-series expansion we rewrite the exact input quantities as $\hat{y}^{[n-1]} = B_{T_y}\big(\hat{\mathbf{S}}^{n-1}; y(t_0), z(t_0)\big)$ and $\hat{z}^{[n-1]} = B_{T_z}\big(\hat{\mathbf{S}}^{n-1}; y(t_0), z(t_0)\big)$ such that

$$\hat{Y}_n = B_{T_y}\big(\hat{\mathbf{v}}; y(t_0), z(t_0)\big), \qquad h\hat{Z}'_n = B_{T_z}\big(\hat{\mathbf{k}}; y(t_0), z(t_0)\big).$$

$\hat{\mathbf{v}}$ and $\hat{\mathbf{k}}$ are defined similar to $\mathbf{v}$, $\mathbf{k}$ but using $\hat{\mathbf{S}}^{n-1}$ instead of $\mathbf{S}^{n-1}$.

Since the global error has order p for the y and z component, the two starting procedures agree for trees of order less than p, i.e. $\mathbf{S}^{n-1}(\tau) = \hat{\mathbf{S}}^{n-1}(\tau)$ for all trees $\tau \in T_y \cup T_z$ with $|\tau| < p$. This immediately implies

$$\mathbf{v}(\tau) = \hat{\mathbf{v}}(\tau) \quad \text{for } |\tau| < p \quad \text{and} \quad \mathbf{k}(\sigma) = \hat{\mathbf{k}}(\sigma) \quad \text{for } |\sigma| < p.$$

(c) Putting the results of (a) and (b) together we find

$$Y_i = \hat{Y}_i + \mathcal{O}(h^p) = y(t_{n-1} + c_i h) + \mathcal{O}(h^{q+1}) + \mathcal{O}(h^p),$$
$$hZ'_i = h\hat{Z}'_i + \mathcal{O}(h^p) = h\, z'(t_{n-1} + c_i h) + \mathcal{O}(h^{q+1}) + \mathcal{O}(h^p). \qquad \square$$

In general the stage derivatives Z'_n are calculated with global error of magnitude $\Delta Z'_n = \mathcal{O}(h^{\min(p-1,q)})$ as was shown in the previous lemma. In spite of this Figure Figure 8.5 (d) shows that from the third step onwards higher accuracy $\Delta Z'_n = \mathcal{O}(h^{\min(p,q)})$ is realised. It was pointed out earlier that the stability matrix at infinity, M_∞, plays a crucial role in explaining this behaviour.

Lemma 8.30. *Let the assumptions of Lemma 8.29 be satisfied. In particular the initial input vectors are assumed to satisfy $y^{[0]} = \hat{y}^{[0]} + \mathcal{O}(h^p)$ and $z^{[0]} = \hat{z}^{[0]} + \mathcal{O}(h^p)$. If the matrix $M_\infty = \mathcal{V} - \mathcal{B}\mathcal{A}^{-1}\mathcal{U}$ is nilpotent with nilpotency index k_0, then $\Delta Z'_n = \mathcal{O}(h^{\min(p,q)})$ after $k_0 + 1$ steps.*

Proof. Define $\mathbf{y}^0 = \mathbf{S}^0$ and $\mathbf{z}^0 = \mathbf{S}^0$ where the elementary weight function $\mathbf{y}^0 : T_y \to \mathbb{R}^r$ is defined on T_y but $\mathbf{z}^0 : T_z \to \mathbb{R}^r$ operates on T_z. For $n \geq 1$ consider

$$\mathbf{v}^n = \mathcal{A}\,\mathbf{l}^n + \mathcal{U}\,\mathbf{y}^{n-1}, \qquad\qquad \mathbf{v}^n D = \mathcal{A}\,\mathbf{k}^n + \mathcal{U}\,\mathbf{z}^{n-1},$$
$$\mathbf{y}^n = \mathcal{B}\,\mathbf{l}^n + \mathcal{V}\,\mathbf{y}^{n-1}, \qquad\qquad \mathbf{z}^n = \mathcal{B}\,\mathbf{k}^n + \mathcal{V}\,\mathbf{z}^{n-1}.$$

Similarly introduce

$$\hat{\mathbf{v}}^n = \mathcal{A}\,\hat{\mathbf{l}}^n + \mathcal{U}\,\hat{\mathbf{S}}^{n-1}, \qquad\qquad \hat{\mathbf{v}}^n D = \mathcal{A}\,\hat{\mathbf{k}}^n + \mathcal{U}\,\hat{\mathbf{S}}^{n-1}$$
$$\hat{\mathbf{y}}^n = \mathcal{B}\,\hat{\mathbf{l}}^n + \mathcal{V}\,\hat{\mathbf{S}}^{n-1}, \qquad\qquad \hat{\mathbf{z}}^n = \mathcal{B}\,\hat{\mathbf{k}}^n + \mathcal{V}\,\hat{\mathbf{S}}^{n-1}$$

as was already done in the previous proof. From Lemma 8.29 it is clear that $\mathbf{k}(\sigma) = \hat{\mathbf{k}}(\sigma)$ for all trees $\sigma \in T_z$ with $|\sigma| < p$. In order to prove the relation

$\Delta Z_n' = \mathcal{O}(h^{\min(p,q)})$ we need to show that $\mathbf{k}(\sigma) = \hat{\mathbf{k}}(\sigma)$ holds for trees of order p as well.

Using the above notation it turns out that $\mathbf{z}^n = \mathcal{B}\mathcal{A}^{-1}(\mathbf{v}^n D) + M_\infty \mathbf{z}^{n-1}$ and therefore

$$\mathcal{A}\mathbf{k}^n = \mathbf{v}^n D - \mathcal{U}\mathbf{z}^{n-1} = \mathbf{v}^n D - \mathcal{U}M_\infty \mathbf{z}^{n-2} - \mathcal{U}\mathcal{B}\mathcal{A}^{-1}(\mathbf{v}^{n-1}D)$$

$$= \cdots = \mathbf{v}^n D - \mathcal{U}M_\infty^{n-1}\mathbf{z}^0 - \sum_{i=1}^{n-1} \mathcal{U}M_\infty^{i-1}\mathcal{B}\mathcal{A}^{-1}(\mathbf{v}^{n-i}D).$$

Given that $M_\infty^{k_0} = 0$ this simplifies to

$$\mathcal{A}\mathbf{k}^n = \mathbf{v}^n D - \sum_{i=1}^{k_0} \mathcal{U}M_\infty^{i-1}\mathcal{B}\mathcal{A}^{-1}(\mathbf{v}^{n-i}D) \tag{8.37a}$$

for $n \geq k_0 + 1$. As the local error is of order $\mathcal{O}(h^{p+1})$, we have $\hat{\mathbf{z}}^n(\sigma) = \hat{\mathbf{S}}^n(\sigma)$ for every tree σ with $|\sigma| \leq p$. Thus a similar argument as above shows that

$$\mathcal{A}\hat{\mathbf{k}}^n =_p \hat{\mathbf{v}}^n D - \sum_{i=1}^{k_0} \mathcal{U}M_\infty^{i-1}\mathcal{B}\mathcal{A}^{-1}(\hat{\mathbf{v}}^{n-i}D). \tag{8.37b}$$

In contrast to (8.37a) equality holds only for trees with $|\sigma| \leq p$. This is indicated using the symbol $=_p$.

Let $\sigma \in T_z$ be an arbitrary tree with order $|\sigma| = p$. If the root of σ has ramifications, i.e. $\sigma = \overset{\tau_1 \cdots \tau_k}{\smile}$ with $k \geq 2$, then each subtree τ_i has order strictly lower than p. Therefore

$$(\mathbf{v}^k D)(\sigma) = \mathbf{v}^k(\tau_1)\cdots\mathbf{v}^k(\tau_k) = \hat{\mathbf{v}}^k(\tau_1)\cdots\hat{\mathbf{v}}^k(\tau_k) = (\hat{\mathbf{v}}^k D)(\sigma)$$

for every $k \geq 1$ and the recursions (8.37) imply $\mathbf{k}^n(\sigma) = \hat{\mathbf{k}}^n(\sigma)$ for $n \geq k_0 + 1$.

It remains to prove that this relationships holds for trees of the form $\sigma = [\tau]_z$, too. For such a tree $\mathbf{k}^n(\sigma) = \mathbf{l}^n(\tau)$ holds (Lemma 8.16). Recall that τ has the form $\tau = \overset{\tau_k \; \sigma_l}{\underset{}{\curlyvee}}\overset{\sigma_l}{}$ with $(k,l) \neq (0,1)$.

Since $|\tau_i| < |\tau|$ and $|\sigma_j| < |\tau|$ for $i = 1, \ldots, k$ and $j = 1, \ldots, l$, the calculation of $\mathbf{l}^n$ requires the evaluation of $\mathbf{v}^n$ and $\mathbf{k}^n$ only for trees of order less than p. This means $\mathbf{l}^n(\tau) = \hat{\mathbf{l}}^n(\tau)$ and therefore also $\mathbf{k}^n(\sigma) = \hat{\mathbf{k}}^n(\sigma)$. $\qquad\square$

Consider a stiffly accurate Runge-Kutta method, i.e. $\mathcal{M} = [\mathcal{A}, e, b^\top, 1]$ with $e_s^\top \mathcal{A} = b^\top$ such that the last row of $\mathcal{A}$ coincides with the vector $b^\top$. The stability function reads $R(z) = 1 + z\,b^\top(I - z\,\mathcal{A})^{-1}e$ (cf. Section 2.2) such that $R_\infty = \lim_{z \to \infty} R(z) = 1 - b^\top \mathcal{A}^{-1}e = 0$. Therefore the error in the stage derivatives can be estimated by $\Delta Z_n' = \mathcal{O}(h^{\min(p,q)})$ already for $n \geq 2$.

For a general linear method with $r > 1$ stiff accuracy no longer implies $M_\infty = 0$ but only the first row is zero. Hence, nilpotency of the stability matrix M_∞ becomes an additional requirement.

A statement similar to Lemma 8.30 was already proved in [10] for BDF methods. Since the BDF schemes are included within the framework of general linear methods a result such as Lemma 8.30 had to be expected.

Example 8.31. Recall from Example 7.2 that the BDF$_3$ method written in Nordsieck form reads

$$
\mathcal{M} = \left[\begin{array}{c|c} \mathcal{A} & \mathcal{U} \\ \hline \mathcal{B} & \mathcal{V} \end{array} \right] = \left[\begin{array}{c|cccc} \frac{6}{11} & 1 & \frac{5}{11} & -\frac{1}{22} & -\frac{7}{66} \\ \hline \frac{6}{11} & 1 & \frac{5}{11} & -\frac{1}{22} & -\frac{7}{66} \\ 1 & 0 & 0 & 0 & 0 \\ \frac{12}{11} & 0 & -\frac{12}{11} & -\frac{1}{11} & \frac{5}{11} \\ \frac{6}{11} & 0 & -\frac{6}{11} & -\frac{6}{11} & \frac{8}{11} \end{array} \right]
$$

such that the stability matrix at infinity is given by

$$
M_\infty = \mathcal{V} - \mathcal{B}\mathcal{A}^{-1}\mathcal{U} = \left[\begin{array}{cccc} 0 & 0 & 0 & 0 \\ -\frac{11}{6} & -\frac{5}{6} & \frac{1}{12} & \frac{7}{36} \\ -2 & -2 & 0 & \frac{2}{3} \\ -1 & -1 & -\frac{1}{2} & \frac{5}{6} \end{array} \right].
$$

It can be verified easily that

$$
M_\infty^2 = \left[\begin{array}{cccc} 0 & 0 & 0 & 0 \\ \frac{7}{6} & \frac{1}{3} & -\frac{1}{6} & \frac{1}{18} \\ 3 & 1 & -\frac{1}{2} & \frac{1}{6} \\ 2 & 1 & -\frac{1}{2} & \frac{1}{6} \end{array} \right], \quad M_\infty^3 = \left[\begin{array}{cccc} 0 & 0 & 0 & 0 \\ -\frac{1}{3} & 0 & 0 & 0 \\ -1 & 0 & 0 & 0 \\ -1 & 0 & 0 & 0 \end{array} \right], \quad M_\infty^4 = 0. \qquad \square
$$

9

Properly Stated Index-2 DAEs

In this chapter we return to studying nonlinear index-2 DAEs

$$A(t)\big[D(t)x(t)\big]' + b\big(x(t),t\big) = 0 \qquad (9.1)$$

with a properly stated leading term. In Chapter 6 a decoupling procedure was derived showing that the components $u = D\bar{P}_1 x$ and $v = D\bar{Q}_1 x$ satisfy the inherent index-1 equation

$$u' = \mathbb{f}\big(u, \mathbb{w}(u, v', t), t\big), \qquad v = D(t)\mathbb{z}(u, t). \qquad (9.2)$$

When studying numerical methods for (9.1) it is seminal that the inherent index-1 DAE (9.2) is treated correctly. This work was carried out in the previous section. General linear methods were studied for (9.2). Order conditions and a convergence result have been derived.

We will investigate how these results can be transferred to the general index-2 formulation (9.1) using the decoupling procedure from Chapter 6.

By ensuring that a general linear method, when applied to (9.1), correctly discretises the inherent system (9.2) the results of [90] are extended in two directions: Not only nonlinear DAEs are treated but also the large class of general linear methods is considered.

9.1 Discretisation and Decoupling

Assume that we wanted to employ a general linear method $\mathcal{M} = [\mathcal{A}, \mathcal{U}, \mathcal{B}, \mathcal{V}]$ in order to solve the initial value problem

$$A[Dx]' + b(x, \cdot) = 0, \qquad D(t_0)P_1(x^0, t_0)\big(x(t_0) - x^0\big) = 0. \qquad (9.3)$$

The vector x_0 originates from an initialisation $A(t_0)y^0 + b(x^0, t_0) = 0$. Observe that x_0 will be inconsistent in general, but the formulation (9.3) ensures that there is a unique local solution as was shown in Theorem 6.7.

From Chapter 6 the exact solution x_* is known to satisfy the decoupled system

$$u' = \mathfrak{f}\big(u, \mathsf{w}(u, v', \cdot), \cdot\big) = f(u, v', \cdot), \qquad v = D\mathsf{z}(u, \cdot) = g(u, \cdot), \qquad (9.4a)$$

$$x_* = D^-u + \mathsf{z}(u, \cdot) + \mathsf{w}(u, v', \cdot). \qquad (9.4b)$$

Recall that $u = D\bar{P}_1 x$ and $v = D\bar{Q}_1 x$ are components of the solution determined by the projector functions $\bar{P}_1(t) = P_1\big(\bar{x}(t), t\big)$ and $\bar{Q}_1(t) = Q_1\big(\bar{x}(t), t\big)$. The function $\bar{x} \in C^1(\mathcal{I}, \mathbb{R}^m)$ is assumed to satisfy

$$\bar{x}(t_0) = x^0, \qquad \big(\bar{x}(t), t\big) \in \mathcal{D} \times \mathcal{I} \qquad \forall\ t \in \mathcal{I}, \qquad (9.5)$$

where $\mathcal{D} \times \mathcal{I} \subset \mathbb{R}^m \times \mathbb{R}$ is the domain where the DAE (9.3) is defined.

As the decoupled system (9.4) is known, it is straightforward to apply the general linear method $\mathcal{M}$ directly to the inherent index-1 system (9.4a),

$$U = h\,\mathcal{A}\,F(U, V') + \mathcal{U}\,u^{[n]}, \qquad G(U) = h\,\mathcal{A}\,V' + \mathcal{U}\,v^{[n]}, \qquad (9.6a)$$

$$u^{[n+1]} = h\,\mathcal{B}\,F(U, V') + \mathcal{V}\,u^{[n]}, \qquad v^{[n+1]} = h\,\mathcal{B}\,V' + \mathcal{V}\,v^{[n]}. \qquad (9.6b)$$

The numerical solution is then given by

$$x_n = D^-(t_n)u_n + \mathsf{z}(u_n, t_n) + \mathsf{w}(u_n, v'_n, t_n) \qquad (9.6c)$$

where u_n and v'_n are numerical approximations to $u(t_n)$ and $v'(t_n)$ respectively. For stiffly accurate methods these quantities are given by the last stages, i.e. $u_n = U_s$ and $v'_n = V'_s$. As in Chapter 7 the shorthand notation

$$F(U, V') = \begin{bmatrix} f(U_1, V'_1, t_n + c_1 h) \\ \vdots \\ f(U_s, V'_s, t_n + c_s h) \end{bmatrix}, \qquad G(U) = \begin{bmatrix} g(U_1, t_n + c_1 h) \\ \vdots \\ g(U_s, t_n + c_s h) \end{bmatrix}$$

has been used.

Unfortunately the system (9.4) is not available for direct computation. Even though the implicit function theorem guarantees the existence of the functions

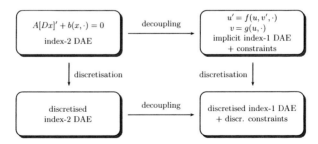

Figure 9.1: The relationship between discretisation and decoupling

$\mathfrak{f}$, $\mathbf{w}$ and $\mathbf{z}$, these mappings are not accessible for practical applications. The decoupling procedure is only a theoretical tool to study the behaviour of the exact/numerical solution. The decoupling is *not* intended to be carried out in practice.

However, we will derive conditions on the DAE (9.3) which guarantee that a general linear method, when applied to the original formulation (9.3), behaves as if it was applied to the inherent system (9.4a). Thus the aim of this section is to derive conditions on the DAE (9.3) such that the diagram in Figure 9.1 commutes.

We have to compare the following two approaches:

(a) The DAE (9.3) is decoupled first and the method $\mathcal{M}$ is applied to the decoupled system (9.4a).

(b) The general linear method is applied directly to the original formulation (9.3) and the decoupling procedure is applied to the resulting discretised scheme afterwards.

The decoupling procedure is explained in detail in Chapter 6. General linear methods for implicit index-1 DAEs have been studied in Chapter 8. Thus approach (a) has received considerable attention so far in this thesis.

For practical computations only the approach (b) is feasible. Additional requirements on the DAE (9.3) will ensure that the numerical quantities obtained using approach (b) coincide with those originating from approach (a). Hence the numerical result will correctly reflect qualitative properties of the inherent index-1 system.

9.1.1 Decoupling the Discretised Equations

The general linear method's discretisation of (9.3) is given by

$$A(t_n + c_i\,h)[DX]_i' + b(X_i, t_n + c_i h) = 0, \quad i = 1, \ldots, s, \tag{9.7a}$$

$$[DX] = h\,\mathcal{A}\,[DX]' + \mathcal{U}\,[Dx]^{[n]}, \tag{9.7b}$$

$$[Dx]^{[n+1]} = h\,\mathcal{B}\,[DX]' + \mathcal{V}\,[Dx]^{[n]}. \tag{9.7c}$$

The stages $X_i \approx x(t_n + c_i h)$ are approximations to the exact solution at intermediate timepoints. The super vectors $[DX]$ and $[DX]'$ are given by

$$[DX] = \begin{bmatrix} D(t_n + c_1\,h)X_1 \\ \vdots \\ D(t_n + c_s\,h)X_s \end{bmatrix}, \qquad [DX]' = \begin{bmatrix} [DX]_1' \\ \vdots \\ [DX]_s' \end{bmatrix},$$

where $[DX]_i' \approx \frac{\mathrm{d}}{\mathrm{d}t}(Dx)(t_n + c_i h)$ represents stage derivatives for the component Dx. The input vector

$$[Dx]^{[n]} \approx \begin{bmatrix} (Dx)\,(t_n) \\ h\,(Dx)'(t_n) \\ \vdots \\ h^{r-1}(Dx)^{(r-1)}(t_n) \end{bmatrix}$$

is assumed to be a Nordsieck vector. Observe that only information about the solution's D-component is passed on from step to step. Hence errors in this component are the only ones that are possibly propagated.

As a consequence the numerical result x_{n+1} at $t_{n+1} = t_n + h$ has to be obtained as a linear combination of the stages X_i. Every stage X_i satisfies the obvious constraints, such that

$$X_i \in \mathcal{M}_0(t_n + c_i h), \qquad \mathcal{M}_0(t) = \{\, z \in \mathcal{D} \,|\, b(z,t) \in \mathrm{im}\, A(t) \,\}. \qquad (9.8)$$

The set $\mathcal{M}_0(t)$ was already introduced in Section 4.2 on page 74. It is a desirable feature for the numerical solution to satisfy $x_{n+1} \in \mathcal{M}_0(t_{n+1})$ as well. Thus stiffly accurate methods, where x_{n+1} coincides with the last stage X_s, guarantee this situation. We will therefore restrict attention to stiffly accurate general linear methods, i.e.

$$\mathcal{M} = \left[\begin{array}{c|c} \mathcal{A} & \mathcal{U} \\ \hline \mathcal{B} & \mathcal{V} \end{array} \right] \qquad \text{with} \qquad e_s^\top \mathcal{A} = e_1^\top \mathcal{B}, \qquad e_s^\top \mathcal{U} = e_1^\top \mathcal{V}, \qquad c_s = 1,$$

such that the last row of $[\mathcal{A}, \mathcal{U}]$ coincides with the first row of $[\mathcal{B}, \mathcal{V}]$.

In order to decouple the discretised system (9.7) according to the approach (b) described above, the vector $[Dx]^{[n]}$ of incoming approximations needs to be split according to its $D\bar{P}_1$ and $D\bar{Q}_1$ parts. For the moment assume that

$$[Dx]_i^{[n]} = \mathbf{u}_i^{[n-1]} + \mathbf{v}_i^{[n-1]} \in \mathrm{im}(D\bar{P}_1)(t_{n-1}) \oplus \mathrm{im}(D\bar{Q}_1)(t_{n-1}) \qquad (9.9)$$

for $i = 1, \ldots, r$. We will justify this assumption later on.

The stages X_i are split according to

$$\mathbf{U}_i = D(t_n + c_i h)\bar{P}_1(t_n + c_i h)X_i, \qquad \mathbf{Z}_i = \bar{Z}(t_n + c_i h)X_i, \qquad (9.10\mathrm{a})$$
$$\mathbf{V}_i = D(t_n + c_i h)\bar{Q}_1(t_n + c_i h)X_i, \qquad \mathbf{W}_i = T(t_n + c_i h)X_i, \qquad (9.10\mathrm{b})$$

such that

$$X_i = D^-(t_n + c_i h)\mathbf{U}_i + \mathbf{Z}_i + \mathbf{W}_i, \qquad \mathbf{V}_i = D(t_n + c_i h)\mathbf{Z}_i. \qquad (9.10\mathrm{c})$$

Recall from Section 5.1 that T is a projector function with $\mathrm{im}\, T = N_0 \cap S_0$. As we are always assuming that the structural condition

(A1) $N_0(t) \cap S_0(x,t)$ *does not depend on x*

holds, T can be chosen to depend on time only. Using $U = I - T$ the projector function

$$\bar{Z}(t) = P_0(t)\bar{Q}_1(t) + U(t)Q_0(t)$$

was introduced in Section 5.1 as well. Finally, for a nonsingular matrix $\mathcal{A}$ stage derivatives $\mathbf{U}'_i$ and $\mathbf{V}'_i$ can be defined using the relations

$$\mathbf{U} = h\,\mathcal{A}\,\mathbf{U}' + \mathcal{U}\,\mathbf{u}^{[n]}, \qquad\qquad \mathbf{V} = h\,\mathcal{A}\,\mathbf{V}' + \mathcal{U}\,\mathbf{v}^{[n]}.$$

A quick comparison with (9.7b), (9.9) and (9.10) shows that $[DX]' = \mathbf{U}' + \mathbf{V}'$ holds by construction.

Rewriting (9.7a) in terms of the new variables introduced so far yields

$$\begin{aligned} F\big(\mathbf{U}_i, \mathbf{W}_i, & \mathbf{Z}_i, \mathbf{U}'_i, \mathbf{V}'_i, t_n + c_i h\big) \\ & = A_{ni}(\mathbf{U}'_i + \mathbf{V}'_i) + b\big(D_{ni}^-\mathbf{U}_i + \mathbf{Z}_i + \mathbf{W}_i,\, t_n + c_i h\big) = 0, \end{aligned} \tag{9.7a'}$$

where $A_{ni} = A(t_n + c_i h)$ was used to shorten notations. D_{ni} has a similar interpretation. The mapping $F(u, w, z, \eta, \zeta, t)$ was first defined in (6.14b) on page 93. Later, in Section 6.3, F was split into three different parts, $\hat{F}_1$, $\hat{F}_2$ and $\hat{F}_3$ used to determine z, w and f, respectively. This procedure will be repeated for the discretised system (9.7a').

Lemma 9.1. *Let (9.3) be a regular index-2 DAE with a properly stated leading term. Assume that the structural condition* (A1) *is satisfied and that (9.5) holds for some function $\bar{x} \in C^1(\mathcal{I}, \mathbb{R}^m)$. If a general linear method $\mathcal{M} = [\mathcal{A}, \mathcal{U}, \mathcal{B}, \mathcal{V}]$ with nonsingular $\mathcal{A}$ and sufficiently small h is used for the numerical scheme (9.7), then the following relations hold:*

$$\mathbf{Z}_i = \mathsf{z}(\mathbf{U}_i, t_n + c_i h), \tag{9.11a}$$

$$\mathbf{V}_i = D_{ni}\,\mathsf{z}(\mathbf{U}_i, t_n + c_i h), \tag{9.11b}$$

$$\mathbf{U}'_i = \mathbf{T}^i_1 - \mathbf{T}^i_2 + \mathsf{f}(\mathbf{U}_i, \mathbf{W}_i, t_i), \tag{9.11c}$$

$$0 = \hat{F}_2\big(\mathbf{U}_i, \mathbf{W}_i, \mathbf{V}'_i, t_n + c_i h\big) + (Q_0\bar{Q}_1 D^-)_{ni}\mathbf{T}^i_2 + \mathbf{T}^i_3. \tag{9.11d}$$

z *is the mapping from Lemma 6.4. The functions f and $\hat{F}_2$ are defined in (6.16) and Lemma 6.5, respectively. The terms $\mathbf{T}^i_j$ are given by*

$$\begin{aligned} \mathbf{T}^i_1 &= \big(I - (D\bar{P}_1 D^-)_{ni}\big)\mathbf{U}'_i - (D\bar{P}_1 D^-)'_{ni}\mathbf{U}_i, \\ \mathbf{T}^i_2 &= (D\bar{P}_1 D^-)_{ni}\mathbf{V}'_i + (D\bar{P}_1 D^-)'_{ni}\mathbf{V}_i, \\ \mathbf{T}^i_3 &= (Q_0\bar{Q}_1 D^-)_{ni}(D\bar{P}_1 D^-)'_{ni}\mathbf{U}_i - (Q_0\bar{Q}_1 D^-)_{ni}\mathbf{U}'_i. \end{aligned}$$

Proof. Lemma 5.2 and 6.9 imply $\bar{Z}\bar{G}_2^{-1}AD = 0$ such that the representation (9.7a') leads to $\hat{F}_1(\mathbf{U}_i, \mathbf{Z}_i, t_n + c_i h) = 0$ for the function

$$\hat{F}_1(u, z, \cdot) = \bar{Z}\bar{G}_2^{-1} b(D^- u + \bar{Z}z, \cdot) + (I - \bar{Z})z$$

(see Section 6.3 for more details). As the mapping z is implicitly defined by $\hat{F}_1(u, z, t) = 0$, the stages $\mathbf{Z}_i$ are given by (9.11a). The relation (9.11b) follows from (9.11a) due to $\mathbf{V}_i = D_{ni}\bar{Q}_1 X_i = D_{ni}\mathbf{Z}_i$.

To be precise, it is assumed that the stepsize h is small enough to guarantee that $(\mathbf{U}_i, t_n + c_i h)$ remains in the domain where z is defined. This will not be stated every time, but all arguments are meant locally in that sense.

As a next step, (9.7a') is multiplied by $D\bar{P}_1\bar{G}_2^{-1}$. The leading term can be written as

$$(D\bar{P}_1\bar{G}_2^{-1}A)_{ni}[DX]_i' = (D\bar{P}_1 D^-)_{ni}(\mathbf{U}_i' + \mathbf{V}_i')$$
$$= \mathbf{U}_i' - \left(I - (D\bar{P}_1 D^-)_{ni}\right)\mathbf{U}_i' + (D\bar{P}_1 D^-)_{ni}\mathbf{V}_i'$$

such that

$$\mathbf{U}_i' = \left(I - (D\bar{P}_1 D^-)_{ni}\right)\mathbf{U}_i' - (D\bar{P}_1 D^-)_{ni}\mathbf{V}_i' - D\bar{P}_1\bar{G}_2^{-1}b(X_i, t_n + c_i h)$$

and therefore (9.11c) follows. Finally, (9.7a) shows that

$$(T\bar{G}_2^{-1}A)_{ni}(\mathbf{U}_i' + \mathbf{V}_i') + (T\bar{G}_2^{-1})_i\, b(D_{ni}^-\mathbf{U}_i + \mathbf{Z}_i + \mathbf{W}_i,\, t_n + c_i h) = 0.$$

Using (9.11a) and (9.11c) this equation can be written as

$$0 = \hat{F}_2(\mathbf{U}_i, \mathbf{W}_i, \mathbf{V}_i', t_n + c_i h) + (T\bar{G}_2^{-1}A)_{ni}(\mathbf{T}_1^i - \mathbf{T}_2^i). \tag{9.12}$$

It remains to take $T\bar{G}_2^{-1}A = -Q_0\bar{Q}_1 D^-$ into account to see that (9.12) is identical to (9.11d). □

The terms $\mathbf{T}_j^i$ appearing in the previous lemma are similar to corresponding quantities obtained in [90] when studying Runge-Kutta methods for linear DAEs. There we also find practical criteria for $\mathbf{T}_j^i = 0$.

Lemma 9.2.   (a) *If the space* $\operatorname{im} D\bar{P}_1 D^-$ *is constant, then* $\mathbf{T}_1^i = 0$, $\mathbf{T}_3^i = 0$.
 (b) *If the space* $\operatorname{im} D\bar{Q}_1 D^-$ *is constant, then* $\mathbf{T}_2^i = 0$.

Proof. The proof is essentially the same as in [90, Lemma 7]. □

9.1.2 Comparing Approach (a) and (b)

We are now in a position to prove commutativity for the diagram in Figure 9.1. Recall that we wanted to compare the following two approaches:

(a) The DAE (9.3) is decoupled first and the method $\mathcal{M}$ is applied to the decoupled system (9.4a).

This approach leads to the numerical scheme

$$U = h\,\mathcal{A}\,F(U, V') + \mathcal{U}\,u^{[n]}, \qquad G(U) = h\,\mathcal{A}\,V' + \mathcal{U}\,v^{[n]}, \qquad (9.6\text{a})$$
$$u^{[n+1]} = h\,\mathcal{B}\,F(U, V') + \mathcal{V}\,u^{[n]}, \qquad v^{[n+1]} = h\,\mathcal{B}\,V' + \mathcal{V}\,v^{[n]}, \qquad (9.6\text{b})$$

with

$$F(U, V')_i = \mathbb{f}\big(U_i, \mathbb{w}(U_i, V'_i, t_n + c_i h), t_n + c_i h\big),$$
$$G(U)_i = D_{ni}\,\mathbb{z}(U_i, t_n + c_i h).$$

For stiffly accurate methods the numerical solution is given by

$$x_{n+1} = D^-(t_{n+1})U_s + \mathbb{z}(U_s, t_{n+1}) + \mathbb{w}(U_s, V'_s, t_{n+1}). \qquad (9.6\text{c})$$

(b) The general linear method is applied directly to the original formulation (9.3) and the decoupling procedure is applied to the resulting discretised scheme afterwards.

We saw above that this approach leads to

$$\mathbf{U} = h\,\mathcal{A}\,\mathbf{U}' + \mathcal{U}\,\mathbf{u}^{[n]}, \qquad \mathbf{V} = h\,\mathcal{A}\,\mathbf{V}' + \mathcal{U}\,\mathbf{v}^{[n]}, \qquad (9.13\text{a})$$
$$\mathbf{u}^{[n+1]} = h\,\mathcal{B}\,\mathbf{U} + \mathcal{V}\,\mathbf{u}^{[n]}, \qquad \mathbf{v}^{[n+1]} = h\,\mathcal{B}\,\mathbf{V}' + \mathcal{V}\,\mathbf{v}^{[n]}. \qquad (9.13\text{b})$$

Lemma 9.1 ensures that

$$\mathbf{V}_i = D_{ni}\,\mathbb{z}(\mathbf{U}_i, t_n + c_i h), \qquad \mathbf{U}'_i = \mathbf{T}_1^i - \mathbf{T}_2^i + \mathbb{f}(\mathbf{U}_i, \mathbf{W}_i, t_i),$$
$$0 = \hat{F}_2\big(\mathbf{U}_i, \mathbf{W}_i, \mathbf{V}'_i, t_n + c_i h\big) + (Q_0 \bar{Q}_1 D^-)_{ni}\mathbf{T}_2^i + \mathbf{T}_3^i.$$

Finally the numerical result can be represented in the form

$$x_{n+1} = X_s = D_{ns}\mathbf{U}_s + \mathbb{z}(\mathbf{U}_s, t_{n+1}) + \mathbf{W}_s. \qquad (9.13\text{c})$$

Obviously, the two schemes (9.6) and (9.13) coincide provided that

- the same initial vectors $u^{[0]} = \mathbf{u}^{[0]}$, $v^{[0]} = \mathbf{v}^{[0]}$ are used
- the terms $\mathbf{T}_j^i = 0$ vanish,
- $\mathbf{W}_i = \mathbb{w}(\mathbf{U}_i, \mathbf{V}'_i, t_n + c_i h)$ holds.

Theorem 9.3. *Let the assumptions of Lemma 9.1 be satisfied and assume that the subspaces* $\operatorname{im} D\bar{P}_1 D^-$ *and* $\operatorname{im} D\bar{Q}_1 D^-$ *are constant. If the initial input vector satisfies*

$$[Dx]_i^{[0]} = \mathbf{u}_i^{[0]} + \mathbf{v}_i^{[0]} \in \operatorname{im}(D\bar{P}_1)(t_0) \oplus \operatorname{im}(D\bar{Q}_1)(t_0) \qquad (9.14)$$

for $i = 1, \ldots, r$, *then the diagram in Figure 9.1 commutes.*

Proof. Due to Lemma 9.2 the terms $\mathbf{T}_j^i$ vanish such that $\mathbf{T}_1^i = 0$, $\mathbf{T}_3^i = 0$ and $\mathbf{T}_2^i = 0$. Therefore (9.11d) reduces to

$$\hat{F}_2(\mathbf{U}_i, \mathbf{W}_i, \mathbf{V}_i', t_n + c_i h) = 0.$$

This equation determines $\mathbf{W}_i$ as a function of $\mathbf{U}_i$, $\mathbf{V}_i'$ and the timepoint $t_n + c_i h$ (cf. Lemma 6.5). Thus $\mathbf{W}_i = \mathsf{w}(\mathbf{U}_i, \mathbf{V}_i', t_n + c_i h)$, since the implicitly defined function w is uniquely determined. From (9.11c) we find

$$\mathbf{U}_i' = \mathsf{f}\big(\mathbf{U}_i, \mathsf{w}(\mathbf{U}_i, \mathbf{V}_i', t_n + c_i h), t_n + c_i h\big) = F(\mathbf{U}, \mathbf{V}')_i \qquad (9.15\text{a})$$

and (9.11b) implies

$$\mathbf{V}_i = D_{ni}\mathbf{Z}_i = D_{ni}\mathsf{z}(\mathbf{U}_i, t_n + c_i h) \qquad = G(\mathbf{U})_i. \qquad (9.15\text{b})$$

As a consequence, (9.6) and (9.13) yield the same numerical scheme. Note that (9.7b) and (9.7c) split into

$$\mathbf{U} = h\,\mathcal{A}\,\mathbf{U}' + \mathcal{U}\,\mathbf{u}^{[n]}, \qquad\qquad \mathbf{V} = h\,\mathcal{A}\,\mathbf{V}' + \mathcal{U}\,\mathbf{v}^{[n]}$$
$$\mathbf{u}^{[n+1]} = h\,\mathcal{B}\,\mathbf{U}' + \mathcal{V}\,\mathbf{u}^{[n]}, \qquad\qquad \mathbf{v}^{[n+1]} = h\,\mathcal{B}\,\mathbf{V}' + \mathcal{V}\,\mathbf{v}^{[n]}.$$

This justifies (9.9) whenever we have a similar splitting (9.14) for the initial input vector. □

The case of index-1 equations is again included in the results presented here. If (9.3) was an index-1 DAE, $G_2 = G_1$ would be non-singular. Hence $Q_1 = 0$, $P_1 = I$ is an obvious choice for the projectors in the sequence (6.7). We have $DQ_1D^- = 0$ and $DP_1D^- = DD^- = R$, where R is the projector function related to the properly stated leading term (see Definition 3.2). For Theorem 9.3 the only remaining requirement is a constant subspace $\mathrm{im}\, R = \mathrm{im}\, D$.

As for linear DAEs [89, 90] we will say that a given equation is numerically qualified whenever Theorem 9.3 is applicable. More precisely we consider

Definition 9.4.   • *A regular index-1 DAE* (9.3) *with a properly stated leading term is said to be numerically qualified if* $\mathrm{im}\, D$ *is constant.*

• *A regular index-2 DAE* (9.3) *with a properly stated leading term is said to be numerically qualified if the two subspaces* $DS_1 = \mathrm{im}\, DP_1D^-$ *and* $DN_1 = \mathrm{im}\, DQ_1D^-$ *are constant.*

For a general linear method to be suited for integrating nonlinear index-2 DAEs with properly stated leading terms it has to be designed for dealing with the implicit index-1 systems but at the same time the method has to discretise the constraints correctly. The commutativity of diagram 9.1 guarantees that both properties are combined correctly.

When doing practical computations the quantities $\mathbf{U}$, $\mathbf{V}$, $\mathbf{u}^{[n]}$, $\mathbf{v}^{[n]}$ and so on are never used explicitly. All calculations are based on X, $[DX]'$ and $[Dx]^{[n]}$, since the numerical scheme has to be applied directly to the index-2 formulation. Nevertheless Theorem 9.3 ensures that these quantities are interrelated by (9.13). Thus for numerically qualified DAEs the method behaves as if it was integrating the inherent index-1 equation.

It remains to prove convergence for general linear methods applied to properly stated index-2 DAEs. To this point we now turn.

9.2 Convergence

In this section we again consider properly stated differential algebraic equations

$$A(t)\big[D(t)x(t)\big]' + b\big(x(t),t\big) = 0 \tag{9.16}$$

with index 1 or 2 and assume that (9.16) is numerically qualified in the sense of Definition 9.4. Recall from Chapter 6 that the exact solution $x(t)$ can be written as

$$x(t) = D^-(t)u(t) + z(t) + w(t). \tag{9.17a}$$

The functions u and $v = Dz$ satisfy the inherent index-1 system equation

$$u'(t) = f\big(u(t), v'(t), t\big), \qquad\qquad v(t) = g\big(u(t), t\big), \tag{9.17b}$$

while the components w and z are given by

$$w(t) = \mathsf{w}\big(u(t), v'(t), t\big), \qquad\qquad z(t) = \mathsf{z}\big(u(t), t\big). \tag{9.17c}$$

In the previous section a general linear method $\mathcal{M} = [\mathcal{A}, \mathcal{U}, \mathcal{B}, \mathcal{V}]$ was applied to (9.16) giving rise to the numerical scheme

$$A_{ni}[DX]'_i + b(X_i, t_n + c_i h) = 0, \qquad\qquad i = 1, \dots, s, \tag{9.18a}$$

$$[DX] = h\,\mathcal{A}\,[DX]' + \mathcal{U}\,[Dx]^{[n]}, \tag{9.18b}$$

$$[Dx]^{[n+1]} = h\,\mathcal{B}\,[DX]' + \mathcal{V}\,[Dx]^{[n]}. \tag{9.18c}$$

As the DAE (9.16) is assumed to be numerically qualified, Theorem 9.3 ensures that the stages X_i exhibit a structure quite similar to (9.17a). In particular

$$X_i = D_{ni}^-\,\mathbf{U}_i + \mathbf{Z}_i + \mathbf{W}_i, \tag{9.19a}$$

$$\left.\begin{array}{ll} \mathbf{U} = h\,\mathcal{A}\,F(\mathbf{U}, \mathbf{V}') + \mathcal{U}\,\mathbf{u}^{[n]}, & g(\mathbf{U}) = h\,\mathcal{A}\,\mathbf{V}' + \mathcal{U}\,\mathbf{v}^{[n]}, \\[2mm] \mathbf{u}^{[n+1]} = h\,\mathcal{B}\,\mathbf{U}' + \mathcal{V}\,\mathbf{u}^{[n]}, & \mathbf{v}^{[n+1]} = h\,\mathcal{B}\,\mathbf{V}' + \mathcal{V}\,\mathbf{v}^{[n]}, \end{array}\right\} \tag{9.19b}$$

$$\mathbf{W}_i = \mathsf{w}(\mathbf{U}_i, \mathbf{V}'_i, t_n + c_i h), \qquad\qquad \mathbf{Z}_i = \mathsf{z}(\mathbf{U}_i, t_n + c_i h). \tag{9.19c}$$

Observe that (9.19b) is precisely the general linear method's discretisation of the inherent index-1 system (9.17b). As usual the shorthand notations $F(\mathbf{U}, \mathbf{V}')_i = f(\mathbf{U}_i, \mathbf{V}'_i, t_n + c_i h)$ and $G(\mathbf{U})_i = D_{ni}\mathsf{z}(\mathbf{U}_i, t_n + c_i h)$ have been used.

For stiffly accurate methods the numerical result x_{n+1} at time t_{n+1} is given by the last stage such that $x_{n+1} = X_s$. In order to prove convergence we therefore have to compare (9.17a) and (9.19a).

Theorem 9.5. *Let the regular index-μ DAE*

$$A(t)\big[D(t)x(t)\big]' + b\big(x(t),t\big) = 0, \qquad D(t_0)P_1(x^0,t_0)\big(x(t_0) - x^0\big) = 0, \quad (9.20)$$

with a properly stated leading term, $\mu \in \{1,2\}$, satisfy the following conditions:

(A1) $N_0(t) \cap S_0(x,t)$ *does not depend on x*

(A2) $\exists \ (y^0, x^0, t_0) \in \text{im } D(t_0) \times \mathcal{D} \times \mathcal{I}$ *such that* $A(t_0)y^0 + b(x^0, t_0) = 0$,

(A3) $\exists \ \bar{x} \in C^1(\mathcal{I}, \mathbb{R}^m)$ *such that* $\bar{x}(t_0) = x^0$ *and* $\big(\bar{x}(t), t\big) \in \mathcal{D} \times \mathcal{I} \ \ \forall \ t \in \mathcal{I}$,

(A4) *the derivatives* $\frac{\partial b}{\partial x}$, $\frac{dD}{dt}$ *and* $\frac{\partial}{\partial t}\big(\bar{Z}\bar{G}_2^{-1}b\big)$ *exist and are continuous,*

(A5) *(9.16) is numerically qualified in the sense of Definition 9.4.*

Then there exists a unique local solution $x \in C_D^1([t_0, t_{end}], \mathbb{R}^m)$ of the initial value problem (9.20). Let $\mathcal{M} = [\mathcal{A}, \mathcal{U}, \mathcal{B}, \mathcal{V}]$ be a general linear method with nonsingular $\mathcal{A}$. Assume that

(a) $\mathcal{V}$ *is power bounded and M_∞ is nilpotent with $M_\infty^{k_0} = 0$.*

(b) f, v, w *are sufficiently differentiable.*

(c) *The initial input vector $[Dx]^{[0]}$ satisfies $[Dx]^{[0]} = \widehat{[Dx]}^{[0]} + \mathcal{O}(h^p)$, where $\widehat{[Dx]}^{[0]}$ denotes the exact input vector.*

(d) *The method has order p for implicit index-1 DAEs (9.17b) and stage order $q = p$ for ordinary differential equations.*

(e) *M is stiffly accurate.*

Then, after $k_0 + 1$ steps, the numerical result computed using a constant stepsize h converges to the exact solution of the DAE (9.16) with order (at least) p.

Before actually proving this results, some brief remarks are due.

The structural condition (A1) ensures that the decoupled system (9.17) can be constructed. As in Theorem 6.7 the assumptions (A2)-(A4) guarantee the existence of the solution x.

(A5) ensures that Theorem 9.3 is applicable. Since discretisation and decoupling are therefore known to commute, the discrete system (9.18) decouples into (9.19) and we can study (9.19a) and (9.19b) separately.

The assumptions (a)-(d) guarantee convergence of order p for the inherent index-1 system (9.17b). This result was established in Theorem 8.27. The initial input vector $[Dx]^{[0]}$ can be computed using generalised Runge-Kutta methods taking only the initial value x^0 as input but producing r output quantities. In this case but also when building up $[Dx]^{[0]}$ gradually using a variable order implementation, the splitting (9.14) is evident.

Finally, assumption (e) ensures that the numerical result $x_{n+1} = X_s$ coincides with the last stage. Thus the numerical result at t_{n+1} satisfies the obvious constraint (9.8) and x_{n+1} can be expressed using (9.19a).

Observe that the assumption on M_∞ to be nilpotent is a stronger requirement than in Theorem 8.27. There M_∞ was assumed to be power bounded, but the nilpotency of M_∞ allows the application of Lemma 8.29 such that $\mathbf{V}'$ is calculated with order $\min(p, q) = p$.

Proof of Theorem 9.5. Due to stiff accuracy, comparing (9.17a) and (9.19a) yields an estimate for the global error,

$$\|x_{n+1} - x(t_{n+1})\| \le \|D_{ns}^-\big(\mathbf{U}_s - u(t_{n+1})\big)\| + \|\mathbf{z}(\mathbf{U}_s, t_{n+1}) - \mathbf{z}\big(u(t_{n+1}), t_{n+1}\big)\|$$

$$+ \|\mathbf{w}(\mathbf{U}_s, \mathbf{V}'_s, t_{n+1}) - \mathbf{w}\big(u(t_{n+1}), v'(t_{n+1}), t_{n+1}\big)\|$$

$$\le \|D_{ns}^-\big(\mathbf{U}_s - u(t_{n+1})\big)\| + \int_0^1 \|\mathbf{z}_u\big(\varphi(\tau), t_n\big)\| \, d\tau \, \|\mathbf{U}_s - u(t_{n+1})\|$$

$$+ \int_0^1 \|\mathbf{w}_u\big(\varphi(\tau), \psi(\tau), t_n\big)\| d\tau \, \|\mathbf{U}_s - u(t_{n+1})\|$$

$$+ \int_0^1 \|\mathbf{w}_{v'}\big(\varphi(\tau), \psi(\tau), t_n\big)\| d\tau \, \|\mathbf{V}'_s - v'(t_{n+1})\|.$$

The functions φ and ψ are given by

$$\varphi(\tau) = \tau \mathbf{U}_s + (1 - \tau) u(t_{n+1}), \qquad \psi(\tau) = \tau \mathbf{V}'_s + (1 - \tau) v'(t_{n+1}).$$

Since the partial derivatives $\mathbf{z}_u$, $\mathbf{w}_u$ and $\mathbf{w}_{v'}$ are bounded, there are h-independent constants C_1, C_2 such that

$$\|x_{n+1} - x(t_{n+1})\| \le C_1 \|\mathbf{U}_s - u(t_{n+1})\| + C_2 \|\mathbf{V}'_s - v'(t_{n+1})\|.$$

Recall that the general linear method was assumed to be stiffly accurate. Thus Theorem 8.27 implies $\mathbf{U}_s = u(t_{n+1}) + \mathcal{O}(h^p)$. On the other hand, Lemma 8.29 and $M_\infty^{k_0} = 0$ show that after $k_0 + 1$ steps the stage derivatives satisfy $\mathbf{V}'_s = v'(t_{n+1}) + \mathcal{O}(h^p)$ due to the stage order being $q = p$. In summary, the global error can be estimated as $\|x_{n+1} - x(t_{n+1})\| = \mathcal{O}(h^p)$ proving the assertion. $\square$

The proof presented here hinges on three essential ingredients:

- The DAE (9.20) is properly stated and the structural condition (A1) ensures the existence of the decoupled system (9.17).
- Theorem 8.27 guarantees convergence of order p for the inherent index-1 system (9.17b).
- Lemma 8.29 yields the same accuracy for the stage derivatives $\mathbf{V}'_i$.

In view of Theorem 8.22 the second requirement can be expressed in terms of the method's order and stage order for ordinary differential equations.

Corollary 9.6. *Let $\mathcal{M}$ be a general linear method in Nordsieck form with $s = r > 1$. Then the assumption (d) in Theorem 9.5 can be replaced by*

(d') *$\mathcal{M}$ has order and stage order equal to $p = q = s - 1$ for ordinary differential equations.* $\square$

Part IV

Practical General Linear Methods

10

Construction of Methods

The derivation of integration schemes is a well established topic in numerical analysis. Methods later called Runge-Kutta schemes were used as early as in 1901 by Kutta [103]. The construction of Runge-Kutta methods is a complex task as complicated nonlinear expressions, the order conditions, have to be solved. The use of rooted trees by Butcher [15, 16, 18] made a systematic study of order conditions possible.

This work lead to well-known order barriers. Explicit methods with order p, for example, require at least $s \geq p$ stages. In case of $p \geq 5$ the number of stages necessarily satisfies $s > p$ [25]. On the other hand, the maximum attainable order of an s-stage implicit Runge-Kutta scheme is $p = 2s$.

The RadauIIA methods suitable for stiff equations where constructed by John Butcher in [17]. These methods form the basis of the popular code RADAU written by Hairer and Wanner [85, 86]. An extensive overview of Runge-Kutta formulae is given in [22].

Alexander and Nørsett [1, 141] focused on diagonally implicit methods. This work is continued in Trondheim until today. A comprehensive overview is available at [106]. The code SIMPLE written by Nørsett and Thomsen uses an embedded pair of SDIRK methods with order 3(2). Recent research in Trondheim focuses on SDIRK methods with an explicit first stage [105] as well as on multirate methods [5].

Linear multistep schemes were used even earlier than Runge-Kutta methods. Adams and Bashforth [6] used linear multistep methods in 1883 to study capillary action by investigating drop formation. BDF methods were introduced by Curtiss and Hirschfelder [48] and became popular through the work of Gear [70, 71]. DASSL, one of the most successful BDF solvers, was written by Petzold [128]. It was also Gear who promoted the use of Nordsieck vectors for linear multistep methods. The codes DIFSUB of Gear and later LSODE of Hindmarsh [134] use this technique.

With [71, 73, 75, 84, 111] and other references differential algebraic equations came into focus in the 1980's. The numerical methods used for DAEs were the ones that had been successfully applied to stiff ordinary differential equations. The BDF code DASSL and the Runge-Kutta code RADAU are the most prominent examples.

Methods of Rosenbrock type were studied for DAEs as well [137, 140]. Günther took up this approach and adapted Rosenbrock methods to the charge oriented MNA equations [79] (see also Section 1.1). Details about his code CHORAL can be found in [81].

Richardson extrapolation is a further technique that can be efficiently used provided that the global error has a known asymptotic expansion into powers of the stepsize h. For index-1 equations Deuflhard, Hairer and Zugck [54] were able to derive a perturbed asymptotic expansion such that extrapolation methods can be successfully used. The code LIMEX developed at the Konrad-Zuse-Zentrum für Informationstechnik Berlin (ZIB) [159] is based on the extrapolated implicit Euler scheme. LIMEX is capable of solving linear implicit index-1 DAEs $B(y,t)y'(t) = f(y,t)$.

Extrapolation schemes were modified by Gerstberger and Günther [74] to be applicable to integrated circuit design. Their work showed that index-1 circuits are feasible but it is not known how to derive the required asymptotic expansion for index-2 equations.

General linear methods (GLMs) were introduced by Butcher [18] in 1966 in order to provide a unifying framework for both linear multistep and Runge-Kutta methods. General linear methods are addressed in detail in the monograph [22]. There three matrices are used to characterise a method, but the equivalent formulation

$$\mathcal{M} = \left[\begin{array}{c|c} \mathcal{A} & \mathcal{U} \\ \hline \mathcal{B} & \mathcal{V} \end{array} \right], \qquad \begin{aligned} Y &= h\,\mathcal{A}\,F(Y) + \mathcal{U}\,y^{[n]}, \\ y^{[n+1]} &= h\,\mathcal{B}\,F(Y) + \mathcal{V}\,y^{[n]}, \end{aligned} \qquad (10.1)$$

using four matrices later became a standard. This formulation due to Burrage and Butcher [13] was already used in Section 2.3 and Chapter 7. Observe that (10.1) covers ordinary differential equations

$$y' = f(y,t), \qquad\qquad y(t_0) = y_0,$$

where $F(Y)_i = f(Y_i, t_n + c_i h)$ for $i = 1, \ldots, s$. Recall that a general linear method $\mathcal{M}$ is characterised by four integers:

   s: the number of internal stages Y_i,
   r: the number of external stages $y_i^{[n]}$,
   p: the order of the method,
   q: the stage order of the method.

Among the huge class of general linear methods Diagonally Implicit Multistage Integration Methods (DIMSIMs) seem to be most likely to be used for practical applications. A DIMSIM is loosely defined by the quantities s, r, p, and q

all being approximately equal [25]. DIMSIMs were introduced by Butcher in 1993 [23]. It is common practice to require the matrix $\mathcal{A}$ to have lower a triangular structure,

$$\mathcal{A} = \begin{bmatrix} a_{11} & & 0 \\ \vdots & \ddots & \\ a_{s1} & \cdots & a_{ss} \end{bmatrix}. \tag{10.2}$$

Thus the stages Y_i can be evaluated sequentially such that an efficient implementation is possible (see Section 2.2). We will adopt the diagonally implicit structure (10.2) as a standard as well. Whenever possible the diagonal elements a_{ii} will be chosen to be equal. This allows a further reduction of implementation costs.

As stability of a general linear method is directly connected with the matrix $\mathcal{V}$ being power bounded (see Definition 7.3), this matrix is often assumed to be of low rank. The example methods constructed in [23] focus on the structure $\mathcal{U} = I$ and $\mathcal{V} = e \cdot v$ where $e = \begin{bmatrix} 1 & \cdots & 1 \end{bmatrix}^\top$ and $v = \begin{bmatrix} v_1 & \cdots & v_r \end{bmatrix}$. In [33, 34] Butcher and Jackiewicz require the same structure for $\mathcal{U}$ and $\mathcal{V}$. Computer algebra tools are used in [33] for the construction of DIMSIMs with order 1, 2, 3. Order 4 methods are obtained with the aid of homotopy techniques. Higher order methods are constructed in [34] using least square minimisation algorithms from MINPACK [125]. The main reason for using numerical searches is the fact that computer algebra tools such as MAPLE or MATHEMATICA are no longer powerful enough to solve the order conditions associated with high p and q. Methods of even higher order, $p = 7$, $p = 8$, are derived in [39] using a nonlinear optimisation approach.

General linear methods for parallel computing have been developed by Butcher and Chartier [28]. The matrix $\mathcal{A} = \lambda I$ is assumed to be a diagonal matrix. Methods of this kind are sometimes referred to as type-4 DIMSIMs. Type-4 methods with the additional property that the stability matrix $M(z)$ has only one nonzero eigenvalue are called DIMSEMs by Enenkel and Jackson[1] [60]. Using the somewhat unusual choice of all parameters c_i being zero, they succeeded in constructing methods with $p = r - 1$ for any $r \geq 2$.

The concept of Runge-Kutta stability was introduced in Chapter 7. Obviously, DIMSEMs belong to this class of methods. A systematic study of general linear methods having inherent Runge-Kutta stability by Wright [158] lead to the surprising result that DIMSIMs with $s = r = p + 1 = q + 1$ can be constructed using only linear operations. The important concept of inherent Runge-Kutta stability was already introduced in Definition 7.9.

Methods of this type mark the starting point for constructing practical methods for index-2 DAEs in the next two sections. As described above, most examples of general linear methods were derived in the context of (stiff) ODEs. In

---

[1]DIMSEM: Diagonally IMplicit Single-Eigenalue Method.

view of Theorem 9.5 and Corollary 9.6, we know that DAEs require additional properties such as stiff accuracy or improved stability at infinity. More precisely we focus on methods having the following properties:

(P1) $\mathcal{M}$ has order p and stage order $q = p$ for ordinary differential equations.

(P2) $\mathcal{A}$ has a diagonally implicit structure (10.2) with $a_{ii} = \lambda$ for $i = 1, \ldots, s$.

(P3) $\mathcal{V}$ is power bounded and M_∞ is nilpotent. $M_\infty^{k_0} = 0$ for some $k_0 \geq 1$ is a necessary requirement for L-stability ensuring that the stage derivatives are calculated with the desired accuracy (see Lemma 8.30).

(P4) The abscissae c_i satisfy $c_i \in [0, 1]$ and $c_i \neq c_j$ for $i \neq j$.

(P5) $\mathcal{M}$ is stiffly accurate, i.e. $c_s = 1$ and the last row of $[\mathcal{A} \quad \mathcal{U}]$ coincides with the first row of $[\mathcal{B} \quad \mathcal{V}]$.

(P6) $\mathcal{M}$ is given in Nordsieck form such that the stepsize can be varied easily.

Apart from these requirements a small error constant is desirable. The error constant C_p signifies that the local truncation error for an order p method applied to the ordinary differential equation $y' = f(y)$ is given by the term $C_p h^{p+1} y^{(p+1)}(t_n) + \mathcal{O}(h^{p+2})$. Thus a small $|C_p|$ allows larger steps to be taken for a given tolerance. More details on computing error constants will be given in Chapter 11.

Finally, we will pay close attention to stability properties when constructing methods. Recall from the introduction that for electrical circuit simulation the stability behaviour along the imaginary axis is an indicator for the methods damping properties. We saw in Example 2.1 and 2.2 that two kinds of oscillations need to be addressed:

- Oscillations of physical significance should be preserved. Hence for a sampling rate of about $8 - 20$ steps per period, the largest eigenvalue of $M(y\mathrm{i})$ is required to stay close to 1 provided that $y \in [0.1\pi, 0.25\pi]$.

- High frequent oscillations caused by numerical noise, perturbations, inconsistent initial values or errors from the Newton iteration should be damped out quickly. This requires the eigenvalues of $M(y\mathrm{i})$ to decay as y tends towards infinity.

We will first look at methods with inherent Runge-Kutta stability satisfying $s = r = p + 1 = q + 1$. In order to improve stability at infinity, stiffly accurate methods with $M_\infty = 0$ will be constructed as well.

These methods use $s = p + 1$ internal stages. Thus the computational costs per step can be reduced by considering DIMSIMs with $s = r = p = q$. Methods of this type will be considered in Section 10.2.

Recall from Chapter 2 that the MNA equations modelling electrical circuits typically suffer from poor smoothness properties of the semiconductor models being used. Thus we restrict attention to low order methods with $1 \leq p \leq 3$.

10.1 Methods with $s = p + 1$ Stages

General linear methods $\mathcal{M} = [\mathcal{A}, \mathcal{U}, \mathcal{B}, \mathcal{V}]$ with inherent Runge-Kutta stability and $s = r = p + 1 = q + 1$ have been constructed in [38, 40, 158]. Other examples can be found in [93], where implementation details are addressed as well. The construction is based on an algorithm developed by Wright [158].

Recall from Section 7.2 that Runge-Kutta stability,

$$\Phi(w, z) = \det\left(w\,I - M(z)\right) = w^{r-1}\left(w - R(z)\right), \tag{10.3}$$

is ensured by requiring that $\mathcal{V}$ has a single nonzero eigenvalue given by $\mathcal{V}\,e_1 = e_1$ and imposing the conditions[2]

$$\mathcal{B}\mathcal{A} = X\mathcal{B}, \qquad \mathcal{B}\mathcal{U} \equiv X\mathcal{V} - \mathcal{V}X. \tag{10.4}$$

As seen in Lemma 7.10, the conditions (10.4) guarantee that the stability matrix $M(z)$ and $\mathcal{V}$ are related by a similarity transformation

$$(I - zX)\,M(z)\,(I - zX)^{-1} \equiv \mathcal{V}. \tag{10.5}$$

Wright showed that the most general matrix X satisfying (10.4), (10.5) has doubly companion form [158]

$$X = \begin{bmatrix} -\alpha_1 & -\alpha_2 & \cdots & -\alpha_p & -\alpha_{p+1} \\ 1 & 0 & \cdots & 0 & -\beta_p \\ 0 & 1 & \cdots & 0 & -\beta_{p-1} \\ \vdots & \vdots & \ddots & \vdots & \vdots \\ 0 & 0 & \cdots & 1 & -\beta_1 \end{bmatrix}. \tag{10.6}$$

The vector $\beta = \begin{bmatrix} \beta_1 & \cdots & \beta_p \end{bmatrix}$ is considered as a free parameter. The entries of $\alpha = \begin{bmatrix} \alpha_1 & \cdots & \alpha_{p+1} \end{bmatrix}$ are determined by $\det(wI - X) = (w - \lambda)^s$ since the matrix $X = \mathcal{B}\mathcal{A}\mathcal{B}^{-1}$ has a single s-fold eigenvalue λ due to (10.4) and the requirement (P2) from the previous page. It will be always assumed that $\mathcal{B}$ is nonsingular (see [158] for more details).

Observe that in (10.4) the matrix $\mathcal{A}$ is determined by $\mathcal{B}$. Similarly, for methods in Nordsieck form, the conditions for order $p = q = s - 1$ fix the matrices $\mathcal{U}$ and $\mathcal{V}$ in terms of $\mathcal{A}$ and $\mathcal{B}$,

$$\mathcal{U} = C - \mathcal{A}CK, \qquad \mathcal{V} = E - \mathcal{B}CK. \tag{10.7}$$

---

[2]As in Section 7.2, $A \equiv B$ indicates equality for the matrices A and B with the possible exception of the first row.

The matrices

$$
K = \begin{bmatrix} 0 & 1 & 0 & \cdots & 0 & 0 \\ 0 & 0 & 1 & \cdots & 0 & 0 \\ \vdots & \vdots & \vdots & \ddots & \vdots & \vdots \\ 0 & 0 & 0 & \cdots & 0 & 1 \\ 0 & 0 & 0 & \cdots & 0 & 0 \end{bmatrix}, \quad
E = \begin{bmatrix} 1 & \frac{1}{1!} & \frac{1}{2!} & \cdots & \frac{1}{(p-1)!} & \frac{1}{p!} \\ 0 & 1 & \frac{1}{1!} & \cdots & \frac{1}{(p-2)!} & \frac{1}{(p-1)!} \\ \vdots & \vdots & \vdots & \ddots & \vdots & \vdots \\ 0 & 0 & 0 & \cdots & 1 & \frac{1}{1!} \\ 0 & 0 & 0 & \cdots & 0 & 1 \end{bmatrix} = \exp(K)
$$

and $C = \begin{bmatrix} e & c & \frac{c^2}{2!} & \cdots & \frac{c^p}{p!} \end{bmatrix}$ have been introduced in Theorem 7.6 on page 118. Thus, the construction of a general linear method $\mathcal{M} = [\mathcal{A}, \mathcal{U}, \mathcal{B}, \mathcal{V}]$ with inherent Runge-Kutta stability boils down to choosing the coefficients of $\mathcal{B}$.

Instead of the matrix $\mathcal{B}$ itself, Wright considers $\tilde{\mathcal{B}} = \Psi^{-1}\mathcal{B}$. The nonsingular matrix Ψ is known from the transformation of X to Jordan canonical form, i.e. $\Psi^{-1}X\Psi = \lambda I + J$ where $J = K^{\top}$. In order to guarantee a lower triangular structure for $\mathcal{A}$, the matrix $\tilde{\mathcal{B}}$ has to be lower triangular, too.

Stiff accuracy as required by the condition (P5) can be ensured by setting

$$
c_s = 1, \qquad \beta_p = 0, \qquad e_2^{\top}\mathcal{B} = e_s^{\top}. \tag{10.8}
$$

It is easily seen that (10.8) indeed implies stiff accuracy since

$$
e_s^{\top}\mathcal{A} = e_s^{\top}\mathcal{B}^{-1}X\mathcal{B} = e_2^{\top}X\mathcal{B} = e_1^{\top}\mathcal{B},
$$
$$
e_s^{\top}\mathcal{U} = e_s^{\top}(C - \mathcal{A}CK) = e_1^{\top}(E - \mathcal{B}CK) = e_1^{\top}\mathcal{V}.
$$

For the second equation, observe that $c_s = 1$ ensures $e_s^{\top}C = e_1^{\top}E$.

Methods satisfying (10.8) will not only be stiffly accurate but they also satisfy the FSAL property known from the study of Runge-Kutta methods.

The relations

$$
e_2^{\top}\mathcal{B} = \begin{bmatrix} 0 & \cdots & 0 & 1 \end{bmatrix},
$$
$$
e_2^{\top}\mathcal{V} = e_2^{\top}(E - \mathcal{B}CK) = e_2^{\top}E - e_s^{\top}CK = e_2^{\top}E - e_1^{\top}EK = \begin{bmatrix} 0 & \cdots & 0 \end{bmatrix}
$$

show that for ordinary differential equations $y' = f(y, t)$ the exact derivative $Y_s' = f(Y_s, t_n + c_s h)$ of the last stage is passed on to the next step as the second component of the Nordsieck vector. More precisely we have $y_2^{[n+1]} = hY_s'$ such that the FSAL property (*first same as last*) is satisfied: hY_s' could be equally well computed using an explicit zeroth stage in step number $n + 1$. We will see in Section 11 that this property simplifies error estimation considerably (see also [40]).

The construction of general linear methods $\mathcal{M} = [\mathcal{A}, \mathcal{U}, \mathcal{B}, \mathcal{V}]$ with inherent Runge-Kutta stability can now proceed along the following lines (full details are given in [158]):

(1) Choose $s = r = p+1 = q+1$ in advance and fix parameters $\lambda, c_1, \ldots, c_{s-1}$, $\beta_1, \ldots, \beta_{p-1} \in \mathbb{R}$. Set $c_s = 1$, $\beta_p = 0$ and choose a nonsingular matrix $T \in \mathbb{R}^{(s-2) \times (s-2)}$.

(2) Compute α, X and the transformation matrix Ψ.

(3) Compute $\tilde{\mathcal{B}}$ from the following set of linear conditions:

- inherent Runge-Kutta stability (10.4),
- stiff accuracy and FSAL property (10.8),
- zero-stability, i.e. $T^{-1}\hat{\mathcal{V}}T$ is strictly upper triangular for $\mathcal{V} = \begin{bmatrix} 1 & v_{12} & \hat{v}_1^\top \\ 0 & 0 & 0 \\ 0 & \hat{v}_2 & \hat{\mathcal{V}} \end{bmatrix}$ with $\hat{v}_1, \hat{v}_2 \in \mathbb{R}^{s-2}$, $\hat{\mathcal{V}} \in \mathbb{R}^{(s-2)\times(s-2)}$,
- $\mathrm{tr}(M_\infty) = 0$, as a vanishing trace implies that the single nonzero eigenvalue is zero at infinity. The error constant can be prescribed as well [158].

(4) Compute $\mathcal{B} = \Psi\tilde{\mathcal{B}}$, $\mathcal{A} = \lambda I + \tilde{\mathcal{B}}^{-1}J\tilde{\mathcal{B}}$ and $\mathcal{U} = C - \mathcal{A}CK$, $\mathcal{V} = E - \mathcal{B}CK$.

This algorithm was used in [40] to derive the family of methods presented in Table 10.1. Observe that the order-1 method does not use the second component of the Nordsieck vector. Thus, in a variable order implementation, no starting procedure is required.

$$
\left[\begin{array}{c|c} \mathcal{A} & \mathcal{U} \\ \hline \mathcal{B} & \mathcal{V} \end{array}\right] = \left[\begin{array}{cc|cc} \frac{3}{10} & 0 & 1 & 0 \\ \frac{7}{10} & \frac{3}{10} & 1 & 0 \\ \hline \frac{7}{10} & \frac{3}{10} & 1 & 0 \\ 0 & 1 & 0 & 0 \end{array}\right]
$$

(a) Order $p=q=1$, $c = \begin{bmatrix} \frac{3}{10} & 1 \end{bmatrix}^\top$

$$
\left[\begin{array}{ccc|ccc} \frac{4}{9} & 0 & 0 & 1 & -\frac{1}{9} & -\frac{5}{54} \\ \frac{124}{27} & \frac{4}{9} & 0 & 1 & -\frac{118}{27} & -\frac{130}{81} \\ -\frac{40}{81} & -\frac{4}{27} & \frac{4}{9} & 1 & \frac{97}{81} & \frac{155}{486} \\ \hline -\frac{40}{81} & -\frac{4}{27} & \frac{4}{9} & 1 & \frac{97}{81} & \frac{155}{486} \\ 0 & 0 & 1 & 0 & 0 & 0 \\ -\frac{21}{4} & \frac{3}{4} & \frac{9}{4} & 0 & \frac{9}{4} & 0 \end{array}\right]
$$

(b) Order $p=q=2$, $c = \begin{bmatrix} \frac{1}{3} & \frac{2}{3} & 1 \end{bmatrix}^\top$

$$
\left[\begin{array}{cccc|cccc} \frac{9}{40} & 0 & 0 & 0 & 1 & \frac{1}{40} & -\frac{1}{40} & -\frac{17}{3840} \\ \frac{52443}{17200} & \frac{9}{40} & 0 & 0 & 1 & -\frac{47713}{17200} & \frac{51583}{68800} & \frac{169369}{1651200} \\ -\frac{1996641}{1350200} & -\frac{387}{1570} & \frac{9}{40} & 0 & 1 & \frac{759579}{337550} & \frac{3269871}{5400800} & \frac{907929}{10801600} \\ -\frac{59481}{40000} & -\frac{189}{800} & \frac{1413}{8000} & \frac{9}{40} & 1 & \frac{46433}{20000} & \frac{50593}{80000} & \frac{9659}{120000} \\ \hline -\frac{59481}{40000} & -\frac{189}{800} & \frac{1413}{8000} & \frac{9}{40} & 1 & \frac{46433}{20000} & \frac{50593}{80000} & \frac{9659}{120000} \\ 0 & 0 & 0 & 1 & 0 & 0 & 0 & 0 \\ -5 & \frac{38}{45} & -\frac{157}{45} & \frac{40}{9} & 0 & \frac{16}{5} & 0 & -\frac{137}{720} \\ \frac{2672}{81} & -\frac{256}{27} & -\frac{2512}{81} & \frac{1600}{81} & 0 & -\frac{992}{81} & 0 & 0 \end{array}\right]
$$

(c) Order $p=q=3$, $c = \begin{bmatrix} \frac{1}{4} & \frac{1}{2} & \frac{3}{4} & 1 \end{bmatrix}^\top$

Table 10.1: A family of methods constructed by Butcher and Podhaisky in [40] having inherent Runge-Kutta stability.

The methods from Table 10.1 have Runge-Kutta stability by construction. It can be verified that for the method of order p the single nonzero eigenvalue of the stability matrix

$$M(z) = \mathcal{V} + z\mathcal{B}(I - z\mathcal{A})^{-1}\mathcal{U}$$

is given by $R_p(z)$ where

$$R_1(z) = \frac{1+(1-2\lambda)z}{(1-\lambda z)^2}, \qquad\qquad R_2(z) = \frac{1+(1-3\lambda)z+(3\lambda^2-3\lambda+\frac{1}{2})z^2}{(1-\lambda z)^3},$$

$$R_3(z) = \frac{1+(1-4\lambda)z+\frac{1}{2}(1-6\lambda)(1-2\lambda)z^2+(-4\lambda^3+6\lambda^2-2\lambda+\frac{1}{6})z^3}{(1-\lambda z)^4}.$$

The parameter λ denotes the diagonal element. Observe that $R_p(z)$ coincides with the stability function of stiffly accurate SDIRK methods having $s = p+1$ stages and order p [85]. Thus, the methods from Table 10.1 have the same linear stability behaviour as the corresponding SDIRK methods. A-stability (and hence L-stability) is ensured by an appropriate choice of λ.

The stability matrix at infinity

$$M_{\infty,1} = \begin{bmatrix} 0 & 0 \\ \frac{40}{9} & 0 \end{bmatrix}, \; M_{\infty,2} = \begin{bmatrix} 0 & 0 & 0 \\ \frac{9}{4} & 0 & 0 \\ \frac{261}{8} & \frac{51}{8} & 0 \end{bmatrix}, \; M_{\infty,3} = \begin{bmatrix} 0 & 0 & 0 & 0 \\ \frac{16}{5} & 0 & -\frac{137}{720} & 0 \\ -\frac{992}{81} & 0 & 0 & 0 \\ -\frac{355712}{243} & -\frac{69088}{243} & -\frac{56}{3} & 0 \end{bmatrix}$$

plays a decisive role for the methods' applicability to index-2 differential algebraic equations. Although $M_{\infty,p}$ is nilpotent for $1 \le p \le 3$ with $M_{\infty,p}^s = 0$, the entries of $M_{\infty,p}$ tend to become large in magnitude (up to ≈ 1463.84 for $p = 3$).

For differential algebraic equations certain parts of the global error are amplified by M_∞ (see Lemma 8.25). Hence suitability of these methods for DAEs might be improved by requiring $M_{\infty,p} = 0$ for $p = 1, 2, 3$.

Recall from Theorem 7.6 that a method in Nordsieck form with $s = r = p+1 = q + 1$ needs to satisfy $\mathcal{U} = C - \mathcal{A}CK$, $\mathcal{V} = E - \mathcal{B}CK$. Therefore

$$M_\infty = \mathcal{V} - \mathcal{B}\mathcal{A}^{-1}\mathcal{U} = (E - \mathcal{B}CK) - \mathcal{B}\mathcal{A}^{-1}(C - \mathcal{A}CK) = E - \mathcal{B}\mathcal{A}^{-1}C$$

vanishes if and only if

$$\mathcal{B} = E\,C^{-1}\mathcal{A}. \tag{10.9}$$

For (10.9) to make sense we have to restrict attention to non-confluent methods, i.e. $c_i \ne c_j$ for $i \ne j$. In this case C is a Vandermonde matrix and hence nonsingular. The matrix $\mathcal{V}$ can be written as

$$\mathcal{V} = E - \mathcal{B}CK = E - E\,C^{-1}\mathcal{A}CK \tag{10.10}$$

such that $\mathcal{A}$ and c remain the only free parameters of the method.

Given $\mathcal{A}$ and c we consider general linear methods in Nordsieck form with

$$
\mathcal{M}_p = \left[\begin{array}{c|c} \mathcal{A} & C - \mathcal{A}CK \\ \hline E\,C^{-1}\mathcal{A} & E - E\,C^{-1}\mathcal{A}CK \end{array} \right], \quad
\mathcal{A} = \begin{bmatrix} \lambda & 0 & \cdots & 0 & 0 \\ a_{21} & \lambda & \cdots & 0 & 0 \\ \vdots & \vdots & \ddots & \vdots & \vdots \\ a_{s-1,1} & a_{s-1,2} & & \lambda & 0 \\ a_{s1} & a_{s2} & \cdots & a_{s,s-1} & \lambda \end{bmatrix} \tag{10.11}
$$

and $c = \begin{bmatrix} c_1 & \cdots & c_{s-1} & 1 \end{bmatrix}^{\top}$. The structure (10.9) for the matrix $\mathcal{B}$ ensures $M_{\infty,p} = 0$. The particular choice of $\mathcal{U}$ and $\mathcal{V}$ shows that $\mathcal{M}_p$ has order p and stage order $q = p$.

Order $p = 1$

For order 1 methods with 2 stages (10.11) takes the form

$$
\mathcal{M}_1 = \left[\begin{array}{cc|cc} \lambda & 0 & 1 & c_1 - \lambda \\ a_{21} & \lambda & 1 & 1 - a_{21} - \lambda \\ \hline a_{21} & \lambda & 1 & 1 - a_{21} - \lambda \\ \frac{\lambda - a_{21}}{c_1 - 1} & -\frac{\lambda}{c_1 - 1} & 0 & \frac{c_1 + a_{21} - 1}{c_1 - 1} \end{array} \right], \quad
c = \begin{bmatrix} c_1 \\ 1 \end{bmatrix}.
$$

Observe that $e_s^{\top}\mathcal{A} = e_1^{\top}\mathcal{B}$ and $c_s = 1$ guarantee stiff accuracy, but we do not require the FSAL property explicitly. As $\mathcal{M}_1$ will be used to start the integration, picking a zero second column for $\mathcal{U}$ and $\mathcal{V}$ ensures that only the initial value is required for startup. We arrive at

$$
\mathcal{M}_1 = \left[\begin{array}{cc|cc} \lambda & 0 & 1 & 0 \\ 1 - \lambda & \lambda & 1 & 0 \\ \hline 1 - \lambda & \lambda & 1 & 0 \\ \frac{1 - 2\lambda}{1 - \lambda} & \frac{\lambda}{1 - \lambda} & 0 & 0 \end{array} \right], \quad
c = \begin{bmatrix} \lambda \\ 1 \end{bmatrix}. \tag{10.12}
$$

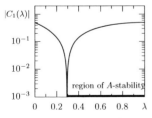

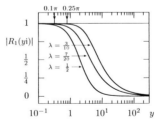

Figure 10.1 (a): $C_1(\lambda) = \lambda^2 - 2\lambda + \frac{1}{2}$.   Figure 10.1 (b): Stab. function $R_1(yi)$.

Figure 10.1: Choosing λ for the method (10.12): The left hand picture shows the modulus $|C_1(\lambda)|$ of the error constant. In Figure 10.1 (b) the stability function $R_1(z)$ is plotted for $z = yi$. Increasing λ leads to poorer preservation of physical oscillations.

This method has the stability function

$$\Phi(w, z) = \det\left(w\,I - M(z)\right) = w^2 - \frac{1+(1-2\lambda)z}{(1-\lambda z)^2}w = w\left(w - R_1(z)\right).$$

It turns out that $\mathcal{M}_1$ has Runge-Kutta stability and hence the same linear stability behaviour as the SDIRK method with $p = 1$ and two stages. The diagonal element λ can be chosen in the interval $[1 - \frac{\sqrt{2}}{2}, 1 + \frac{\sqrt{2}}{2}]$ for A-stability [85]. Figure 10.1 suggests $\lambda = \frac{3}{10}$ such that a small error constant $C_1(\frac{3}{10}) = -\frac{1}{100}$ and good damping behaviour along the imaginary axis is realised.

Order $p = 2$

For $p = s - 1 = 2$ the formulation (10.11) leaves six free parameters: λ, c_1, c_2 and a_{21}, a_{31}, a_{32}. Recall from (10.4) that a necessary condition for *inherent* Runge-Kutta stability is

$$\mathcal{B}\mathcal{A} = X\mathcal{B} \quad \Leftrightarrow \quad E\,C^{-1}\mathcal{A} = XEC^{-1} \quad \Leftrightarrow \quad X = (EC^{-1})\mathcal{A}\,(EC^{-1})^{-1}.$$

Since X has doubly companion form (10.6), the submatrix $X_{2:3,1:2} = I$ yields four conditions on $\mathcal{A}$ and c. These are linear in $\hat{a} = \begin{bmatrix} a_{21} & a_{31} & a_{32} \end{bmatrix}^\top$. In particular, the linear system

$$X_{2,1} = 1, \qquad X_{3,1} = 0, \qquad X_{3,2} = 1,$$

or, equivalently,

$$\begin{bmatrix} \frac{c_1+3}{(c_2-1)(c_1-c_2)} & -\frac{c_1+2+c_2}{(c_2-1)(c_1-1)} & -\frac{c_1+2+c_2}{(c_2-1)(c_1-1)} \\ -\frac{2}{(c_2-1)(c_1-c_2)} & \frac{2}{(c_2-1)(c_1-1)} & \frac{2}{(c_2-1)(c_1-1)} \\ -\frac{2(c_1+1)}{(c_2-1)(c_1-c_2)} & \frac{2(c_1+1)}{(c_2-1)(c_1-1)} & \frac{2(c_2+1)}{(c_2-1)(c_1-1)} \end{bmatrix} \begin{bmatrix} a_{21} \\ a_{31} \\ a_{32} \end{bmatrix} = \begin{bmatrix} 1 \\ 0 \\ 1 \end{bmatrix}$$

can be solved for

$$a_{21} = c_2 - c_1, \qquad a_{31} = \frac{(1-c_1)(2c_1-3c_2+1)}{2(c_1-c_2)}, \qquad a_{32} = \frac{(1-c_1)(c_2-1)}{2(c_1-c_2)}.$$

The remaining condition $X_{2,2} = \frac{1}{2}(c_1 - c_2) + \lambda = 0$ can easily be satisfied by choosing $c_1 = c_2 - 2\lambda$. For simplicity, c_2 is chosen such that the abscissae c_i are spaced equidistantly in $[0, 1]$, i.e. $c_1 = \frac{1}{2} - \lambda$, $c_2 = \frac{1}{2} + \lambda$. This results in the method

$$\begin{bmatrix} \lambda & 0 & 0 & 1 & \frac{1}{2} - 2\lambda & \frac{1}{8} - \lambda + \frac{3}{2}\lambda^2 \\ 2\lambda & \lambda & 0 & 1 & \frac{1}{2} - 2\lambda & \frac{1}{8} - \lambda + \frac{3}{2}\lambda^2 \\ \frac{(2\lambda+1)(10\lambda-1)}{16\lambda} & \frac{1-4\lambda^2}{16\lambda} & \lambda & 1 & \frac{1}{2} - 2\lambda & \frac{1}{8} - \lambda + \frac{3}{2}\lambda^2 \\ \frac{(2\lambda+1)(10\lambda-1)}{16\lambda} & \frac{1-4\lambda^2}{16\lambda} & \lambda & 1 & \frac{1}{2} - 2\lambda & \frac{1}{8} - \lambda + \frac{3}{2}\lambda^2 \\ \frac{4\lambda^2+4\lambda-1}{4\lambda(2\lambda+1)} & \frac{4\lambda^2+4\lambda-1}{4\lambda(2\lambda-1)} & \frac{-4\lambda}{4\lambda^2-1} & 0 & 0 & 0 \\ \frac{2\lambda-1}{2\lambda(2\lambda+1)} & \frac{6\lambda-1}{2\lambda(2\lambda-1)} & \frac{-8\lambda}{4\lambda^2-1} & 0 & 0 & 0 \end{bmatrix}. \qquad (10.13)$$

The diagonal element λ is again a free parameter. The stability function

$$\Phi(w,z) = w^3 - \frac{1+(1-3\lambda)z+(3\lambda^2-3\lambda+\frac{1}{2})z^2}{(1-\lambda z)^3}w^2 = w^2\big(w - R_2(z)\big)$$

shows that for every λ the method (10.13) has Runge-Kutta stability. $R_2(z)$ is the stability function of the SDIRK method with $s = p + 1 = 3$. Therefore, A-stability is achieved for $\lambda \in [0.18042531, 2.18560010]$ (see again [85]). Figure 10.2 shows that again small values of λ lead to smaller error constants and to the desired stability behaviour along the imaginary axis. The diagonal element $\lambda = \frac{2}{11} \approx 0.1818$ might be a good compromise since the corresponding method is A-stable with $C_2(\frac{2}{11}) = -\frac{103}{7986} \approx -0.0129$. Methods with even smaller error constants are possible for $\lambda \approx 0.436 = \frac{109}{250}$, but Figure 10.2 (b) indicates stronger numerical damping for these methods.

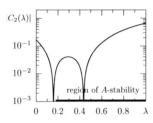
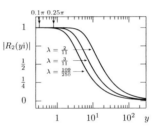

Figure 10.2 (a): $C_2(\lambda) = -\lambda^3 + 3\lambda^2 - \frac{3}{2}\lambda + \frac{1}{6}$.   Figure 10.2 (b): Stab. function $R_2(yi)$.

Figure 10.2: Choosing λ for the method (10.13): $\lambda = \frac{2}{11}$ leads to a small error constant $C_2(\frac{2}{11}) = -\frac{103}{7986}$ and nice stability properties.

Order $p = 3$

Unfortunately the approach taken for order 2 methods cannot be generalised to $p = 3$. In fact, for $p = s - 1 = 3$ and $\lambda \neq 0$ there is no inherently Runge-Kutta stable method of the form (10.11) satisfying all requirements (P1)–(P6) from page 182. This can be seen as follows:

Let X be the doubly companion matrix from (10.6) and assume that

$$\beta_1 = 1 - \delta_1, \qquad \beta_2 = \tfrac{1}{2} - \delta_1 - \delta_2, \qquad \beta_3 = \tfrac{1}{6} - \tfrac{1}{2}\delta_1 - \delta_2 - \delta_3.$$

Then

$$\hat{X} = E^{-1}XE = \begin{bmatrix} -\hat{\alpha}_1 & -\hat{\alpha}_2 & \cdots & -\hat{\alpha}_p & -\hat{\alpha}_{p+1} \\ 1 & 0 & \cdots & 0 & -\delta_p \\ 0 & 1 & \cdots & 0 & -\delta_{p-1} \\ \vdots & \vdots & \ddots & \vdots & \vdots \\ 0 & 0 & \cdots & 1 & -\delta_1 \end{bmatrix}$$

is again a doubly companion matrix. For a method (10.11) to have inherent Runge-Kutta stability the relation

$$\mathcal{B}\mathcal{A} = X\mathcal{B} \qquad \Leftrightarrow \qquad E\,C^{-1}\mathcal{A} = XEC^{-1} \qquad \Leftrightarrow \qquad \mathcal{A}C = C\hat{X}$$

needs to be satisfied. This system of nonlinear equations[3] can be solved for the unknowns λ, c_i, δ_i, a_{ij}. Simple but quite lengthy computations not being reproduced here show that either $\lambda = 0$ or $c_i = c_j$ for some $i \neq j$.

Nevertheless Runge-Kutta stability is still possible. For simplicity assume that $c = \begin{bmatrix} \frac{1}{4} & \frac{1}{2} & \frac{3}{4} & 1 \end{bmatrix}^{\top}$ such that the stability function

$$\Phi(w, z) = \left(p_0 + p_1\,w + p_2\,w^2 + p_3\,w^3 + (1 - \lambda z)^4 w^4\right)\frac{1}{(1-\lambda z)^4}$$

can be written in terms of polynomials

$$p_k = \sum_{l=0}^{k} p_{kl}\,z^l, \qquad k = 0, \dots, 3,$$

where p_{kl} are polynomials depending on the method's remaining parameters λ and a_{ij}. Details on this particular structure are given in [33, 34].

For Runge-Kutta stability we have to ensure $p_0 = 0$, $p_1 = 0$, $p_2 = 0$ such that $\Phi(w, z) = w^3\left(w - R_3(z)\right)$. This yields $1+2+3 = 6$ conditions on the remaining seven parameters. Thus λ is fixed in advance and the system

$$p_{00} = 0, \quad p_{10} = 0, \quad p_{11} = 0, \quad p_{20} = 0, \quad p_{21} = 0, \quad p_{22} = 0 \quad (10.14)$$

is solved for $a_{21}, a_{31}, a_{32}, a_{41}, a_{42}, a_{43}$.

The equations (10.14) can easily be generated using computer algebra tools. Consider MAPLE [112] as an example. Appropriate commands read

```
    s := 4:
    A := evalm(matrix(s, s, (i, j) -> 'if'(i < j, 0, cat('a', i, j) )) + lambda*&*() ):
    C := matrix(s, s, (i, j) -> (i/s)^(j-1) / (j-1)! ):
    K := matrix(s, s, (i, j) -> 'if'(j = i+1, 1, 0) ):
 5  E := exponential(K):
    B := evalm(E &* inverse(C) &* A):
    U := evalm(C - A &* C &* K):
    V := evalm(E - B &* C &* K):
    Mz := map(simplify, evalm(V + z*B &* inverse(&*()-z*A) &* U)):
10  Phi := series(det( w*&*() - Mz ), w, s+1):
    eqns := { seq( seq(
            coeff(series(numer(coeff(Phi, w, k)), z, k+1), z, l), l=0..k), k=0..2) };
```

The set eqns contains all equations from (10.14). However, solving this system symbolically,

```
    solve(eqns, indets(eqns) minus {lambda} );
```

was not possible on an Intel® Pentium® M processor with 1.4 GHz, 512 MB RAM, using MAPLE 9.5. Hence it was decided to solve (10.14) numerically using routines from MINPACK.

---

[3]Recall that c_i appears nonlinearly due to $C = \begin{bmatrix} e & c & \frac{c^2}{2!} & \cdots & \frac{c^p}{p!} \end{bmatrix}$.

The FORTRAN subroutine hybrd from MINPACK [125] solves a system of n non-linear equations in n variables using a modification of the Powell hybrid method. For every iteration the correction is chosen as a convex combination of the Newton and scaled gradient directions. This ensures (under reasonable conditions) global convergence for starting points far from the solution and a fast rate of convergence. The Jacobian is computed using forward differences and rank-1 Broyden updates.

Input for the simplified driver hybrd1 can be generated using MAPLE,

```
fvec := convert(eqns, list):
codegen[fortran](fevec):
```

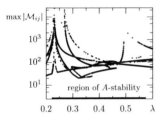

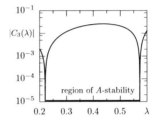

Figure 10.3 (a): Magnitude of $\mathcal{M}$'s largest element. Different initial values a^0 lead to a variety of methods for each λ.

Figure 10.3 (b): Error constant $C_3(\lambda)$. Two methods $\mathcal{M}(\lambda, a^0) \neq \mathcal{M}(\lambda, \bar{a}^0)$ have the same error constant.

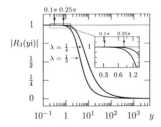

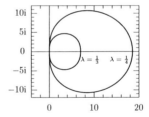

Figure 10.3 (c): Stability behaviour on the imaginary axis. For $\lambda = \frac{1}{3}$ oscillations of physical significance will be damped more rapidly than for $\lambda = \frac{1}{4}$.

Figure 10.3 (d): Comparison of stability regions. The stable region is given by the outside of the kidney shaped area.

Figure 10.3: $\lambda = \frac{1}{4}$ leads to a small error constant $C_3(\frac{1}{4}) \approx 0.00391$. The largest coefficient of the corresponding method has magnitude ≈ 38.7 (see Table 10.2). Although $\lambda = \frac{1}{3}$ leads to smaller coefficients, the stability behaviour is inferior to that of the $\frac{1}{4}$-method. The error constant $C_3(\frac{1}{3}) \approx 0.017$ is larger as well.

For fixed $\lambda \in [0.22364780, 0.57281606]$ this approach yields A-stable methods [85] as the single nonzero eigenvalue is given by the stability function $R_3(z)$ of the SDIRK method with $s = p + 1 = 4$. Different starting points $a^0 = \begin{bmatrix} a_{21}^0 & a_{31}^0 & a_{32}^0 & a_{41}^0 & a_{42}^0 & a_{43}^0 \end{bmatrix}^\top$ lead to methods $\mathcal{M}(\lambda, a^0)$. Figure 10.3 (a) shows that the coefficients of $\mathcal{M}(\lambda, a^0)$ vary enormously in absolute size. Smaller coefficients are preferable over larger ones, as exceedingly large entries will lead to cancellation of significant values due to round-off effects. The order 3 method from Table 10.2 with $\lambda = \frac{1}{4}$ offers a good compromise between reasonably sized coefficients, a small error constant and good damping properties. For $\lambda = \frac{1}{3}$ smaller coefficients are possible, but the error constant is considerably larger and the stability behaviour is no longer acceptable (see Figure 10.3).

Even though the order 3 method in Table 10.2 is reproduced using (truncated) decimal numbers, the matrices

$$\mathcal{B} = EC^{-1}\mathcal{A}, \qquad \mathcal{U} = C - \mathcal{A}CK, \qquad \mathcal{V} = E - \mathcal{B}CK$$

will be computed using (10.7) and (10.9). In an actual implementation the order conditions will thus be satisfied exactly. The errors resulting from truncating $\mathcal{A}$'s coefficients will affect stability only. As reported in [34], these changes in the method's stability behaviour are generally negligible.

$$\left[\begin{array}{c|c} \mathcal{A} & \mathcal{U} \\ \hline \mathcal{B} & \mathcal{V} \end{array} \right] = \left[\begin{array}{cc|cc} \frac{3}{10} & 0 & 1 & 0 \\ \frac{7}{10} & \frac{3}{10} & 1 & 0 \\ \frac{7}{10} & \frac{3}{10} & 1 & 0 \\ \hline \frac{7}{10} & \frac{3}{10} & 0 & 0 \\ \frac{4}{7} & \frac{3}{7} & 0 & 0 \end{array} \right]$$

$$\left[\begin{array}{ccc|ccc} \frac{2}{11} & 0 & 0 & 1 & \frac{3}{22} & -\frac{7}{968} \\ \frac{4}{11} & \frac{2}{11} & 0 & 1 & \frac{3}{22} & -\frac{7}{968} \\ \frac{135}{352} & \frac{105}{352} & \frac{2}{11} & 1 & \frac{3}{22} & -\frac{7}{968} \\ \frac{135}{352} & \frac{105}{352} & \frac{2}{11} & 1 & \frac{3}{22} & -\frac{7}{968} \\ \hline -\frac{17}{120} & \frac{17}{56} & \frac{88}{105} & 0 & 0 & 0 \\ -\frac{77}{60} & -\frac{11}{28} & \frac{176}{105} & 0 & 0 & 0 \end{array} \right]$$

(a) Order $p = q = 1$, $c = \begin{bmatrix} \frac{3}{10} & 1 \end{bmatrix}^\top$ (b) Order $p = q = 2$, $c = \begin{bmatrix} \frac{7}{22} & \frac{15}{22} & 1 \end{bmatrix}^\top$

$$\left[\begin{array}{cccc|cccc} 0.25000000 & 0 & 0 & 0 & 1 & 0 & -0.03125000 & -0.005208333 \\ 3.96644133 & 0.25000000 & 0 & 0 & 1 & -3.71644133 & -0.99161033 & -0.134367958 \\ 7.10874919 & 0.19011989 & 0.25000000 & 0 & 1 & -6.79886908 & -1.77849724 & -0.245913398 \\ 9.22821238 & 0.30032578 & 0.14504057 & 0.25000000 & 1 & -8.92357873 & -2.31599641 & -0.325048352 \\ \hline 9.22821238 & 0.30032578 & 0.14504057 & 0.25000000 & 1 & -8.92357873 & -2.31599641 & -0.325048352 \\ 5.83388174 & 1.42095043 & -1.93636919 & 1.83333333 & 0 & -6.15179631 & -1.55000209 & -0.231990442 \\ -23.5448944 & 10.4008341 & -15.3587019 & 8.00000000 & 0 & 20.5027622 & 5.20483297 & 0.755308606 \\ -28.7175183 & 30.7178318 & -38.7174038 & 16.0000000 & 0 & 20.7170903 & 4.85851648 & 0.946963295 \end{array} \right]$$

(c) Order $p = q = 3$, $c = \begin{bmatrix} \frac{1}{4} & \frac{1}{2} & \frac{3}{4} & 1 \end{bmatrix}^\top$

Table 10.2: A family of methods having Runge-Kutta stability and $M_{\infty,p} = 0$.

10.2 Methods with $s = p$ Stages

The methods from Table 10.1 and 10.2 use $s = p + 1$ internal stages to perform an order p computation. Hence, each step requires at least $p + 1$ function evaluations – one for each stage. Due to the diagonally implicit structure the stages can be evaluated sequentially. Newton's method is used in order to determine the stages Y_i for $i = 1, \ldots, s$. Evaluating the Jacobian and performing the Newton iteration requires additional function calls.

In electrical circuit simulation a function call means the evaluation of all device models. This process is often referred to as 'loading'. For standard applications with up to 10^3 equations the loading requires approximately 85% of the work while the linear solver causes only about 10% [66]. Most time is spent for loading since even standard transistor models are very complex. The overhead required for stepsize and convergence control is usually below 5%.

Consequently, a large number of internal stages may slow down the simulation significantly. Spending more time per step can be justified only if larger steps are taken. This, in turn, requires small error constants and a relatively smooth behaviour of the solution. Discontinuities and breakpoints, as they appear frequently in circuit simulation, will render an efficient stepsize control more difficult. It is not clear for realistic applications whether the steps can be taken large enough to benefit from the small error constants and the good stability properties of the methods constructed above.

In order to judge how much work should be spent per step, it was decided not only to look at methods with $s = p + 1$ stages but also to construct methods with only $s = p$ stages. In this section general linear methods $\mathcal{M} = [\mathcal{A}, \mathcal{U}, \mathcal{B}, \mathcal{V}]$ with $s = r = p = q$ are derived. In other words, we will construct DIMSIMs of type 2, where the matrix $\mathcal{A}$ has the diagonally implicit structure (10.2).

The method $\mathcal{M}$ has to satisfy the properties (P1) – (P6) from page 182 such that it can be used efficiently for solving differential algebraic equations. However, for $s = r = p = q$ fewer coefficients are available for satisfying these requirements. Compared to the previous section little freedom remains. It will turn out that $M_{\infty,p} = 0$ is not possible for $p = 2$ and the property of Runge-Kutta stability has to be dropped for $p = 3$.

Order $p = 1$

For $s = r = 1$ the method's tableau simply reads $\mathcal{M}_1 = \left[\begin{array}{c|c} a & u \\ \hline b & v \end{array} \right]$ with real numbers a, u, b, v. Stiff accuracy and the conditions for $p = q = 1$ imply that $\mathcal{M}_1 = \left[\begin{array}{c|c} 1 & 1 \\ \hline 1 & 1 \end{array} \right]$ is uniquely defined. This method, however, is nothing but the implicit Euler scheme already introduced in Section 2.1.

Order $p = 2$

The construction of general linear methods with $s = r = p = q = 2$ is more interesting. The generic singly diagonally implicit scheme being stiffly accurate reads

$$\mathcal{M}_2 = \left[\begin{array}{c|c} \mathcal{A} & \mathcal{U} \\ \hline \mathcal{B} & \mathcal{V} \end{array}\right] = \left[\begin{array}{cc|cc} \lambda & 0 & u_{11} & u_{12} \\ a_{21} & \lambda & u_{21} & u_{22} \\ a_{21} & \lambda & u_{21} & u_{22} \\ b_{21} & b_{22} & v_{21} & v_{22} \end{array}\right], \qquad c = \begin{bmatrix} c_1 \\ 1 \end{bmatrix}$$

such that eleven parameters can be chosen to satisfy the requirements (P1)–(P6) from page 182. In order to guarantee $p = q = 2$ the method's coefficients have to satisfy

$$\mathcal{U}W = C - \mathcal{A}CK, \qquad \mathcal{V}W = WE - \mathcal{B}CK. \tag{10.15}$$

These relations were derived in Theorem 7.6 for $C = \begin{bmatrix} e & c & \frac{c^2}{2!} & \cdots & \frac{c^p}{p!} \end{bmatrix}$, $K = \begin{bmatrix} 0 & I \\ 0 & 0 \end{bmatrix}$ and $E = \exp(K)$. More details are given in Section 7.1. Recall that methods in Nordsieck form are characterised by $W_{ij} = \begin{cases} 1 , i = j \\ 0 , i \neq j \end{cases}$ such that $W = \begin{bmatrix} 1 & 0 & 0 \\ 0 & 1 & 0 \end{bmatrix}$ for $s = p = 2$. The order and stage order conditions (10.15) can therefore be written as

$$\mathcal{U} = \begin{bmatrix} 1 & c_1 - \lambda \\ 1 & 1 - a_{21} - \lambda \end{bmatrix}, \quad \mathcal{A}c = \frac{1}{2}c^2, \quad \mathcal{V} = \begin{bmatrix} 1 & 1 - a_{21} - \lambda \\ 0 & 1 - b_{21} - b_{22} \end{bmatrix}, \quad \mathcal{B}c = \begin{bmatrix} \frac{1}{2} \\ 1 \end{bmatrix}.$$

As usual, the exponentiation c^k has to be understood in a component by component sense. Solving these equations for c_1, a_{21} and b_{22} the method $\mathcal{M}_2$ takes the form[4]

$$\mathcal{M}_2 = \left[\begin{array}{cc|cc} \lambda & 0 & 1 & \lambda \\ \frac{1}{4\lambda} - \frac{1}{2} & \lambda & 1 & \frac{3}{2} - \lambda - \frac{1}{4\lambda} \\ \frac{1}{4\lambda} - \frac{1}{2} & \lambda & 1 & \frac{3}{2} - \lambda - \frac{1}{4\lambda} \\ b_{21} & 1 - 2b_{21}\lambda & 0 & b_{21}(2\lambda - 1) \end{array}\right], \qquad c = \begin{bmatrix} 2\lambda \\ 1 \end{bmatrix}.$$

Computing the stability matrix $M(z)$ and determining the limit for $z \to \infty$ yields the matrix

$$M_{\infty,2} = \begin{bmatrix} 0 & 0 \\ \frac{2b_{21}\lambda(4\lambda^2 - 1) + 1 - 2\lambda - 4\lambda^2}{2\lambda^3} & \frac{2b_{21}\lambda(2\lambda - 1) + 1 - 4\lambda + 2\lambda^2}{2\lambda^3} \end{bmatrix} = \begin{bmatrix} 0 & 0 \\ m_{21} & m_{22} \end{bmatrix}.$$

It can be checked easily that the system $m_{21} = 0$, $m_{22} = 0$ has no solution. Hence, $M_{\infty,2} = 0$ is not possible. Nevertheless, it is easy to achieve Runge-Kutta stability by requiring $b_{21} = 0$. This can be seen by computing the stability function

$$\Phi(w, z) = \det\big(wI - M(z)\big) = \frac{1}{(1-\lambda z)^2}\big(p_0 + p_1 w + (1 - \lambda z)^2 w^2\big).$$

---

[4]For simplicity it is assumed that $c_1 \neq 0$.

The polynomial p_0 is given by

$$p_0 = b_{21}(2\lambda - 1)(2 - 2\lambda z + z).$$

Thus p_0 vanishes for $b_{21} = 0$. Assuming that $b_{21} = 0$ holds, the stability function reads

$$\Phi(w, z) = w\left(w - \frac{2z^2\lambda^2 - 4z^2\lambda + z^2 - 4z\lambda + 2z + 2}{2(1 - \lambda z)^2}\right) = w\left(w - \tilde{R}_2(\lambda, z)\right).$$

The single nonzero eigenvalue $\tilde{R}_2(\lambda, z)$ is given by the stability function of the corresponding SDIRK method with two stages and order 2. It is well known that in this case A-stability can be achieved if and only if $\lambda = 1 \pm \frac{\sqrt{2}}{2}$ (see again [85]). Observe that $\lambda = 1 - \frac{\sqrt{2}}{2}$ leads to $c_1 = 2\lambda \approx 0.586 \in [0, 1]$ such that the final method reads

$$\mathcal{M}_2 = \left[\begin{array}{cc|cc} 1 - \frac{1}{2}\sqrt{2} & 0 & 1 & 1 - \frac{1}{2}\sqrt{2} \\ \frac{(-1+\sqrt{2})(2+\sqrt{2})}{4} & 1 - \frac{1}{2}\sqrt{2} & 1 & \frac{(-1+\sqrt{2})(2+\sqrt{2})}{4} \\ \hline \frac{(-1+\sqrt{2})(2+\sqrt{2})}{4} & 1 - \frac{1}{2}\sqrt{2} & 1 & \frac{(-1+\sqrt{2})(2+\sqrt{2})}{4} \\ 0 & 1 & 0 & 0 \end{array}\right], \quad c = \left[\begin{array}{c} 2 - \sqrt{2} \\ 1 \end{array}\right].$$

This scheme is A- and L-stable. The stability matrix at infinity is given by

$$M_{\infty,2} = \left[\begin{array}{cc} 0 & 0 \\ \frac{(-4+3\sqrt{2})(3+2\sqrt{2})}{2} & 0 \end{array}\right] \approx \left[\begin{array}{cc} 0 & 0 \\ 0.707106778 & 0 \end{array}\right]. \tag{10.16}$$

Notice that $\mathcal{M}_2$ exhibits the FSAL property. This method could therefore be interpreted equally well as a diagonally implicit Runge-Kutta scheme with an explicit first stage. Using this slightly different point of view, the method $\mathcal{M}_2$ was also constructed by Kværnø in [105].

It is interesting to note that a strictly lower triangular structure (10.16) for $M_{\infty,2}$ can be obtained by fixing $\mathcal{A}$'s first diagonal element only. Repeating the above computations for $\mathcal{A} = \left[\begin{smallmatrix} a_{11} & 0 \\ a_{21} & a_{22} \end{smallmatrix}\right]$ leads to the method

$$\left[\begin{array}{cc|cc} a_{11} & 0 & 1 & a_{11} \\ \frac{1-2a_{22}}{4a_{11}} & a_{22} & 1 & \frac{4a_{11}(1-a_{22})+2a_{22}-1}{4a_{11}} \\ \hline \frac{1-2a_{22}}{4a_{11}} & a_{22} & 1 & \frac{4a_{11}(1-a_{22})+2a_{22}-1}{4a_{11}} \\ 0 & 1 & 0 & 0 \end{array}\right], \quad c^* = \left[\begin{array}{c} 2a_{11} \\ 1 \end{array}\right].$$

Straightforward computations show that the stability matrix at infinity has the form

$$\left[\begin{array}{cc} 0 & 0 \\ \frac{\frac{1}{4} - a_{11}^2 - \frac{1}{2}a_{22}}{a_{11}^2 a_{22}} & \frac{\frac{1}{2} - a_{22} - a_{11}(1 - a_{22})}{a_{11}a_{22}} \end{array}\right]$$

such that a strictly lower triangular structure can be ensured by choosing $a_{11} = \frac{2a_{22}-1}{2(a_{22}-1)}$. The resulting method

$$
\mathcal{M}_2^*(a_{22}) =
\left[
\begin{array}{cccc|cc}
\frac{2a_{22}-1}{2a_{22}-2} & 0 & 1 & \frac{2a_{22}-1}{2a_{22}-2} \\
\frac{1}{2} - \frac{1}{2}a_{22} & a_{22} & 1 & \frac{1}{2} - \frac{1}{2}a_{22} \\
\frac{1}{2} - \frac{1}{2}a_{22} & a_{22} & 1 & \frac{1}{2} - \frac{1}{2}a_{22} \\
0 & 1 & 0 & 0
\end{array}
\right],
\qquad
c^*(a_{22}) =
\left[
\begin{array}{c}
\frac{2a_{22}-1}{a_{22}-1} \\
1
\end{array}
\right],
$$

has Runge-Kutta stability due to

$$
\Phi^*(w, z) = w\left(w - \frac{2a_{22}(1+z(1-a_{22}))-z-2}{(2a_{22}(z-1)-z+2)(za_{22}-1)} \right) = w\left(w - \tilde{R}_2^*(z) \right),
$$

but a_{22} is left as a free parameter. In order to check for A-stability we write $\tilde{R}_2^*(z) = \frac{P(z)}{Q(z)}$ and compute the poles by solving

$$
Q(z) = 0 \qquad \Leftrightarrow \qquad z \in \left\{ \frac{1}{a_{22}},\ \frac{2a_{22}-2}{2a_{22}-1} \right\}.
$$

For $0 < a_{22} < \frac{1}{2}$ all poles are positive numbers such that A-stability is equivalent to requiring

$$
E(y) = Q(\mathrm{i}y)\, Q(-\mathrm{i}y) - P(\mathrm{i}y)\, P(-\mathrm{i}y) \geq 0 \qquad \forall\ y \in \mathbb{R}.
$$

This mapping, the E-polynomial, is widely used e.g. in [25, 85]. The property $E(y) \geq 0$ ensures that the nonzero eigenvalue satisfies $|R_2^*(yi)| \leq 1$ on the imaginary axis. Since all poles belong to the right half of the complex plane, the maximum principle shows that $|R_2^*(z)| \leq 1$ holds whenever z has negative real part. This, in turn, is nothing but A-stability.

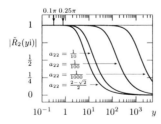

Figure 10.4 (a): Stability behaviour on the imaginary axis for different a_{22}.

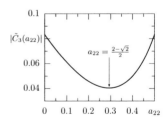

Figure 10.4 (b): Error constant $\tilde{C}_2(a_{22})$ for the method $\mathcal{M}_2^*$.

Figure 10.4: The parameter a_{22} controls the damping behaviour of the method $\mathcal{M}_2^*$. For $a_{22} \to 0$ the damping behaviour of the trapezoidal rule is approximated. The error constant is minimal for $a_{22} = 1 - \frac{\sqrt{2}}{2}$ where $\mathcal{M}_2^*(a_{22}) = \mathcal{M}_2$.

For the special case of $\mathcal{M}_2^*(a_{22})$ the E-polynomial is easy to compute,

$$E(y) = y^4 a_{22}^2 (2a_{22} - 1)^2.$$

Thus $\mathcal{M}_2^*(a_{22})$ is A-stable (and hence also L-stable) for every $0 < a_{22} < \frac{1}{2}$.

The significance of the parameter a_{22} is visualised in Figure 10.4 (a). Varying a_{22} allows to control the damping behaviour of the method. For $a_{22} \to 0$ the damping behaviour of the trapezoidal rule is approximated. In fact, for $a_{22} = 0$ the method

$$\mathcal{M}_2^*(0) = \left[\begin{array}{cc|cc} \frac{1}{2} & 0 & 1 & \frac{1}{2} \\ \frac{1}{2} & 0 & 1 & \frac{1}{2} \\ \frac{1}{2} & 0 & 1 & \frac{1}{2} \\ 0 & 1 & 0 & 0 \end{array}\right], \qquad c^*(0) = \begin{bmatrix} 1 \\ 1 \end{bmatrix},$$

is nothing but a complicated representation of the trapezoidal rule itself.

For $a_{22} = 1 - \frac{\sqrt{2}}{2}$ the methods $\mathcal{M}_2^*\left(1 - \frac{\sqrt{2}}{2}\right) = \mathcal{M}_2$ coincide.

Example 10.1. In Chapter 2 the damping behaviour of BDF methods was investigated. Recall from Example 2.1 on page 27 that the BDF_2 scheme showed strong artificial damping for the circuit in Figure 10.5 (a) below. Plotting the node potential u_3 and the current through the voltage source i_V in phase space this becomes clearly visible as the numerical solution spirals inwards. The general linear method $\mathcal{M}_2^*$, by contrast, shows little numerical damping for $a_{22} = 1 - \frac{\sqrt{2}}{2}$. For $a_{22} = \frac{1}{100}$ the solution is almost a perfect circle in phase space. $\qquad \square$

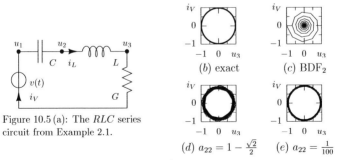

Figure 10.5 (a): The RLC series circuit from Example 2.1.

(b) exact

(c) BDF_2

(d) $a_{22} = 1 - \frac{\sqrt{2}}{2}$

(e) $a_{22} = \frac{1}{100}$

Figure 10.5: For a small resistance $R = \frac{1}{G} = 1\,\mathrm{m\Omega}$ the exact solution is nearly undamped such that u_3 and i_V trace out a circle in phase space (b). The BDF_2 (c) shows strong numerical damping while the general linear method, (d) and (e), allows a control of the damping behaviour via the parameter a_{22}.

The possible control of the damping behaviour comes at the price of the diagonal elements a_{11}, a_{22} not being equal. This is far from optimal as different iteration matrices will be necessary for each of the two stages. The additional evaluations of the Jacobian may slow down the simulation significantly.

For electrical circuit simulation this is not a severe restriction. Although function evaluations are extremely expensive, information about the Jacobian can be obtained at little additional costs. ROW methods have been adapted to a cheap Jacobian by Günther [82]. For general linear methods exploiting cheap Jacobians may lead to an adaptive control of the damping behaviour. However, these investigations are beyond the scope of this thesis and, for the time being, it is recommended to use $a_{22} = 1 - \frac{\sqrt{2}}{2}$ such that $\mathcal{M}_2^\star(a_{22}) = \mathcal{M}_2$ becomes singly diagonally implicit and the error constant is minimised (see Figure 10.4 (b)).

Order $p = 3$

The generic stiffly accurate method with $s = r = 3$ is given by the tableau

$$
\mathcal{M}_3 = \left[\begin{array}{c|c} \mathcal{A} & \mathcal{U} \\ \hline \mathcal{B} & \mathcal{V} \end{array} \right] =
\left[\begin{array}{ccc|ccc}
\lambda & 0 & 0 & u_{11} & u_{12} & u_{13} \\
a_{21} & \lambda & 0 & u_{21} & u_{22} & u_{23} \\
a_{31} & a_{32} & \lambda & u_{31} & u_{32} & u_{33} \\
\hline
a_{31} & a_{32} & \lambda & u_{31} & u_{32} & u_{33} \\
b_{21} & b_{22} & b_{23} & v_{21} & v_{22} & v_{23} \\
b_{31} & b_{32} & b_{33} & v_{31} & v_{32} & v_{33}
\end{array} \right], \qquad
c = \left[\begin{array}{c} c_1 \\ c_2 \\ 1 \end{array} \right].
$$

$$(10.17)$$

Solving the order and stage order conditions

$$
\mathcal{U}W = C - \mathcal{A}CK, \qquad \mathcal{V}W = WE - \mathcal{B}CK
$$

from Theorem 7.6 shows that for $s = r = p = q = 3$ and $c_1 \neq 0$ the method $\mathcal{M}_3$ has the form

$$
\mathcal{A} = \left[\begin{array}{ccc}
\lambda & 0 & 0 \\
\frac{c_2{}^2(c_2 - 3\lambda)}{27\lambda^2} & \lambda & 0 \\
\frac{1 - 3a_{32}c_2{}^2 - 3\lambda}{27\lambda^2} & a_{32} & \lambda
\end{array} \right], \qquad
c = \left[\begin{array}{c} 3\lambda \\ c_2 \\ 1 \end{array} \right],
$$

$$
\mathcal{U} = \left[\begin{array}{ccc}
1 & 2\lambda & \frac{3}{2}\lambda^2 \\
1 & \frac{27\lambda^2 c_2 - c_2^3 + 3\lambda c_2^2 - 27\lambda^3}{27\lambda^2} & \frac{c_2\left(15\lambda c_2 - 2c_2^2 - 18\lambda^2\right)}{18\lambda} \\
1 & \frac{27\lambda^2 - 1 + 3a_{32}c_2^2 + 3\lambda - 27a_{32}\lambda^2 - 27\lambda^3}{27\lambda^2} & \frac{15\lambda - 2 + 6a_{32}c_2^2 - 18a_{32}c_2\lambda - 18\lambda^2}{18\lambda}
\end{array} \right],
$$

$$
\mathcal{B} = \left[\begin{array}{ccc}
\frac{1 - 3a_{32}c_2^2 - 3\lambda}{27\lambda^2} & a_{32} & \lambda \\
b_{21} & b_{22} & 1 - 9b_{21}\lambda^2 - b_{22}c_2^2 \\
b_{31} & b_{32} & 2 - 9b_{31}\lambda^2 - b_{32}c_2^2
\end{array} \right],
$$

$$
\mathcal{V} = \left[\begin{array}{ccc}
1 & \frac{27\lambda^2 - 1 + 3a_{32}c_2^2 + 3\lambda - 27a_{32}\lambda^2 - 27\lambda^3}{27\lambda^2} & \frac{15\lambda - 2 + 6a_{32}c_2^2 - 18a_{32}c_2\lambda - 18\lambda^2}{18\lambda} \\
0 & b_{22}c_2^2 - b_{21} - b_{22} + 9b_{21}\lambda^2 & b_{22}c_2^2 - 3b_{21}\lambda - b_{22}c_2 + 9b_{21}\lambda^2 \\
0 & b_{32}c_2^2 - b_{31} - b_{32} - 2 + 9b_{31}\lambda^2 & b_{32}c_2^2 - 1 - 3b_{31}\lambda - b_{32}c_2 + 9b_{31}\lambda^2
\end{array} \right].
$$

Observe that the fourth column of the stage order conditions $\mathcal{U}W = C - \mathcal{A}CK$ reads $0 = \frac{1}{6}c^3 - \frac{1}{2}\mathcal{A}c^2$. Thus stage order $q = 3$ requires

$$0 = \frac{1}{3}c^3 - \mathcal{A}c^2 = \begin{bmatrix} \frac{1}{3}c_1^3 - \lambda c_1^2 \\ \frac{1}{3}c_2^3 - a_{21}c_1^2 - \lambda c_2^2 \\ \frac{1}{3} - a_{31}c_1^2 - a_{32}c_2^2 - \lambda \end{bmatrix}$$

to hold. For $c_1 \neq 0$ the first component implies $c_1 = 3\lambda$ as is already indicated by the above formulae.

On the other hand, if $\mathcal{M}_3$ had Runge-Kutta stability,

$$\Phi(w, z) = w^2\big(w - \tilde{R}_3(\lambda, z)\big),$$

the single nonzero eigenvalue $\tilde{R}_3(\lambda, z)$ would be given by the stability function of the SDIRK method with $s = p = 3$. Thus, A-stability would be possible only for $\lambda = 0.43586652$ which is a root of $\lambda^3 - 3\lambda^2 + \frac{3}{2}\lambda - \frac{1}{6} = 0$ (see [85]).

As a consequence, there is no A-stable singly diagonally implicit general linear method in Nordsieck form having stiff accuracy, $s = r = p = q = 3$ and satisfying $0 < c_1 < 1$.

Sacrificing A-stability in favour of a value $c_1 \in [0, 1]$ will inevitably lead to a stability matrix M_∞ having a nonzero eigenvalue. Hence, nilpotency of M_∞ would be impossible for $0 < c_1 < 1$, but this was one of the requirements (P1) – (P6) on page 182.

The difficulty of finding L-stable methods with $s = r + 1 = p = q = 3$ was already observed in [33]. Although Butcher and Jackiewicz succeed in constructing A- and L-stable DIMSIMs with $s = r = p = q = 3$ in [33], these methods are neither in Nordsieck form nor stiffly accurate. In [27] it is conjectured that no A-stable type-4 DIMSIM[5] with $M_\infty = 0$ exists for $s = r = p = q = 3$.

In order to cope with these difficulties, we will drop the requirement of Runge-Kutta stability from now on.

In view of Theorem 9.5 good stability at zero and at infinity will be ensured first. The remaining free parameters can be used to achieve A-stability, or at least large regions of $A(\alpha)$-stability.

Recall from Chapter 7 that a general linear method $\mathcal{M} = [\mathcal{A}, \mathcal{U}, \mathcal{B}, \mathcal{V}]$ applied to the ordinary differential equation $y' = f(y, t)$ reads

$$Y = h_n \mathcal{A}\, F(Y) + \mathcal{U}\, y^{[n]}, \qquad y^{[n+1]} = h_n \mathcal{B}\, F(Y) + \mathcal{V}\, y^{[n]}, \qquad (10.18)$$

where $F(Y) = \big[f(Y_1, t_n + c_1 h_n)^\top \;\; \cdots \;\; f(Y_s, t_n + c_s h_n)^\top\big]^\top$. Using the scheme (10.18) the method proceeds from t_n to $t_{n+1} = t_n + h_n$ using a stepsize h_n. If the next step is taken with a different stepsize $h_{n+1} \neq h_n$, the output vector

---

[5]A DIMSIM is said to be of type-4 if $\mathcal{A} = \lambda I$.

$y^{[n+1]}$ has to be modified. For methods in Nordsieck form the appropriate modification is given by simply multiplying the Nordsieck vector with the diagonal matrix $D(\sigma_n) = \text{diag}(1, \sigma_n, \ldots, \sigma_n^{r-1})$,

$$y^{[n+1]} \approx \begin{bmatrix} y(t_{n+1}) \\ h_n y'(t_{n+1}) \\ \vdots \\ h_n^{r-1} y^{(r-1)}(t_{n+1}) \end{bmatrix} \quad \Rightarrow \quad D(\sigma_n)\, y^{[n+1]} \approx \begin{bmatrix} y(t_{n+1}) \\ h_{n+1} y'(t_{n+1}) \\ \vdots \\ h_{n+1}^{r-1} y^{(r-1)}(t_{n+1}) \end{bmatrix}.$$

Here, $\sigma_n = \frac{h_{n+1}}{h_n}$ denotes the stepsize ratio[6]. In case of the linear scalar test equation $y' = \lambda y$, (10.18) turns into

$$D(\sigma_n)\, y^{[n+1]} = D(\sigma_n) M(\lambda h_n) y^{[n]} = \prod_{i=0}^{n} D(\sigma_i) M(\lambda h_i)\, y^{[0]}.$$

In a variable stepsize implementation, stability is thus related to products of matrices $M(\sigma_i, z_i) = D(\sigma_i) M(z_i)$. To ensure that $y^{[n+1]}$ remains bounded as $n \to \infty$, it is not sufficient to require a spectral radius $\varrho(M(\sigma_i, z_i)) < 1$ for each factor, but we have to guarantee a spectral radius less than 1 for the product.

This goal can be achieved by limiting the stepsize ratio σ. Corresponding results for low order BDF schemes and a two-stage order 1 general linear method were derived by Butcher and Heard [31] using appropriate norms. Guglielmi and Zennaro use the theory of a joint spectral radius for a family of matrices and the concept of polytope norms [77, 76]. Results for two-step W-methods are given in [95].

Butcher and Jackiewicz use a different approach in [38]. They show that general linear methods being unconditionally zero-stable can be constructed using special error estimators and appropriate modifications of the Nordsieck vector.

For the order-3 method (10.17), by contrast, we will try to guarantee stability at zero and at infinity by construction. If the matrices $\mathcal{V}$ and M_∞ were of the form

$$\mathcal{V} = \begin{bmatrix} 1 & v_{12} & v_{13} \\ 0 & 0 & 0 \\ 0 & v_{32} & 0 \end{bmatrix}, \qquad M_\infty = \begin{bmatrix} 0 & 0 & 0 \\ m_{21} & 0 & 0 \\ m_{31} & m_{32} & 0 \end{bmatrix}, \tag{10.19}$$

then $M(\sigma, 0) = D(\sigma)\mathcal{V}$ and $M(\sigma, \infty) = D(\sigma)M_\infty$ would have the same structure. Thus, multiplication by $D(\sigma)$ does not affect stability since the special zero-pattern ensures that the spectra $\sigma(\mathcal{V}) = \{1, 0\}$ and $\sigma(M_\infty) = \{0\}$ do not depend on the particular values of v_{ij}, m_{ij}.

---

[6] A more refined adjustment of the Nordsieck vector that preserves the first order error terms will be discussed in Chapter 11.

In order for the method $\mathcal{M}_3$ to be of type (10.19), the system

$$
\begin{aligned}
v_{22} &= b_{22}c_2^2 - b_{21} - b_{22} + 9b_{21}\lambda^2 &= 0, \\
v_{23} &= b_{22}c_2^2 - 3b_{21}\lambda - b_{22}c_2 + 9b_{21}\lambda^2 &= 0, \\
v_{33} &= b_{32}c_2^2 - 1 - 3b_{31}\lambda - b_{32}c_2 + 9b_{31}\lambda^2 &= 0
\end{aligned}
$$

is solved first. One particularly simple solution is given by

$$
b_{21} = 0, \qquad b_{22} = 0, \qquad b_{32} = \frac{-1-3b_{31}\lambda+9b_{31}\lambda^2}{c_2(1-c_2)}, \tag{10.20}
$$

where it is assumed that $c_2 \neq 0$ and $c_2 \neq 1$. Observe that (10.20) ensures that $\mathcal{M}_3$ has the FSAL property since $e_2^\top \mathcal{B} = \begin{bmatrix} 0 & 0 & 1 \end{bmatrix}$ and $e_2^\top \mathcal{V} = \begin{bmatrix} 0 & 0 & 0 \end{bmatrix}$. For the structure (10.19) it remains to compute $M_\infty = (m_{ij})$ and to solve the system $m_{22} = 0$, $m_{23} = 0$, $m_{33} = 0$. The first row $\begin{bmatrix} m_{11} & m_{12} & m_{13} \end{bmatrix} = 0$ vanishes due to stiff accuracy. Using e.g. MAPLE the solution is found to be

$$
c_2 = \frac{1-6\lambda+6\lambda^2}{3\lambda^2-5\lambda+1},
$$

$$
a_{32} = \frac{\left(3\lambda^2-5\lambda+1\right)^3 \lambda}{(1-9\lambda+21\lambda^2-9\lambda^3)(1-6\lambda+6\lambda^2)},
$$

$$
b_{31} = \frac{1+510\lambda^4-288\lambda^5+120\lambda^2+54\lambda^6-366\lambda^3-18\lambda}{3\lambda(6\lambda-1)(-1+3\lambda)(3\lambda^2-6\lambda+1)(3\lambda^3-9\lambda^2+6\lambda-1)}
$$

and the diagonal element λ is left as a free parameter. Of course, we have to require $\lambda \notin \left\{0, \frac{1}{6}, \frac{1}{3}, 1 \pm \frac{\sqrt{6}}{3}, \frac{1}{2} \pm \frac{\sqrt{3}}{6}, \frac{5}{6} \pm \frac{\sqrt{13}}{6}\right\}$ and that λ is not a root of $3\lambda^3 - 9\lambda^2 + 6\lambda - 1 = 0$. This avoids a zero denominator. The coefficients of the method $\mathcal{M}_3(\lambda)$ are reproduced in Table 10.3.

Due to the special structure (10.19) for the matrices $\mathcal{V}$ and M_∞ this method is stable at zero and at infinity for every stepsize pattern $(\sigma_n)_{n \in \mathbb{N}}$. Hence we have ensured unconditional stability at zero and at infinity by construction.

Figure 10.6 (a) suggests to choose $0 < \lambda < 1 - \frac{\sqrt{6}}{3}$ such that $0 < c_1 < c_2 < 1$. Unfortunately, the corresponding methods are not A-stable.

Information on possible angles for $A(\alpha)$-stability is given in Figure 10.6 (b). For each $\lambda \in [0,1]$ the method $\mathcal{M}_3(\lambda)$ has been computed. The angle of $A(\alpha)$-stability was computed from the corresponding stability region. This angle $\alpha(\lambda)$ is plotted in Figure 10.6 (b) as a function of the diagonal element λ. $\mathcal{M}(\lambda)$ is A-stable provided that $\alpha(\lambda) = 90°$, i.e. for $\lambda \geq \lambda_* \approx 0.3558$.

For $\lambda \geq \lambda_*$ the abscissa c_1 lies outside the unit interval $[0,1]$. As we chose the condition $c_1, c_2 \in [0,1]$ as a design criterion, the optimal angle of $A(\alpha)$ stability is achieved for $\lambda \approx 0.158$ with $\alpha \approx 75.8053°$. For $\lambda = \frac{4}{25} = 0.16$ and $\alpha(\frac{4}{25}) \approx 73.7535°$ much simpler rational coefficients can be obtained (see Table 10.4). The entry of largest magnitude is $b_{32} \approx -13.4165$ and the error constant is given by $C_3(\frac{4}{25}) = -\frac{35404169}{6805687500} \approx -0.0052$.

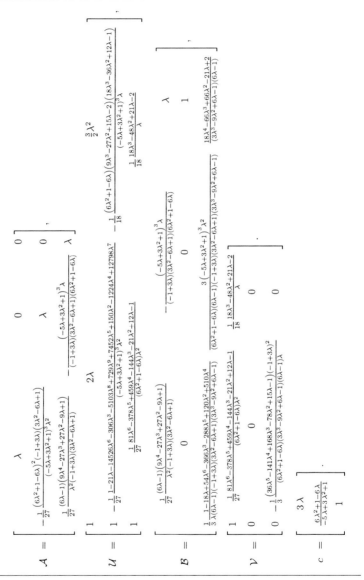

Table 10.3: The method $\mathcal{M}_3(\lambda)$ satisfies $s = r = p = q = 3$ and is unconditionally stable at zero and at infinity.

Although the region of $A(\alpha)$-stability is considerably smaller than for the BDF scheme with the same order (see Table 2.2 on page 32), we will see in Chapter 12 that the family from Table 10.4 performs surprisingly well for most test examples.

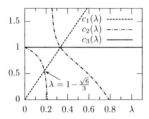

Figure 10.6 (a): Values for the abscissae c_i at different λ.

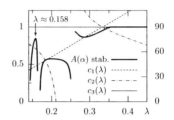

Figure 10.6 (b): Possible angles for $A(\alpha)$-stability.

Figure 10.6: Choosing λ for the method $\mathcal{M}_3$. Figure 10.6 (a) shows that $0 < \lambda < 1 - \frac{\sqrt{6}}{3}$ guarantees $0 < c_1 < c_2 < 1$. For $\lambda > \frac{1}{2} - \frac{\sqrt{3}}{6} \approx 0.211$ at least one c_i lies outside the unit interval. Figure 10.6 (b) contains an enlarged copy of the left hand picture. Additionally the angle α is plotted (rightmost axis) indicating $\mathcal{M}_3$'s $A(\alpha)$ stability.

$$\left[\begin{array}{c|c} \mathcal{A} & \mathcal{U} \\ \hline \mathcal{B} & \mathcal{V} \end{array}\right] = \left[\begin{array}{c|c} 1 & 1 \\ \hline 1 & 1 \end{array}\right]$$

$$\left[\begin{array}{cc|cc} 1-\frac{1}{2}\sqrt{2} & 0 & 1 & 1-\frac{1}{2}\sqrt{2} \\ \frac{(\sqrt{2}-1)(2+\sqrt{2})}{4} & 1-\frac{1}{2}\sqrt{2} & 1 & \frac{(\sqrt{2}-1)(2+\sqrt{2})}{4} \\ \hline \frac{(\sqrt{2}-1)(2+\sqrt{2})}{4} & 1-\frac{1}{2}\sqrt{2} & 1 & \frac{(\sqrt{2}-1)(2+\sqrt{2})}{4} \\ 0 & 1 & 0 & 0 \end{array}\right]$$

(a) Order $p = q = 1$, $c = \begin{bmatrix} 1 \end{bmatrix}$ (b) Order $p = q = 2$, $c = \begin{bmatrix} 2 - \sqrt{2} & 1 \end{bmatrix}^{\top}$

$$\left[\begin{array}{ccc|ccc} \frac{4}{25} & 0 & 0 & 1 & \frac{8}{25} & \frac{24}{625} \\ \frac{347357725}{2236773744} & \frac{4}{25} & 0 & 1 & \frac{21480179099}{55919343600} & \frac{270961229}{4659945300} \\ \frac{57229}{409968} & \frac{20710868}{71768125} & \frac{4}{25} & 1 & \frac{13454351}{32670000} & \frac{1601}{22500} \\ \hline \frac{57229}{409968} & \frac{20710868}{71768125} & \frac{4}{25} & 1 & \frac{13454351}{32670000} & \frac{1601}{22500} \\ 0 & 0 & 1 & 0 & 0 & 0 \\ \frac{334994375}{45927804} & -\frac{6213260400}{463105357} & \frac{27758}{4033} & 0 & -\frac{4451291}{5855916} & 0 \end{array}\right]$$

(c) Order $p = q = 3$, $c = \begin{bmatrix} \frac{1}{4} & \frac{1}{2} & \frac{3}{4} & 1 \end{bmatrix}^{\top}$

Table 10.4: A family of DIMSIMS with $s = r = q = p$

Larger regions of $A(\alpha)$-stability are possible by weakening the requirement (10.19). As an example assume that $\mathcal{V}$ and M_∞ satisfy

$$\mathcal{V} = \begin{bmatrix} 1 & v_{12} & v_{13} \\ 0 & 0 & 0 \\ 0 & v_{32} & 0 \end{bmatrix}, \qquad M_\infty^2 = \begin{bmatrix} 0 & 0 & 0 \\ m_{21} & 0 & 0 \\ m_{31} & m_{32} & 0 \end{bmatrix}, \tag{10.21}$$

where M_∞^2, in contrast to M_∞, is required to have a strictly lower triangular form. Solving $m_{22} = 0$, $m_{23} = 0$, $m_{33} = 0$ yields many different solutions. Unfortunately it was not possible to find A-stable methods within these families. One example of a method with a large region of $A(\alpha)$-stability is given in Table 10.5. There the angle α satisfies $\alpha \approx 88.1°$ such that the method is more stable than the BDF_3 scheme. Unconditional stability at infinity is, however, lost.

$$\mathcal{A} = \begin{bmatrix} \frac{7}{40} & 0 & 0 \\ \frac{343}{1458} & \frac{7}{40} & 0 \\ \frac{5225432851}{18627840000} + \frac{\beta}{126720000} & \frac{2346585921}{19317760000} - \frac{9\beta}{2759680000} & \frac{7}{40} \end{bmatrix},$$

$$\mathcal{U} = \begin{bmatrix} 1 & \frac{7}{20} & \frac{147}{3200} \\ 1 & \frac{11851}{29160} & \frac{6517}{97200} \\ 1 & \frac{44126632861}{104315904000} - \frac{23\beta}{4967424000} & \frac{159211901}{2027520000} - \frac{\beta}{675840000} \end{bmatrix},$$

$$\mathcal{B} = \begin{bmatrix} \frac{5225432851}{18627840000} + \frac{\beta}{126720000} & \frac{2346585921}{19317760000} - \frac{9\beta}{2759680000} & \frac{7}{40} \\ 0 & 0 & 1 \\ \frac{\beta}{670320} + \frac{3022207}{670320} & -\frac{17130621}{1207360} - \frac{3\beta}{1207360} & \frac{8202367}{802560} + \frac{\beta}{802560} \end{bmatrix},$$

$$\mathcal{V} = \begin{bmatrix} 1 & \frac{44126632861}{104315904000} - \frac{23\beta}{4967424000} & \frac{159211901}{2027520000} - \frac{\beta}{675840000} \\ 0 & 0 & 0 \\ 0 & -\frac{2135167}{3951360} - \frac{\beta}{3951360} & 0 \end{bmatrix}.$$

Table 10.5: A DIMSIM with $s = r = p = q = 3$ and $A(\alpha)$-stability for $\alpha \approx 88.1°$. In the tableau β represents $\sqrt{11831751737089}$ and the vector c is given by $c = \left[\frac{21}{40}, \frac{49}{60}, 1\right]$.

10.3 Summary

In the literature [23, 28, 33, 34, 93, 158] general linear methods have been constructed mainly for (stiff) ordinary differential equations. The investigations of Theorem 9.5 and Corollary 9.6 showed that differential algebraic equations can be treated as well, but additional requirements need to be satisfied.

In particular additional order conditions and stronger stability requirements have to be met. It was thus decided to find methods $\mathcal{M}$ having the following properties:

(P1) $\mathcal{M}$ uses $s = r$ stages to achieve order p and stage order $q = p$ for ordinary differential equations.

(P2) $\mathcal{A}$ has a diagonally implicit structure (10.2) with $a_{ii} = \lambda$ for $i = 1, \ldots, s$.

(P3) $\mathcal{V}$ is power bounded and M_∞ is nilpotent[7].

(P4) The abscissae c_i satisfy $c_i \in [0, 1]$ and $c_i \neq c_j$ for $i \neq j$.

(P5) $\mathcal{M}$ is stiffly accurate, i.e. $c_s = 1$ and the last row of $[\mathcal{A} \quad \mathcal{U}]$ coincides with the first row of $[\mathcal{B} \quad \mathcal{V}]$.

(P6) $\mathcal{M}$ is given in Nordsieck form such that the stepsize can be varied easily.

Three families of methods have been considered so far. The methods from Table 10.1 constructed by Butcher and Podhaisky and those in Table 10.2 with $M_{\infty,p} = 0$ use $s = p+1$ stages while the DIMSIMs constructed in the previous section have only $s = p$ stages.

The methods in Table 10.1 were constructed using the algorithm of Wright [158] assuming $\mathcal{B}\mathcal{A} = X\mathcal{B}$ and $\mathcal{B}\mathcal{U} \equiv X\mathcal{V} - \mathcal{V}X$. Thus these methods have inherent Runge-Kutta stability (IRKS). One of the many advantages of this approach is the fact, that only linear operations are necessary to derive a method.

In an attempt to improve stability at infinity, the methods from Table 10.2 were constructed such that $M_{\infty,p} = 0$. For $p = 1$ and $p = 2$ IRKS was still possible. For $p = 3$ *inherent* Runge-Kutta stability could not be realised any more, but Runge-Kutta stability was still possible. Starting from the stability function

$$\Phi(w, z) = \left(p_0 + p_1 w + p_2 w^2 + p_3 w^3 + (1 - \lambda z)^4 w^4\right) \frac{1}{(1-\lambda z)^4}$$

the FORTRAN routine hybrd from MINPACK [125] was used to solve the system $p_0 = 0$, $p_1 = 0$, $p_2 = 0$ numerically. Thus the resulting method has, once again, only one nonzero eigenvalue.

Solving the complicated nonlinear system leads only to isolated methods. The IRKS approach of Wright, by contrast, yields families of methods.

The requirement of having $M_{\infty,p} = 0$ is a strong restriction. Some of the method's coefficients seem to get a bit too large. Also, the methods of Butcher and Podhaisky have smaller error constants. We have to check numerically whether there is any benefit due to the improved stability at infinity. The corresponding numerical experiments will be performed in Chapter 12.

In Section 10.2 DIMSIMs with $s = r = p = q$ have been considered. These methods seem of particular interest as only $s = p$ stages are used per step. The reduced computational work may lead to a clear increase in performance.

---

[7]Recall from Lemma 8.30 that $M_\infty^{k_0} = 0$ for some $k_0 \geq 1$ ensures that the stage derivatives are calculated with the desired accuracy.

However, compared to $s = p + 1$ fewer parameters are available for satisfying the requirements (P1) – (P6).

For $p = 1$ the corresponding method is uniquely determined. Forcing Runge-Kutta stability for $p = 2$ by setting $b_{21} = 0$ leads to an order 2 method that can be interpreted as a singly diagonally implicit Runge-Kutta scheme with an explicit first stage.

If the entries a_{11} and a_{22} on the diagonal are allowed to take different values, a family $\mathcal{M}_2^*(a_{22})$ of A- and L-stable diagonally implicit general linear methods was constructed. Varying $a_{22} \in [0, \frac{1}{2}]$ this family allows a control of the method's damping behaviour.

Unfortunately the search for A-stable methods with $p = 3$ was not successful. If unconditional stability at zero and at infinity is guaranteed by construction, the maximal angle for $A(\alpha)$-stability is $\alpha \approx 75.8053°$ (see Figure 10.7 (c)). There are, however, methods that are more stable than their BDF-counterparts. The method from Table 10.5 with $\alpha \approx 88.1°$ served as an example.

The methods with $s = r = p = q$ from Table 10.4 will be the basis of the variable-order variable-stepsize implementation GLIMDA. The acronym 'GLIMDA' is used to abbreviate General LInear Methods for Differential Algebraic equations.

Although the order 3 method is not A-stable, there is some hope for a competitive code as only $s = p$ stages are used per step. This code will also benefit from the unconditional stability of the order 3 method.

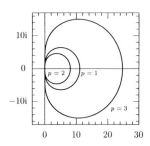

Figure 10.7 (a): Table 10.1 (IRKS).

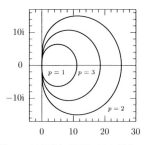

Figure 10.7 (b): Table 10.2 ($M_{\infty,p} = 0$).

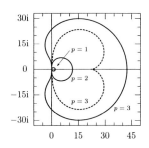

Figure 10.7 (c): Table 10.4 ($s = p$). The method from Table 10.5 is plotted as a dashed line.

Figure 10.7: Stability regions of the three families of methods constructed in the previous sections. The stable region is always given by the outside of the encircled areas.

Since the methods from Table 10.1 and 10.2 use an incremented number of stages, the corresponding code will be denoted as GLIMDA++. The family of methods being used for a particular computation can be chosen easily by changing the method's coefficients.

In Chapter 12 a large number of numerical experiments will be performed in order to assess the potential of these different families of methods. The codes based on general linear methods will not only be compared with each other but also with DASSL and RADAU. It will turn out that general linear methods are competitive and often superior to classical linear multistep or Runge-Kutta codes. In particular GLIMDA will prove a tough competitor.

Before proceeding to the numerical experiments, implementation issues such as error estimation, stepsize selection and strategies for changing the order have to be addressed. These considerations will be the topic of the next chapter.

11

Implementation Issues

Implementation specifics for general linear methods (GLMs) have been discussed in several papers [26, 37, 40, 94] mainly by Butcher and Jackiewicz. The focus is always on (stiff) ordinary differential equations

$$y' = f(y,t), \qquad t \in [t_0, t_{end}], \qquad y(t_0) = y_0. \tag{11.1}$$

For stiff equations singly diagonally implicit methods with high stage order are most appropriate. Recall that in the stiff regime implicit methods have to be used, but due to the diagonally implicit structure the stages can be evaluated sequentially. Thus the computational costs are considerably reduced.

The resulting nonlinear systems are solved using Newton's method. General linear schemes allow the construction of highly accurate predictors for the initial guess. The predictors are based on incoming approximations and on stage derivatives already evaluated within the current step. Thus the initial guess can be obtained at no additional costs. A case study for possible predictors is given in [93].

A distinct feature of general linear methods is the possibility to have diagonally implicit schemes with high stage order. High stage order is not only exploited in deriving predictors for Newton's method but also, as we will see later, for constructing error estimates. In the context of stiff equations and DAEs it is of particular importance to recall that high stage order avoids the order reduction phenomenon discussed in Section 2.2. Finally, dense output comes practically for free (see again [93]).

Solving (11.1) numerically requires discretisation. Starting from the initial condition $y(t_0) = y_0$ approximations $y_n \approx y(t_n)$ to the exact solution are computed on the nonuniform grid

$$t_0 < t_1 < t_2 < \cdots < t_N = t_{end}.$$

After computing y_n at timepoint t_n a stepsize h_n has to be predicted for the next step from t_n to $t_{n+1} = t_n + h_n$. The suggestion for the stepsize h_n is based on an estimate of the local truncation error in step number n. In [29] the error estimator is obtained from information computed in the previous two steps. This is not always convenient in a variable order implementation. Thus, we will use ideas from [37, 40] where the error estimate is based on information from the current step only. As indicated above, error estimation benefits from high stage order. Methods with $s = p + 1$ stages will have clear advantages over those with $s = p$ stages.

Finally, an efficient implementation has to be capable of changing the order. Low order methods are preferred for loose tolerances. Tight tolerances and very smooth solutions are most efficiently dealt with by higher order schemes. Adaptivity, i.e. monitoring how the solution evolves and changing the order accordingly, is the key for an efficient simulation.

Variable-order implementations for general linear methods are discussed in [30, 40, 94]. While [30, 94] use ratios of appropriate error estimates to decide on the new order, the approach taken in [40] is more elegant. There it is shown how to obtain an estimate for $h^{p+2}y^{(p+2)}(t_n)$. As the method currently used is assumed to have order p, this allows to estimate the local truncation error for the method of order $p + 1$.

For methods with $s = p$ stages this approach is not feasible, but techniques of Hairer and Wanner [86] can be adapted to general linear methods. The decision on the new order will be based on the convergence rate of Newton's method.

These topics – Newton's method, error estimation and order control – will now be addressed in more detail. We will restrict attention to methods in Nordsieck form where the input vector

$$y^{[n]} \approx \begin{bmatrix} y(t_n) \\ h\,y'(t_n) \\ \vdots \\ h^{r-1}y^{(r-1)}(t_n) \end{bmatrix}$$

contains approximations to scaled derivatives of the exact solution.

Recall that we are concerned with the numerical solution of differential algebraic equations

$$A(t)\big[d\big(x(t),t\big)\big]' + b\big(x(t),t\big) = 0.$$

The material presented here will be based on the more general structure

$$f\Big(\dot{q}\big(x(t),t\big),\, x(t),\, t\Big) = 0 \tag{11.2}$$

such that the codes GLIMDA and GLIMDA++ will eventually be capable of solving DAEs of the form (11.2).

11.1 Newton Iteration

Let $\mathcal{M} = [\mathcal{A}, \mathcal{U}, \mathcal{B}, \mathcal{V}]$ be a general linear method with the matrix $\mathcal{A}$ being nonsingular. Similar to the approach taken in Chapter 9 the method $\mathcal{M}$ can be applied to the DAE (11.2) by introducing internal stages X_i and stage derivative Q_i' representing $X_i \approx x(t_n + c_i h)$ and $Q_i' \approx \dot{q}\big(x(t_n + c_i h), t_n + c_i h\big)$, respectively, for $i = 1, \ldots, s$.

X_i and Q_i' are related by the differential equation,

$$f\big(Q_i', X_i, t_n + c_i h \big) = 0, \qquad i = 1, \ldots, s, \tag{11.3a}$$

$$Q(X) = h\, \mathcal{A}\, Q' + \mathcal{U}\, q^{[n]}, \tag{11.3b}$$

$$q^{[n+1]} = h\, \mathcal{B}\, Q' + \mathcal{V}\, q^{[n]}. \tag{11.3c}$$

In (11.3) the following abbreviations have been used:

$$X = \begin{bmatrix} X_1 \\ \vdots \\ X_s \end{bmatrix}, \qquad Q' = \begin{bmatrix} Q_1' \\ \vdots \\ Q_s' \end{bmatrix}, \qquad Q(X) = \begin{bmatrix} q(X_1, t_n + c_1 h) \\ \vdots \\ q(X_s, t_n + c_s h) \end{bmatrix}.$$

The input vectors $q_{i+1}^{[n]} \approx h^i \frac{\mathrm{d}}{\mathrm{d}t} q\big(x(t_n), t_n\big)$, $i = 1, \ldots, r$, approximate scaled derivatives of the function q appearing in (11.2). Observe that only information about $q(x, t)$ is propagated from step to step. As a consequence, the numerical solution x_n might not be recoverable from $q^{[n]}$. For stiffly accurate methods this is no restriction at all since the numerical result in step number n coincides with the last stage, i.e. $x_n = X_s$.

Observe that $q^{[n]} \in \mathbb{R}^{r \cdot l}$ for

$$f : \mathbb{R}^l \times \mathbb{R}^m \times \mathbb{R} \to \mathbb{R}^m.$$

In general the relation $l \leq m$ holds such that for $l < m$ the memory required for storing the Nordsieck vector is lower as compared to the case of ordinary differential equations where $q^{[n]}$ satisfies $q^{[n]} \in \mathbb{R}^{r \cdot m}$.

Given that the matrix $\mathcal{A}$ has singly diagonally implicit structure,

$$\mathcal{A} = \begin{bmatrix} \lambda & 0 & \cdots & 0 & 0 \\ a_{21} & \lambda & \cdots & 0 & 0 \\ \vdots & \vdots & \ddots & \vdots & \vdots \\ a_{s-1,1} & a_{s-1,2} & \cdots & \lambda & 0 \\ a_{s1} & a_{s2} & \cdots & a_{s,s-1} & \lambda \end{bmatrix},$$

the relation (11.3b) can be written as

$$q(X_i, t_n + c_i h) = h\, \lambda\, Q_i' + \omega_i \tag{11.4}$$

for $i = 1, \ldots, s$, where, at stage number i, the vector

$$\omega_i = h \sum_{j=1}^{i-1} a_{ij} Q'_j + \sum_{j=1}^{r} u_{ij} q_j^{[n]}$$

is already known from previous results. Thus, in order to determine the value of X_i we need to solve the nonlinear system

$$\mathcal{F}_i(X_i) = h \lambda \, f\Big(\tfrac{1}{h\lambda}[q(X_i, t_n + c_i h) - \omega_i], X_i, t_n + c_i h\Big) = 0.$$

Starting from an initial guess X_i^0 Newton's method can be applied such that

$$X_i^{k+1} = X_i^k + \Delta X_i^k \qquad \text{with} \qquad \mathcal{J}_i(X_i^k)\Delta X_i^k = -\mathcal{F}_i(X_i^k). \qquad (11.5)$$

The Jacobian matrix

$$\mathcal{J}_i(X_i^k) = \frac{\mathrm{d}}{\mathrm{d}X_i}\mathcal{F}_i(X_i^k) = \frac{\partial f}{\partial \dot{q}} \frac{\partial q}{\partial x} + h \lambda \frac{\partial f}{\partial x}$$

is computed using difference approximations. Of course, $\mathcal{J}_i(X_i^k)$ can be assembled analytically, given that the user provides functions $\frac{\partial f}{\partial \dot{q}}$, $\frac{\partial q}{\partial x}$ and $\frac{\partial f}{\partial x}$.

Notice that the stages X_i can indeed be evaluated sequentially. Hence in order to determine the stages from (11.3), s nonlinear systems of dimension m have to be solved. The integer m denotes the problem size.

The diagonal element λ stays the same for every stage X_i. Hence there is some hope that the same Jacobian $\mathcal{J}_1(X_1^k)$ can be used for all stages. The Jacobian information may even be kept constant over several steps. In case of convergence problems the Jacobian is updated as needed. Full details on the iteration scheme are given in [93].

For solving the linear system (11.5), the Jacobian $\mathcal{J}_i(X_i^k) = L\,U$ is decomposed into its $L\,U$ factors using dgetrf from LAPACK [109]. The Newton process is stopped provided that the residuum $\mathcal{F}_i(X_i^k)$ and the correction ΔX_i^k satisfy the mixed criterion

$$\|\mathcal{F}_i(X_i^k)\|_{sc} \leq \mathsf{tolf}, \qquad \|\Delta X_i^k\|_{sc} \leq \mathsf{tolx},$$

where the norm $\|\cdot\|_{sc}$ is defined by

$$\|\eta\|_{sc} = \max_{j=1}^{m} \Big(|\eta_j| / (\mathsf{atol}_j + \mathsf{rtol}_j \cdot |x_{n,j}|) \Big).$$

The vectors atol and rtol are given by the user and contain the absolute and relative accuracy requirements, respectively. The numerical result of the previous step from t_{n-1} to t_n is denoted by $x_n \in \mathbb{R}^m$. This vector is used as a reference when assessing the relative accuracy. The tolerances tolf and tolx can be chosen by the user. Default values are $\mathsf{tolf} = \mathsf{tolx} = \frac{1}{10}$.

It remains to chose an appropriate initial guess X_i^0 for the Newton iteration. Many different approaches are discussed in [93]. These ideas use quantities based on the Nordsieck vector in order to predict a stage value. In case of the DAE (11.2) this leads to an approximation $Q_i^0 \approx q\big(x(t_n + c_i h), t_n + c_i h\big)$, but it was mentioned earlier that X_i^0 cannot be computed from Q_i^0 in general. A similar remark applies when using explicit methods of order p to compute a prediction. Hence a different approach is required.

Extensive numerical tests have shown that extrapolation based on previous stage values gives satisfactory results. Let $\bar{X}$ be a queue of κ past (accepted) stages and corresponding timepoints, i.e.

$$\bar{X} = \big[(\bar{X}_1, \bar{t}_1), (\bar{X}_2, \bar{t}_2), \dots, (\bar{X}_\kappa, \bar{t}_\kappa)\big],$$

and let P denote the unique polynomial satisfying $P(\bar{t}_j) = \bar{X}_j$ for $j = 1, \dots, \kappa$ and $P(t_n + c_j h) = X_j$ for $j = 1, \dots, i-1$. A prediction X_i^0 is found by evaluating $X_i^0 = P(t_n + c_i h)$. After completing step number n, the κ most recent values from the list

$$\big[(\bar{X}_1, \bar{t}_1), (\bar{X}_2, \bar{t}_2), \dots, (\bar{X}_\kappa, \bar{t}_\kappa), (X_1, t_n + c_1 h), \dots, (X_s, t_n + c_s h)\big]$$

are saved in $\bar{X}$ for use in step number $n + 1$. For extrapolation Neville's algorithm from [130] is used. Obviously, the development of more sophisticated stage predictors for DAEs should be a topic for further research.

From the stage values X_i the corresponding stage derivatives Q_i' can be computed using (11.4), i.e.

$$Q_i' = \tfrac{1}{h\lambda}\big(q(X_i, t_n + c_i h) - \omega_i\big).$$

Finally the output vector $q^{[n+1]}$ of step number n is obtained from (11.3c).

11.2 Error Estimation and Stepsize Prediction

After computing the output vector at the end of the step we have to decide whether to accept or reject the result. This decision is based on an estimation of the local truncation error. Appropriate estimators have been developed in [26, 37, 40] in the context of stiff ODEs (11.1). For completeness the key points will be briefly reviewed here.

We assume that $\mathcal{M} = [\mathcal{A}, \mathcal{U}, \mathcal{B}, \mathcal{V}]$ is a general linear method in Nordsieck form with s internal and r external stages. The stage order $q = p$ agrees with the order. Recall that $\mathcal{M}$ is applied to the ODE (11.1) according to

$$Y = h\,\mathcal{A}\,Y' + \mathcal{U}\,y^{[n]},$$
$$y^{[n+1]} = h\,\mathcal{B}\,Y' + \mathcal{V}\,y^{[n]}$$

with $Y_i' = f(Y_i, t_n + c_i h)$ being the stage derivatives.

The exact Nordsieck vector will be denoted by

$$
\hat{y}^{[n]} = \begin{bmatrix} y(t_n) \\ h\,y'(t_n) \\ \vdots \\ h^{r-1}y^{(r-1)}(t_n) \end{bmatrix}.
$$

It is assumed that the input vector $y^{[n]}$ used for numerical computations comes with a first order error term $\beta\,h^{p+1}y^{(p+1)}(t_n)$, i.e

$$
y_i^{[n]} = \hat{y}_i^{[n]} - \beta_i\,h^{p+1}y^{(p+1)}(t_n) + \mathcal{O}(h^{p+2}) \tag{11.6}
$$

for $i = 1, \dots, r$. As we are interested in the local error, the first component is assumed to be exact, i.e. $\beta_1 = 0$. As in [26, 37], using Taylor series expansion yields

$$
Y_i = y(t_n + c_i h) - \varepsilon_i\,h^{p+1}y^{(p+1)}(t_n) \qquad + \mathcal{O}(h^{p+2}), \tag{11.7a}
$$
$$
h\,Y_i' = h\,y'(t_n + c_i h) + \varepsilon_i\,\tfrac{\partial f}{\partial y}\big(y(t_n)\big)\,h^{p+2}y^{(p+1)}(t_n)\mathcal{O}(h^{p+3}), \tag{11.7b}
$$
$$
y_i^{[n+1]} = \hat{y}_i^{[n+1]} \qquad - \gamma_i\,h^{p+1}y^{(p+1)}(t_{n+1}) \qquad + \mathcal{O}(h^{p+2}), \tag{11.7c}
$$

The coefficients ε and γ are given by

$$
\varepsilon = \tfrac{1}{(p+1)!}\,c^{p+1} - \tfrac{1}{p!}\,\mathcal{A}\,c^p + \mathcal{U}\,\beta,
$$
$$
\gamma = \alpha(p,r) - \tfrac{1}{p!}\,\mathcal{B}\,c^p + \mathcal{V}\,\beta,
$$
$$
c^k = \begin{bmatrix} c_1^k \\ c_2^k \\ \vdots \\ c_s^k \end{bmatrix}, \quad \alpha(k,l) = \begin{bmatrix} \tfrac{1}{(k+1)!} \\ \tfrac{1}{k!} \\ \vdots \\ \tfrac{1}{(k+2-l)!} \end{bmatrix}.
$$

The matrix $\mathcal{V}$ plays a crucial role in propagating the error. For methods with Runge-Kutta stability but also for those from Table 10.4 this matrix is given by $\mathcal{V} = \begin{bmatrix} 1 & v^\top \\ 0 & \tilde{\mathcal{V}} \end{bmatrix}$ where the submatrix $\tilde{\mathcal{V}}$ has zero spectral radius. Hence the errors in the first component of the Nordsieck vector are propagated quite differently from those in the remaining $r-1$ components. Indeed, writing

$$
\beta = \begin{bmatrix} 0 \\ \tilde{\beta} \end{bmatrix}, \qquad \mathcal{B} = \begin{bmatrix} b^\top \\ \tilde{\mathcal{B}} \end{bmatrix}, \qquad \alpha(p,r) = \begin{bmatrix} \tfrac{1}{(p+1)!} \\ \tilde{\alpha} \end{bmatrix},
$$

it turns out that

$$
\tilde{\beta} \mapsto \tilde{\alpha} - \tfrac{1}{p!}\,\tilde{\mathcal{B}}\,c^p + \tilde{\mathcal{V}}\,\tilde{\beta}
$$

is a fixed-point mapping. Thus, after a series of steps using a constant stepsize, the values $\tilde{\beta}$ will settle down to a fixed value that is not changed from step to step. This fixed point can be computed by solving

$$
\gamma = \beta + C_p e_1 \quad \Leftrightarrow \quad \big[\,e_1 e_1^\top + (I - \mathcal{V})\,\big] \begin{bmatrix} C_p \\ \tilde{\beta} \end{bmatrix} = \alpha(p,r) - \tfrac{1}{p!}\,\mathcal{B}\,c^p. \tag{11.8}
$$

If C_p and $\tilde{\beta}$ form a solution of (11.8), then input values

$$y_1^{[n]} = \hat{y}_1^{[n]}$$
$$y_i^{[n]} = \hat{y}_i^{[n]} \quad - \tilde{\beta}_i\, h^{p+1} y^{(p+1)}(t_n) \quad + \mathcal{O}(h^{p+2}), \qquad i = 2, \ldots, r,$$

imply that the output at the end of the step is given by

$$y_1^{[n+1]} = \hat{y}_1^{[n+1]} - C_p\, h^{p+1} y^{(p+1)}(t_{n+1}) + \mathcal{O}(h^{p+2}),$$
$$y_i^{[n+1]} = \hat{y}_i^{[n+1]} - \tilde{\beta}_i\, h^{p+1} y^{(p+1)}(t_{n+1}) + \mathcal{O}(h^{p+2}), \qquad i = 2, \ldots, r.$$

Thus, the number C_p is called the method's error constant and

$$\hat{y}_1^{[n+1]} - y_1^{[n+1]} = C_p\, h^{p+1} y^{(p+1)}(t_{n+1}) + \mathcal{O}(h^{p+2}) \tag{11.9}$$

is the local truncation error.

Example 11.1. Consider the order 2 method from Table 10.2 on page 192,

$$\mathcal{M} = \begin{bmatrix} \mathcal{A} & \mathcal{U} \\ \hline \mathcal{B} & \mathcal{V} \end{bmatrix} = \left[\begin{array}{ccc|ccc} \frac{2}{11} & 0 & 0 & 1 & \frac{3}{22} & -\frac{7}{968} \\ \frac{4}{11} & \frac{2}{11} & 0 & 1 & \frac{3}{22} & -\frac{7}{968} \\ \frac{135}{352} & \frac{105}{352} & \frac{2}{11} & 1 & \frac{3}{22} & -\frac{7}{968} \\ \frac{135}{352} & \frac{105}{352} & \frac{2}{11} & 1 & \frac{3}{22} & -\frac{7}{968} \\ -\frac{17}{120} & \frac{17}{56} & \frac{88}{105} & 0 & 0 & 0 \\ -\frac{77}{60} & -\frac{11}{28} & \frac{176}{105} & 0 & 0 & 0 \end{array} \right], \quad c = \begin{bmatrix} \frac{7}{22} \\ \frac{15}{22} \\ 1 \end{bmatrix}.$$

For this method $p = 2$, $s = r = 3$ implies $\alpha(2,3) = \begin{bmatrix} \frac{1}{6} & \frac{1}{2} & 1 \end{bmatrix}^{\top}$ such that the linear system (11.8) reads

$$\begin{bmatrix} 1 & -\frac{3}{22} & \frac{7}{968} \\ 0 & 1 & 0 \\ 0 & 0 & 1 \end{bmatrix} \begin{bmatrix} C_2 \\ \beta_2 \\ \beta_3 \end{bmatrix} = \begin{bmatrix} -\frac{415}{31944} \\ \frac{17}{968} \\ \frac{7}{22} \end{bmatrix} \quad \Leftrightarrow \quad \begin{bmatrix} C_2 \\ \beta_2 \\ \beta_3 \end{bmatrix} = \begin{bmatrix} -\frac{103}{7986} \\ \frac{17}{968} \\ \frac{7}{22} \end{bmatrix}.$$

Hence the error constant is given by $C_2 = -\frac{103}{7986}$. $\qquad\qquad\square$

If a method $\mathcal{M}$ is used with a constant stepsize h, the above computations are valid for all n as the coefficients $\tilde{\beta}$ won't change from step to step. In general, however, an efficient numerical solution of any given problem requires an adaptive change of the stepsize in order to react on how the solution evolves.

If the current step is computed with stepsize h_n and another stepsize h_{n+1} will be used for the next step, the Nordsieck vector needs modification. This was discussed briefly in Section 10.2.

Let $\sigma_n = \frac{h_{n+1}}{h_n}$ be the stepsize ratio. Then the Nordsieck vector $y^{[n+1]}$ needs to be multiplied by the diagonal matrix $D(\sigma_n) = \mathrm{diag}(1, \sigma_n, \ldots, \sigma_n^{r-1})$ in order to provide the correct scaled derivatives for the next step.

This approach, however, distorts the coefficients β as can be seen from the following computation:

$$
y^{[n+1]} = \begin{bmatrix} y(t_{n+1}) & - & C_p\,h_n^{p+1}y^{(p+1)}(t_{n+1}) + \mathcal{O}(h_n^{p+2}) \\ h_n\;\; y'(t_{n+1}) & - & \beta_2\,h_n^{p+1}y^{(p+1)}(t_{n+1}) + \mathcal{O}(h_n^{p+2}) \\ \vdots & & \\ h_n^{r-1}\;\; y^{(r-1)}(t_{n+1}) - & & \beta_r\,h_n^{p+1}y^{(p+1)}(t_{n+1}) + \mathcal{O}(h_n^{p+2}) \end{bmatrix},
$$

$$
D(\sigma_n)\,y^{[n+1]} = \begin{bmatrix} y(t_{n+1}) & - & C_p\,h_n^{p+1}y^{(p+1)}(t_{n+1}) + \mathcal{O}(h_{n+1}^{p+2}) \\ h_{n+1}\;\; y'(t_{n+1}) & - & \sigma_n\beta_2\,h_n^{p+1}y^{(p+1)}(t_{n+1}) + \mathcal{O}(h_{n+1}^{p+2}) \\ \vdots & & \\ h_{n+1}^{r-1}\;\; y^{(r-1)}(t_{n+1}) - & & \sigma_n^{r-1}\beta_r\,h_n^{p+1}y^{(p+1)}(t_{n+1}) + \mathcal{O}(h_{n+1}^{p+2}) \end{bmatrix}.
$$

The desired output, by contrast, reads

$$
\begin{aligned}
\bar{y}_i^{[n+1]} &= h_{n+1}^{i-1}y^{(i-1)}(t_{n+1}) - \quad \beta_i\,h_{n+1}^{p+1}y^{(p+1)}(t_{n+1}) + \mathcal{O}(h_{n+1}^{p+2}) \\
&= h_{n+1}^{i-1}y^{(i-1)}(t_{n+1}) - \sigma_n^{p+1}\beta_i\,h_n^{p+1}y^{(p+1)}(t_{n+1}) + \mathcal{O}(h_{n+1}^{p+2}).
\end{aligned}
$$

Hence the appropriate modification of the Nordsieck vector has to be twofold. First $y^{[n+1]}$ needs to be scaled with the stepsize ratio and then the correction $(\sigma_n^{i-1} - \sigma_n^{p+1})\beta_i\,h_n^{p+1}y^{(p+1)}(t_{n+1})$ has to be added. In other words, $\bar{y}_i^{[n+1]}$ can be computed as

$$
\bar{y}_i^{[n+1]} = \sigma_n^{i-1}\,y^{[n+1]} + (\sigma_n^{i-1} - \sigma_n^{p+1})\beta_i\,h_n^{p+1}y^{(p+1)}(t_{n+1}). \tag{11.10}
$$

This approach is called 'scale and modify' [26, 40]. It ensures that the coefficients $\tilde{\beta}$ are maintained correctly also in case of a variable-stepsize implementation.

In order to apply the scale and modify technique, an estimate

$$
\mathrm{scdrv}_{p+1}(t_{n+1}) \approx h^{p+1}y^{(p+1)}(t_{n+1})
$$

for the scaled derivative is required. This value will not only be used for (11.10), but also in order to estimate the local truncation error (11.9) via

$$
\mathrm{est}_{p+1}(t_{n+1}) = C_p\,\mathrm{scdrv}_{p+1}(t_{n+1}).
$$

The error constant C_p can be computed from the method's coefficients as a solution of the linear system (11.8).

Recall from (11.7) that due to the method's high stage order the (scaled) stage derivatives $h\,Y_i'$ are exact to within $\mathcal{O}(h^{p+2})$. Thus, $\mathrm{scdrv}_{p+1}(t_{n+1})$ will be computed as a linear combination of stage derivatives.

In the previous chapters we restricted attention to stiffly accurate methods. There the first component of the input vector $y_1^{[n]} = \bar{Y}_s$ is computed from the last stage $\bar{Y}_s$ of the previous step. The corresponding scaled derivative $h\bar{Y}_s'$ is thus available for further exploitation as well. Since in case of local error estimation $y_1^{[n]}$ is assumed to be exact, $h\bar{Y}_s'$ constitutes the exact scaled derivative.

For methods having the FSAL property, or Property F as it is called in [26], $h\bar{Y}_s'$ is passed on to the next step as the second component of the Nordsieck vector (see page 184). For methods with $y_2^{[n]} \neq h\bar{Y}_s'$ the quantity $h\bar{Y}_s'$ can be made available using an appropriate implementation.

As in [40] we consider the estimation of $h^{p+1}y^{(p+1)}$ using a linear combination

$$\mathrm{scdrv}_{p+1}(\delta_0, \delta, t_{n+1}) = \delta_0\, h\, \bar{Y}_s' + \sum_{i=1}^{s} \delta_i\, h\, Y_i'. \tag{11.11}$$

Expanding the right hand side of (11.11) into a Taylor series leads to

$$\begin{aligned}
\mathrm{scdrv}_{p+1}&(\delta_0, \delta, t_{n+1}) \\
&= \delta_0\, h\, y'(t_n) + \sum_{i=1}^{s} \delta_i\, h\, y'(t_n + c_i h) \\
&\qquad\qquad - \delta^\top \varepsilon \frac{\partial f}{\partial y} h^{p+2} y^{(p+1)}(t_n) + \mathcal{O}(h^{p+3}) \\
&= \delta_0\, h\, y'(t_n) + \delta^\top \hat{C}\, Z_n - \delta^\top \varepsilon \frac{\partial f}{\partial y} h^{p+2} y^{(p+1)}(t_n) + \mathcal{O}(h^{p+3})
\end{aligned} \tag{11.12}$$

with

$$\delta = \begin{bmatrix} \delta_1 \\ \vdots \\ \delta_s \end{bmatrix}, \quad
\hat{C} = \begin{bmatrix} 1 & \frac{c_1}{1!} & \cdots & \frac{c_1^{p+1}}{(p+1)!} \\ \vdots & \vdots & \ddots & \vdots \\ 1 & \frac{c_s}{1!} & \cdots & \frac{c_s^{p+1}}{(p+1)!} \end{bmatrix} = \begin{bmatrix} C & \frac{c^{p+1}}{(p+1)!} \end{bmatrix}, \quad
Z_n = \begin{bmatrix} h\, y'(t_n) \\ \vdots \\ h^{p+2} y^{(p+2)}(t_n) \end{bmatrix}.$$

The matrix C was introduced in Theorem 7.6 and extensively used in the previous Chapter. Compared to C, the matrix $\hat{C}$ uses an additional column. Observe that the multiplication $\delta^\top \hat{C}\, Z_n = [(\delta^\top \hat{C}) \otimes I_m]\, Z_n$ involves Kronecker products.

For (11.11) to be an estimator satisfying

$$\begin{aligned}
\mathrm{scdrv}_{p+1}(\delta_0, \delta, t_{n+1}) &= h^{p+1} y^{(p+1)}(t_{n+1}) + \mathcal{O}(h^{p+2}) \\
&= h^{p+1} y^{(p+1)}(t_n) \quad + \mathcal{O}(h^{p+2})
\end{aligned}$$

we need to solve the linear system

$$\begin{bmatrix} \delta_0 & \delta_1 & \cdots & \delta_s \end{bmatrix}
\begin{bmatrix} 1 & 0 & \cdots & 0 \\ 1 & \frac{c_1}{1!} & \cdots & \frac{c_1^p}{p!} \\ \vdots & \vdots & \ddots & \vdots \\ 1 & \frac{c_s}{1!} & \cdots & \frac{c_s^p}{p!} \end{bmatrix}
= \begin{bmatrix} 0 & \cdots & 0 & 1 \end{bmatrix} \;\Leftrightarrow\; \begin{bmatrix} \delta_0 \\ \delta \end{bmatrix}^\top \cdot \begin{bmatrix} e_1^\top \\ C \end{bmatrix} = e_{p+1}^\top. \tag{11.13}$$

This system consists of $p + 1$ equations in $s + 1$ unknowns. Observe that the first p equations guarantee that $h^k y^{(k)}(t_n)$ does not appear in the Taylor series expansion (11.12) for $k = 1, \ldots, p$. The last equation ensures coefficient 1 for $h^{p+1} y^{(p+1)}(t_n)$ such that the correct scaled derivative is approximated.

For non-confluent methods with $s = p$, the solution is uniquely defined. In this case the coefficient matrix is a Vandermonde matrix and hence nonsingular. The estimators computed for the methods from Table 10.4 are presented in Table 11.1.

Example 11.2. In order to verify that the estimators derived for methods with $s = r = p = q$ are highly accurate, we want to solve the problem

$$y' = -\tfrac{1}{10}\big(y - g(t)\big) + g'(t), \qquad t \in [0, 20]. \tag{11.14}$$

This equation introduced by Prothero and Robinson [131] was already discussed in Example 2.7. Here, the complex function $g(t) = \exp(x\mathrm{i})$ with $\mathrm{i}^2 = -1$ is used. The exact solution is given by $y(t) = g(t)$. Similar to [38, 40] the periodic stepsize pattern

$$h_{n+1} = \rho^{(-1)^{k(n)} \sin(8\pi t_n/20) \cos(2\pi\, t_n/20)}\, h_n \tag{11.15}$$

is used. The choice $k(n) = 1 - \lfloor \frac{n \bmod 4}{2} \rfloor$ ensures that the stepsize is successively increased twice and then decreased twice such that h_n oscillates around the initial stepsize $h_0 = \frac{20}{N}$, $N = 800$. When changing the stepsize, the scale and modify technique is used.

The integration is started with order $p = 1$. After that a random order change is allowed at every fourth step. More precisely, the following random-variable order strategy is employed,

$$p_{n+1} = \begin{cases} p_n & , n \bmod 4 \neq 0 \\ p_n + 1 & , n \bmod 4 = 0, \ p_n = 2 \text{ and } z \in [\tfrac{2}{3}, 1] \\ p_n + 1 & , n \bmod 4 = 0, \ p_n = 1 \text{ and } z \in [\tfrac{1}{2}, 1] \\ p_n - 1 & , n \bmod 4 = 0, \ p_n = 2 \text{ and } z \in [0, \tfrac{1}{3}] \\ p_n - 1 & , n \bmod 4 = 0, \ p_n = 3 \text{ and } z \in [0, \tfrac{1}{2}] \end{cases}$$

order	δ_0	δ_1	δ_2	δ_3
$p = 1$	-1	1		
$p = 2$	$2 + \sqrt{2}$	$-4 - 3\sqrt{2}$	$2 + 2\sqrt{2}$	
$p = 3$	$-\frac{4325}{242}$	$\frac{2703125}{24674}$	$-\frac{388328775}{2985554}$	$\frac{12975}{338}$

Table 11.1: Error estimators for the methods from Table 10.4 $(s = r = p = q)$ constructed in Chapter 10.

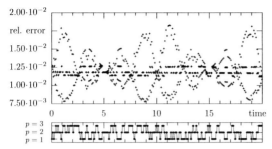

Figure 11.1: Accuracy of the estimators from Table 11.1 for the problem (11.14) using fixed-variable steps (11.15) and random order.

where z is a random number uniformly distributed in $[0, 1]$. Thus, each possibility – changing the order up- or downwards and keeping the same order – has the same probability.

At every timepoint t_n the accuracy of the estimate $\mathrm{scdrv}_{p+1}(t_n)$ is measured by computing the relative error

$$\left| \frac{\mathrm{scdrv}_{p+1}(t_n) - h_n^{p+1} y^{(p+1)}(t_n)}{h_n^{p+1} y^{(p+1)}(t_n)} \right|.$$

This relative error together with the order history is plotted in Figure 11.1.

In spite of the demanding stepsize and order variations used here the estimators perform quite satisfactory. The estimate yields approximately two correct digits. □

Recall that the schemes discussed in the previous example use only $s = r = p$ stages and the estimators in Table 11.1 are uniquely defined by the linear system (11.13). Methods with $s = r = p + 1 = q + 1$ leave more freedom to derive optimised error estimators.

Example 11.3. For illustration consider again the order 2 method from Example 11.1. In order to determine an error estimator (δ_0, δ), the linear system (11.13) needs to be solved, i.e.

$$\begin{bmatrix} \delta_0 & \delta_1 & \delta_2 & \delta_3 \end{bmatrix} \begin{bmatrix} 1 & 0 & 0 \\ 1 & \frac{7}{22} & \frac{49}{968} \\ 1 & \frac{15}{22} & \frac{225}{968} \\ 1 & 1 & \frac{1}{2} \end{bmatrix} = \begin{bmatrix} 0 & 0 & 1 \end{bmatrix}.$$

Obviously the solution is not uniquely defined. One solution is given by the vector

$$\begin{bmatrix} \delta_0 & \delta_1 & \delta_2 & \delta_3 \end{bmatrix} = \begin{bmatrix} \frac{121}{15} & -\frac{847}{60} & \frac{2057}{420} & \frac{121}{105} \end{bmatrix}.$$

This particular solution has advantages over other choices in the sense that $\delta^\top \varepsilon = 0$ holds for

$$\varepsilon = \tfrac{1}{(p+1)!}\, c^{p+1} - \tfrac{1}{p!}\, \mathcal{A}\, c^p + \mathcal{U}\, \beta = \begin{bmatrix} -\tfrac{239}{63888} & -\tfrac{15}{1936} & -\tfrac{103}{7986} \end{bmatrix}^\top.$$

Recall that ε is related to the first order error term of the stages as indicated in (11.7). Satisfying the additional requirement $\delta^\top \varepsilon = 0$ allows to construct estimators of higher accuracy. $\qquad\square$

For general linear methods with $s = r = p + 1 = q + 1$ assume that the (δ_0, δ) is obtained from solving the extended system

$$\begin{bmatrix} \delta_0 & \delta^\top \end{bmatrix} \cdot \begin{bmatrix} e_1^\top & 0 \\ C & \varepsilon \end{bmatrix} = \begin{bmatrix} e_{p+1}^\top & 0 \end{bmatrix}. \tag{11.16}$$

The estimator (11.12) is thus seen to satisfy

$$\mathrm{scdrv}_{p+1}(\delta_0, \delta, t_{n+1}) = h_n^{p+1} y^{(p+1)}(t_n) + \theta\, h_n^{p+2} y^{(p+2)}(t_n) + \mathcal{O}(h_n^{p+3})$$

$$= h_n^{p+1} y^{(p+1)}(t_n + \theta h_n) + \mathcal{O}(h_n^{p+3})$$

for $\theta = \tfrac{1}{(p+1)!}\, \delta^\top c^{p+1}$. The scaled derivative $h_n^{p+1} y^{(p+1)}(t_n + \theta h_n)$ can therefore be estimated with much higher accuracy. The particular values of δ_0, δ and θ corresponding to the methods constructed in the previous chapter are given in Table 11.2 and 11.3.

The increased accuracy of $\mathrm{scdrv}_{p+1}(\delta_0, \delta, t_{n+1})$ offers the possibility to estimate the $(p + 2)$-nd derivative as well. The details of this construction are described in [40]. Since estimating the local truncation error of the method with order

order	δ_0	δ_1	δ_2	δ_3	δ_4	θ
$p = 1$	$-\tfrac{5}{6}$	$-\tfrac{5}{21}$	$\tfrac{15}{14}$			$\tfrac{21}{40}$
$p = 2$	$\tfrac{3906}{629}$	$-\tfrac{6057}{629}$	$\tfrac{396}{629}$	$\tfrac{1755}{629}$		$\tfrac{824}{1887}$
$p = 3$	$-\tfrac{53262347456}{1439714429}$	$\tfrac{120907666368}{1439714429}$	$-\tfrac{43148914368}{1439714429}$	$-\tfrac{63375780544}{1439714429}$	$\tfrac{38879376000}{1439714429}$	$\tfrac{5534123787}{11517715432}$

Table 11.2: Error estimators for the methods with inherent Runge-Kutta stability from Table 10.1

order	δ_0	δ_1	δ_2	δ_3	δ_4	θ
$p = 1$	$-\tfrac{5}{6}$	$-\tfrac{5}{21}$	$\tfrac{15}{14}$			$\tfrac{21}{40}$
$p = 2$	$\tfrac{121}{15}$	$-\tfrac{847}{60}$	$\tfrac{2057}{420}$	$\tfrac{121}{105}$		$\tfrac{3}{8}$
$p = 3$	-5.19608	-43.2157	160.823	-171.216	58.8039	0.604703

Table 11.3: Error estimators for the methods with $M_{\infty,p} = 0$ from Table 10.2

$p + 1$ is seminal for a variable-order implementation, the main ideas will be briefly reviewed.

Denote $\eta_n = \mathrm{scdrv}_{p+1}(\delta_0, \delta, t_{n+1})$ and $\sigma_{n-1} = \frac{h_n}{h_{n-1}}$ such that

$$\eta_n = h_n^{p+1} y^{(p+1)}(t_n + \theta h_n) + \mathcal{O}(h_n^{p+3})$$
$$= h_n^{p+1} y^{(p+1)}(t_n) + \theta\, h_n^{p+2} y^{(p+2)}(t_n) + \mathcal{O}(h_n^{p+3})$$
$$\eta_{n-1} = h_{n-1}^{p+1} y^{(p+1)}(t_{n-1} + \theta h_{n-1}) + \mathcal{O}(h_{n-1}^{p+3})$$
$$= \frac{1}{\sigma_{n-1}^{p+1}} h_n^{p+1} y^{(p+1)}\Big(t_n + (\frac{\theta}{\sigma_{n-1}} - 1)h_n\Big) + \mathcal{O}(h_n^{p+3})$$
$$= \frac{1}{\sigma_{n-1}^{p+1}} h_n^{p+1} y^{(p+1)}(t_n) + \frac{1}{\sigma_{n-1}^{p+1}}(\frac{\theta}{\sigma_{n-1}} - 1)h_n^{p+2} y^{(p+2)}(t_n) + \mathcal{O}(h_n^{p+3}).$$

The aim is to find coefficients μ and λ such that

$$\mu\,\eta_n + \lambda\,\eta_{n-1} = h_n^{p+2} y^{(p+2)}(t_{n+1}) + \mathcal{O}(h_n^{p+3}) = h_n^{p+2} y^{(p+2)}(t_n) + \mathcal{O}(h_n^{p+3}).$$

This goal can be achieved solving the linear system

$$\begin{bmatrix} 1 & \frac{1}{\sigma_{n-1}^{p+1}} \\ \theta & \frac{1}{\sigma_{n-1}^{p+1}}(\frac{\theta}{\sigma_{n-1}} - 1) \end{bmatrix} \begin{bmatrix} \mu \\ \lambda \end{bmatrix} = \begin{bmatrix} 0 \\ 1 \end{bmatrix} \Leftrightarrow \begin{bmatrix} \sigma_{n-1}^{p+1} & 1 \\ \theta\,\sigma_{n-1}^{p+2} & \theta - \sigma_{n-1} \end{bmatrix} \begin{bmatrix} \mu \\ \lambda \end{bmatrix} = \begin{bmatrix} 0 \\ \sigma_{n-1}^{p+2} \end{bmatrix}.$$

Hence an estimator for the $(p + 2)$-nd scaled derivative is given by

$$\mathrm{scdrv}_{p+2}(t_{n+1}) = \frac{\sigma_{n-1}}{\theta\,\sigma_{n-1} - \theta + \sigma_{n-1}}\Big(\eta_n - \sigma_{n-1}^{p+1}\eta_{n-2}\Big)$$
$$= h_n^{p+2} y^{(p+2)}(t_{n+1}) + \mathcal{O}(h_n^{p+3}).$$

Observe that this estimator uses information from the current step and from the step before.

Example 11.4. The estimators for $s = r = p + 1 = q + 1$ were tested using the same numerical experiment as in Example 11.2. Figure 11.2 displays the relative error

$$\left| \frac{\mathrm{scdrv}_{p+1}(t_{n+1}) - h_n^{p+1} y^{(p+1)}(t_n + \theta h_n)}{h_n^{p+1} y^{(p+1)}(t_n + \theta h_n)} \right|$$

and

$$\left| \frac{\mathrm{scdrv}_{p+2}(t_{n+1}) - h_n^{p+2} y^{(p+1)}(t_{n+1})}{h_n^{p+2} y^{(p+2)}(t_{n+1})} \right|$$

for estimating the $(p + 1)$-st and $(p + 2)$-nd scaled derivative, respectively.

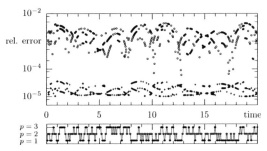

Figure 11.2: Accuracy of the estimators from Table 11.2 (methods with in-herent Runge-Kutta stability, Table 10.1) for the problem (11.14) using fixed-variable steps (11.15) and random order. The relative error in $\mathrm{scdrv}_{p+1}(t_n)$ (+) and $\mathrm{scdrv}_{p+2}(t_n)$ (o) is displayed.

The relative error in the estimator $\mathrm{scdrv}_{p+2}(t_n)$ is displayed only after three consecutive steps with the same order. This allows a settling of the leading error term after changing the order.

Figure 11.2 indeed indicates very high accuracy for these estimators. In case of the scaled $(p+1)$-st derivative, $\mathrm{scdrv}_{p+1}(t_n)$ yields approximately 5 correct digits while the estimator $\mathrm{scdrv}_{p+2}(t_n)$ for the scaled $(p+2)$-nd derivative offers 2–3 significant digits.   □

Using the estimators for scaled derivatives as described above, the local trun-cation error (11.9) can be estimated using

$$\mathrm{est}_{p+1}(t_{n+1}) = C_p \, \mathrm{scdrv}_{p+1}(t_{n+1}) = \hat{q}_1^{[n+1]} - q_1^{[n+1]} + \mathcal{O}(h_n^{p+2}).$$

The error constant C_p is computed from (11.8). Recall that for properly stated DAEs of the form (11.2) the Nordsieck vector $q^{[n]}$ approximates scaled deriv-atives of $q\big(x(t_n), t_n\big) \in \mathbb{R}^l$ such that the estimate satisfies $\mathrm{est}_{p+1}(t_{n+1}) \in \mathbb{R}^l$ as well. The current step is accepted provided that the inequality

$$\| \, \mathrm{est}_{p+1}(t_{n+1}) \|_{scq} \leq 1$$

holds. The scaled norm

$$\|q\|_{scq} = \max_{j=1}^{l} \Big(\, |q_j| \, / \, \big(\mathrm{atolq}_j + \mathrm{rtolq}_j \cdot |q_{n,j}|\big) \Big)$$

depends on absolute and relative tolerances atolq, rtolq for q. These tolerances can be computed from the values atol, rtol that are prescribed by the user (see also Chapter 12). The numerical approximation for $q(x_n, t_n)$ obtained in the previous step is denoted by q_n.

The number $\|\operatorname{est}_{p+1}(t_{n+1})\|_{scq}$ is used to predict a stepsize h_{n+1} for the next step from t_{n+1} to t_{n+2} as well. Recall that

$$\operatorname{est}_{p+1}(t_{n+1}) = C_p \, h_n^{p+1} q^{(p+1)}\big(x(t_{n+1}), t_{n+1}\big) + \mathcal{O}(h_n^{p+2})$$

and similarly we will have

$$\operatorname{est}_{p+1}(t_{n+2}) = C_p \, h_{n+1}^{p+1} q^{(p+1)}\big(x(t_{n+2}), t_{n+2}\big) + \mathcal{O}(h_{n+1}^{p+2})$$

Since optimal performance is realised if the local error coincides with the required tolerance, we will try to choose h_{n+1} such that $\|\operatorname{est}_{p+1}(t_{n+2})\|_{scq} \approx 1$. Assuming that q does not vary too much over the next step, this can be achieved requiring

$$\left(\frac{h_n}{h_{n+1}}\right)^{p+1} = \frac{\|\operatorname{est}_{p+1}(t_{n+1})\|_{scq}}{1},$$

or, equivalently,

$$h_{n+1} = h_n \left(\frac{1}{\|\operatorname{est}_{p+1}(t_{n+1})\|_{scq}}\right)^{p+1}. \tag{11.17}$$

In spite of its simplicity, this standard stepsize controller (11.17) is quite reliable for many problems. The GLM codes described in [93, 94] use this technique. A slightly modified version is used in [40]. The controller (11.17) will also be adopted as standard for the codes GLIMDA and GLIMDA++. As usual, safety factors will be employed to avoid excessively large variations in the stepsize which might destroy stability properties of the methods [85, 93]. Observe that in case of rejected steps, $\|\operatorname{est}_{p+1}(t_{n+1})\|_{scq} > 1$, the controller (11.17) decreases the stepsize.

More sophisticated controllers can be constructed assuming e.g. $\frac{K_{n+1}}{K_n} = \frac{K_n}{K_{n-1}}$ where $K_n = q^{(p+1)}\big(x(t_{n+1}), t_{n+1}\big)$. This leads to the formula

$$h_{n+1} = h_n \left(\frac{1}{\|\operatorname{est}_{p+1}(t_{n+1})\|_{scq}}\right)^{p+1} \frac{h_n}{h_{n-1}} \left(\frac{\|\operatorname{est}_{p+1}(t_n)\|_{scq}}{\|\operatorname{est}_{p+1}(t_{n+1})\|_{scq}}\right)^{p+1}.$$

A control theoretic study of similar estimators can be found in [148, 149].

11.3 Changing the Order

With the error estimator and stepsize control techniques described in the previous section, a variable-stepsize implementation of general linear methods is already possible. It remains to develop a strategy for choosing the appropriate order in an adaptive fashion.

Assume that we are currently using a method with $s = r = p + 1 = q + 1$. In this case, $\text{scdrv}_{p+2}(t_{n+1})$ can be used to estimate the local truncation error

$$\text{est}_{p+2}(t_{n+1}) = C_{p+1}\,\text{scdrv}_{p+2}(t_{n+1})$$

for the method with order $p + 1$. The corresponding estimate

$$\text{est}_p(t_{n+1}) = C_{p-1}q_r^{[n+1]}$$

for the method with order $p - 1$ is trivial to obtain as the last component of the Nordsieck vector approximates the scaled derivative $h^p q^{(p)}\big(x(t_{n+1}), t_{n+1}\big)$.

Thus a stepsize prediction can be computed not only for the method of order p but also for its adjacent methods with order $p + 1$ and $p - 1$, respectively.

The decision on what order should be used for the next step is based on these stepsize predictions $h_{n+1,p-1}$, $h_{n+1,p}$ and $h_{n+1,p+1}$. As methods of different order use a different number of stages, the computational costs per step have to be taken into account as well.

In [40] it is suggested to choose the new order by selecting the maximum of

$$\frac{s-2}{(s-1)^2}\,h_{n+1,p-1}, \qquad \alpha\,\frac{s-1}{s^2}\,h_{n+1,p}, \qquad \frac{s}{(s+1)^2}\,h_{n+1,p+1}.$$

The safety factor $\alpha = 1.2$ is used to prevent frequent order changes.

This strategy is employed for GLIMDA++ as well. Similar to [40] the order is allowed to change only in case of a settled stepsize, i.e. $\left|\frac{h_{n+1,p}-h_{n,p}}{h_{n,p}}\right| < 0.2$.

Example 11.5. This order selection strategy will be tested using the Miller Integrator from Example 1.1. Recall from Chapter 1 that the Miller Integrator circuit leads to an index-1 equation with $m = 5$, $l = 1$. The input signal $v(t) = \sin(\omega\,t^2)$ is used with $\omega = 10^{14}$. The DAE is solved for $t \in [0, 7 \cdot 10^{-7}]$ using $x(0) = 0$ as initial value.

For many different tolerances $\mathsf{rtol} = 10^{-j}$, $j = 0, \ldots, 11$, $\mathsf{atol} = 10^{-3} \cdot \mathsf{rtol}$, Figure 11.3 shows the number of functions evaluations required for solving the problem together with the achieved accuracy, i.e. the relative error at the end

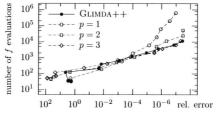

Figure 11.3: Testing the order selection strategy for the Miller Integrator using the methods from Table 10.1.

of the integration interval. The figure contains four curves. Three correspond to solving the Miller Integrator using constant order $p \in \{1, 2, 3\}$ while the fourth curve shows the results of the variable-order code GLIMDA++. For the computations the methods of Butcher and Podhaisky from Table 10.1 were used.

The order control described above shows almost optimal performance: For high accuracy demands high order methods are used while low order methods are chosen for loose tolerances.   □

This order selection strategy is very elegant and seems to work reliably. However, for methods with $s = r = p = q$ we cannot proceed in a similar fashion since no estimator $\mathrm{scdrv}_{p+2}(t_{n+2})$ is available for methods of this type. In order to overcome these difficulties there are at least three possible approaches.

(a) Information from previous steps could be used in order to compute the required estimate $\mathrm{scdrv}_{p+2}(t_{n+2})$. Keeping track of the backward informations is a complex task in a variable-stepsize variable-order setting and a related theory establishing the error expansions quickly becomes difficult to manage (see e.g. [29]).

(b) In [30, 94] the decision on a new order is based on monitoring the ratio $\rho = \frac{\mathrm{est}_{p+1}(t_{n+1})}{\mathrm{est}_p(t_{n+1})}$. If $\rho < \rho_{\min} = 0.9$ and $p < p_{\max}$, the order is increased. On the other hand, the order is decreased provided that $\rho > \rho_{\max} = 1.1$ and $p > p_{\min}$.

(c) Hairer and Wanner [86] discovered that high order methods perform poorly for loose tolerances due to a slowly converging Newton iteration. Hence, order control for their code RADAU is based on the convergence rate of Newton's method.

The approaches (b) and (c) have been tested. It was found that in case of (b) the order is built up quickly, but often it is not decreased sufficiently fast. Hence, most computations are performed using the method of highest order.

The approach (c), by contrast, resulted in more efficient and more reliable computations such that (c) was implemented for GLIMDA.

The technique of Hairer and Wanner is well documented in [85, 86]. It can be used for general linear methods without any modifications being necessary.

For $k \geq 1$ let $\theta_k = \|\Delta X_i^k\| / \|\Delta X_i^{k-1}\|$ denote the quotient of two consecutive Newton corrections in the iterative procedure (11.5). Based on θ_k the convergence rate is measured using

$$\psi_1 = \theta_1, \qquad \psi_k = \sqrt{\theta_m \cdot \theta_{k-1}}, \qquad k \geq 2.$$

For any given problem the computation is started using the lowest order $p = 1$. The order is allowed to change only after computing **same** p consecutive steps with the current order and if the stepsize has settled, i.e. $\left| \frac{h_{n+1} - h_n}{h_n} \right| < 0.2$.

In case of a possible order change, the order is increased provided that $\psi <$ upth and $p < p_{\max}$. Here, ψ denotes the final value ψ_k of the Newton iteration for the current step. Similarly the order is decreased if $\psi >$ downth and $p > p_{\min}$. Convergence failures give rise to an order decrease as well.

The parameters samep and upth, downth can be set by the user. A value samep = 5 is used by default. For the threshold values upth $= 10^{-3}$ and downth $= 0.8$ has been found to be appropriate.

Example 11.6. The order selection strategy is tested using the same numerical experiment as in Example 11.5.

For the Miller Integrator Figure 11.4 (a) shows the desired behaviour. For high accuracy demands the solution is computed using order $p = 3$ but for loose tolerances order $p = 1$ is used.

One of the test problems used most frequently for comparing integrators for stiff ODEs is given by the Robertson Reaction. This problem consists of a system of stiff ordinary differential equation modelling an autocatalytic chemical reaction. The reaction is described in [139]. The equations became popular in numerical analysis through the work of Hairer and Wanner. In [86] it is used for deriving and testing their order selection strategy.

Usually the equations are posed for $t \in [0, 10^{11}]$. Here, the tolerances rtol $= 10^{-j}$, $j = 2, 3, \ldots, 9$, atol $= 10^{-3} \cdot$ rtol have been used.

The results obtained for the code GLIMDA based on the general linear methods from Table 10.4 are presented in Figure 11.4 (b). Obviously the order 3 method is most efficient for almost all tolerances. This method is chosen correctly by the order control mechanism. □

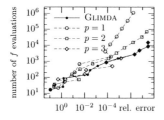
Figure 11.4 (a): Miller Integrator.

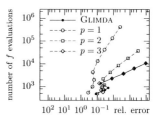
Figure 11.4 (b): Robertson Reaction.

Figure 11.4: Testing the order selection strategy for the methods from Table 10.4.

12

Numerical Experiments

Using techniques from the previous chapter the general linear methods from Table 10.1, 10.2 and 10.4 have been implemented using FORTRAN77. This programming language offers fast performance and high accuracy. Linear algebra can be handled conveniently using routines from BLAS and LAPACK [8, 109].

The resulting codes GLIMDA and GLIMDA++ are capable of solving index-2 differential algebraic equations

$$f\Big(\dot{q}\big(x(t),t\big),\, x(t),\, t \Big) = 0. \tag{12.1}$$

Notice that this formulation is more general than the structure

$$A(t)\big[d\big(x(t),t\big)\big]' + b\big(x(t),t\big) = 0 \tag{12.2}$$

covered by the results of this thesis.

Both GLIMDA and GLIMDA++ are variable-stepsize variable-order codes based on singly diagonally implicit general linear methods. The order p satisfies $1 \leq p \leq 3$ and the stage order $q = p$ coincides with the order. The quantities passed on from step to step are given by the Nordsieck vector

$$q^{[n]} \approx \begin{bmatrix} q\big(x(t_n),t_n\big) \\ h\, q'\big(x(t_n),t_n\big) \\ \vdots \\ h^{r-1} q^{(r-1)}\big(x(t_n),t_n\big) \end{bmatrix}.$$

Our aim is to compare the following methods:

- GLIMDA: $s = r = p$ stages are used. For $p = 1, 2$ the methods from Table 10.4 are A- and L-stable. The order 3 method is $A(\alpha)$-stable with $\alpha \approx 73.7535°$. Unfortunately there was no success in finding an A-stable order 3 method satisfying all requirements (P1)–(P6) from page 182. The method from Table 10.5 with $\alpha \approx 88.1°$ might be an alternative.

  As described in the previous chapter, order control is based on the convergence rate of Newton's method. Fast convergence signals a possible order increase. If the iteration process converges slowly or does not converge at all, the order is switched downwards.

- GLIMDA++: this solver uses an additional flag to distinguish between the two families of methods from Table 10.1 and 10.2. The former methods have been constructed with inherent Runge-Kutta stability while the latter family satisfies $M_{\infty,p} = 0$.

  In all cases $s = r = p + 1$ stages are used and the stage order is $q = p$. The methods are A- and L-stable. Order control is based on estimates of the quantities $h^{p+1} \frac{\mathrm{d}^{p+1}}{\mathrm{d}t^{p+1}} q\big(x(t), t\big)$ and $h^{p+2} \frac{\mathrm{d}^{p+2}}{\mathrm{d}t^{p+2}} q\big(x(t), t\big)$.

The codes GLIMDA and GLIMDA++ are available from the author.

Comparing different codes is a notoriously difficult task. The results of the comparison will clearly depend on the problem, on the prescribed accuracy or on the machine that runs the tests.

Since we are studying general linear methods for integrated circuit design, this chapter focuses on applications from electrical circuit engineering. Thus most problems are realistic circuits. The Test Set for Initial Value Problem Solvers [124] of Bari University (formerly maintained by CWI Amsterdam) offers a wealth of suitable test problems.

In case of electrical circuits, the vector x of the DAE (12.1) contains node potentials as well as the currents through inductors and voltage sources. The vector $q(x, t)$ comprises charges and fluxes. Observe that $x \in \mathbb{R}^m$ and $q(x, t) \in \mathbb{R}^n$ might be of different size. We have seen in Chapter 11 that error estimation is most naturally based on q rather than on x. The user, however, will prescribe a desired tolerance for x, such that this tolerance needs modification in order to be applicable to charges and fluxes. The computation of an appropriate tolerance is performed automatically based on the matrix $\frac{\partial q}{\partial x}(x, t)$. BDF schemes used for electrical circuit simulation give rise to similar difficulties. More details can be found in [147].

In order to achieve a comparison as fair as possible, each code has to solve every test problem for many different tolerances. The results are visualised in work-precision diagrams showing the relation between the required work and the achieved accuracy. A code performs optimal if high accuracy is achieved investing little work.

It is difficult to find a reasonable measure for the work necessary to solve a given problem. The easiest option is to measure the time required by the solver. This approach, however, is likely to give false results as the processor of the test machine might be busy with other tasks at the same time. Hence the results will often be un-reproducible.

For electrical circuit simulation most time is spent evaluating the complex transistor models and setting up the equations. This process called 'loading' requires approximately 85% of the work for standard applications with up to 10^3 devices [66]. Thus a fair comparison is possible when counting the number of function evaluations.

As one code may update the Jacobian much more frequently than another, the number of Jacobian evaluations has to be taken into account as well. Extensive tests using the simulator TITAN of Infineon Technologies AG have shown that for electrical circuit simulation the Jacobian can be obtained rather cheaply. Evaluating the Jacobian *together* with a function call amounts to approximately 1.4 to 1.5 times the work of a *single* function evaluation [64, 82]. To be on the safe side, each Jacobian evaluation will be counted as an additional function call.

The accuracy is measured using scd numbers. The scd value denotes the minimum number of significant correct digits in the numerical solution at the end of the integration interval, i.e

$$\mathsf{scd} = -\log_{10}\left(\|\text{relative error at the end of the integration interval}\|_\infty\right).$$

We will first use GLIMDA++ to compare the two families from Table 10.1 and 10.2. Here we want to find out whether there is any benefit in using methods with $M_{\infty,p} = 0$. Afterwards GLIMDA++ will be judged against GLIMDA. It will turn out that the methods with only $s = p$ stages indeed show a better performance.

Two of the most successful linear multistep and Runge-Kutta codes are DASSL and RADAU. Hence, to assess the potential of general linear methods for realistic applications we have to compare GLIMDA to these highly tuned codes. Additionally the code MEBDFI by Cash [45] will be used for comparison in Section 12.3. MEBDFI is based on modified BDF formulae with improved stability properties. We will see that GLIMDA is indeed competitive. In particular for many electrical circuits it will often be superior to classical codes.

All computations were performed on an Intel® Pentium® M processor with 1.4 GHz and 512 MB RAM running SuSE Linux 9.1 [150]. The GNU FORTRAN compiler was used for compilation and linking. Each code was used with its corresponding standard options. Except for adjusting the tolerances and the initial stepsize there was no tweaking of parameters to improve performance.

12.1 GLIMDA++: Methods with $s = p + 1$ Stages

The techniques for implementing DIMSIMs with $s = r = p+1 = q+1$ presented in Chapter 11 do not depend on the particular coefficients of the method. Hence the two families from Table 10.1 and 10.2 can be used within the same code GLIMDA++.

Four test problems will be used to compare the different methods.

• miller – the Miller Integrator (index 1, $m = 5$, $n = 1$)

This example was used in Chapter 1 to introduce the modified nodal analysis. It is described in detail in [66]. Using a properly stated leading term it consists of one differential and four algebraic equations.

For the miller problem relative tolerances rtol $= 10^{-j}$, $j = 1, \ldots, 9$, were chosen. The absolute tolerance satisfied atol $= 10^{-3} \cdot$ rtol. The initial stepsize was chosen as $h_0 = 10^{-10}$. As in Example 11.5 miller was solved on the interval $[0, 7 \cdot 10^{-7}]$ using $x(0) = 0$ and the input signal $v(t) = \sin(\omega\, t^2)$ with $\omega = 10^{14}$.

Figure 12.1 (a) shows that both families perform similarly for loose tolerances. For tight tolerances the IRKS methods are clearly superior.

• nand – NAND gate model (index 1, $m = n = 14$)

The NAND gate was discussed in Chapter 4 and 5. It served as an example where both functions q and b are nonlinear in the formulation 12.2. Here the NAND gate model from the Bari Testset for Initial Value Problem Solvers is used [124].

A	B	$\neg(A \wedge B)$
true	*true*	*false*
true	*false*	*true*
false	*true*	*true*
false	*false*	*true*

The problem has been contributed by Günther and Rentrop [83]. The equations model a basic logical element expressing the NAND relation as indicated on the right. Due to a simplified transistor model the equations have index 1.

The test set uses the implicit formulation $\hat{f}(x', x, t) = 0$ and the derivative $\dot{q}(x, t) = C(x, t)\, x'$ is calculated explicitly. Since the new solver GLIMDA++ is capable of handling properly stated leading terms, the equations were modified such that the formulation $f\big(\dot{q}(x, t), x, t\big) = 0$ is used directly.

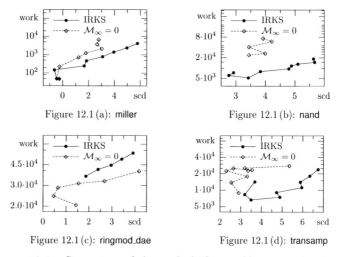

Figure 12.1 (a): miller

Figure 12.1 (b): nand

Figure 12.1 (c): ringmod_dae

Figure 12.1 (d): transamp

Figure 12.1: Comparison of the methods from Table 10.1 with inherent Runge-Kutta stability (IRKS) and those from Table 10.2 with $M_{\infty,p} = 0$. Although the latter family is superior for the index-2 DAE (c), the overall performance is far from optimal. The IRKS methods are generally superior.

The following tolerances have been used: rtol $= 10^{-j/2}$, $j = 2, 3, \ldots, 10$ and atol $= 10^{-3} \cdot$ rtol. As for miller the initial stepsize was $h_0 = 10^{-10}$.

For the NAND gate the methods with $M_{\infty,p} = 0$ show a poor performance. Spending more work does not lead to a significant increase in accuracy. Almost half of the runs failed with stepsize too small (see Table 12.1). The methods with inherent Runge-Kutta stability, by contrast, show the desired behaviour.

- ringmod_dae – the Ring Modulator as a DAE (index 2, $m = 15$, $n = 11$)

The Ring Modulator circuit was introduced in Chapter 2. The circuit diagram from Example 2.6 corresponds to the index-2 formulation used here. Recall that the Ring Modulator mixes a low-frequent input signal U_{in_1} with a high-frequent input signal U_{in_2}.

For this problem, rtol $= 10^{-j/2}$, $j = 4, 6, \ldots, 10$ and atol $= 10^{-2} \cdot$ rtol have been used. Again, the initial stepsize was prescribed as $h_0 = 10^{-10}$.

For this index-2 problem the schemes with $M_{\infty,p} = 0$ are clearly superior to the IRKS methods. For all tolerance requirements higher accuracy is achieved with fewer function evaluations.

- transamp – the transistor amplifier circuit (index 1, $m = 8$, $n = 5$)

The transistor amplifier circuit from Figure 12.2 is one of most frequently used benchmark circuits. Given an input signal U_{in}, the amplified output is delivered at node 8. Amplification is realised using two transistors. Each transistor is modelled as

$$I_G = (1 - \alpha)\, g(u_G - u_S), \qquad I_D = \alpha\, g(u_G - u_S), \qquad I_S = g(u_G - u_S),$$

where I_G, I_D and I_S denotes the current through the gate, drain and source contact, respectively. The node potentials at gate and source are denoted by u_G, u_S. The function g is given by $g(u) = \beta \left(\exp(u/u_F) - 1 \right)$ with $\beta = 10^{-6}$,

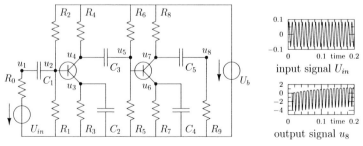

Figure 12.2: The transistor amplifier circuit. The input signal U_{in} is converted into an amplified signal at node 8.

$u_F = 0.026$ and $\alpha = 0.99$. The transistor amplifier problem was published by Rentrop in [137]. It can be found in [84, 124] as well.

For the numerical simulation $\mathsf{rtol} = 10^{-j/2}$, $j = 0, 1, \ldots, 8$, $\mathsf{atol} = 10^{-6} \cdot \mathsf{rtol}$ and $h_0 = 10^{-10}$ has been used.

The results in Figure 12.1 (d) once again show a clear superiority of the IRKS methods. The methods with $M_{\infty,p} = 0$ are competitive only for $\mathsf{rtol} = 1$, $\mathsf{atol} = 10^{-6}$. Tighter tolerances do not lead to an increase in accuracy. Higher scd values seem possible for $\mathsf{rtol} = 10^{-4}$, $\mathsf{atol} = 10^{-10}$. On the other hand the IRKS methods show a nice increase of accuracy when tightening the tolerances. There is, however, a need for improvement in particular for loose tolerances. The same accuracy $\mathsf{scd} \approx 3.5$ is reached with a completely different amount of work. Behaviour of this kind has to be avoided.

The above numerical experiments show a clear difference in performance for the two families of methods considered here. A summary of failed runs is given in Table 12.1. As expected the methods with $M_{\infty,p} = 0$ perform best for nonlinear index-2 problems. Here the improved stability at infinity seems to be a benefit. However, in all other cases the results are far from satisfactory, in particular for nand and transamp.

The methods with inherent Runge-Kutta stability constructed by Butcher and Podhaisky [40] show a good behaviour for all test examples. These methods will therefore be used as the standard option for the solver GLIMDA++.

It becomes clear that the construction of good methods is not a trivial task. Although both families of methods have Runge-Kutta stability and satisfy the same order and stage order conditions, their performance is quite different. Thus the construction of methods should always be accompanied with a corresponding test implementation in order to allow a realistic assessment of the method's potential.

problem	solver	rtol	reason for failing
nand	GLIMDA++ $(M_{\infty,p} = 0)$	10^{-1}, 10^{-4}, $10^{-5/2}$, 10^{-5}	stepsize too small
ringmod_dae	GLIMDA++ (IRKS)	10^{-2}, $10^{-5/2}$	stepsize too small
	GLIMDA++ $(M_{\infty,p} = 0)$	10^{-2}	stepsize too small

Table 12.1: Summary of failed runs for GLIMDA++ applied to miller, nand, ringmod_dae and transamp.

12.2 GLIMDA vs. GLIMDA++

In the previous section it was decided to adopt the methods from Table 10.1 as a standard for GLIMDA++. We are now going to compare this code with the implementation GLIMDA based on the methods from Table 10.4. Recall that GLIMDA not only uses different methods with only $s = p$ stages, but also employs different implementation specifics. In particular, order control is based on the convergence rate of Newton's method (see Chapter 11 for more details).

The same test problems miller, nand, ringmod_dae and transamp as in Section 12.1 have been used.

Figure 12.3 shows a surprisingly well performance for GLIMDA. The code GLIMDA++ is outperformed for all four test examples. Notice in particular that for GLIMDA the relation between the achieved accuracy and the required work is almost given by straight line. This behaviour ensures that GLIMDA is clearly superior not only for high accuracy demands but in particular for loose tolerances (see e.g. Figure 12.3 (d)).

In order to allow a more complete comparison, four additional test examples will be considered next.

• ringmod – the Ring Modulator circuit formulated as an ordinary differential equation (index 0, $m = 15$, $n = 15$)

As described in Example 2.6 the Ring Modulator can be regularised by introducing artificial capacitances C_s. The resulting model is an ordinary differential equation of dimension $m = 15$. For $C_s = 10^{-12}$ simulation results can be found in [124]. The same values are used here. Due to the capacitances C_s, artificial oscillations of high frequency are introduced. Thus the numerical simulation requires an excessively large number of timesteps.

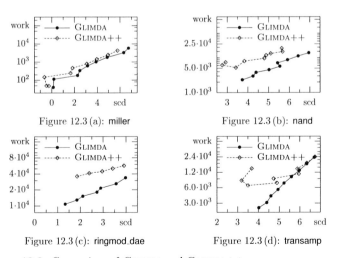

Figure 12.3 (a): miller

Figure 12.3 (b): nand

Figure 12.3 (c): ringmod_dae

Figure 12.3 (d): transamp

Figure 12.3: Comparison of GLIMDA and GLIMDA++

For the comparison of GLIMDA and GLIMDA++ the tolerances $\mathsf{rtol} = 10^{-j/2}$, $j = 4, 5, \ldots, 8$, $\mathsf{atol} = 10^{-3} \cdot \mathsf{rtol}$ have been used. The initial stepsize was chosen as $h_0 = 10^{-6}$.

GLIMDA produces accurate solutions but shows little reaction to different tolerances. GLIMDA++, by contrast, is able to efficiently compute solutions that satisfy low accuracy demands. As this is the required behaviour, GLIMDA++ is superior for this example as well.

Nevertheless the ODE formulation of the Ring Modulator circuit should not be used for practical applications. A comparison with the results for the index-2 formulation in Figure 12.3 (c) shows that there GLIMDA achieves four times the accuracy with only half the effort.

- **rober** – the Robertson reaction (index 0, $m = n = 3$)

This stiff ordinary differential equation models an autocatalytic chemical reaction described in [139]. The problem is used frequently for numerical studies [85]. It is used in particular for comparing integrators for stiff ODEs.

For the **rober** problem stiffness is caused by reaction rates of different magnitude. The exact solution shows a quick initial transient. Afterwards the components vary slowly such that a large stepsize is appropriate. As reported in [85, 124] many codes fail if t becomes very large. Problems arise if components of the solution become accidentally negative, which eventually causes overflow. Thus the integration interval is chosen sufficiently large as $[0, 10^{11}]$.

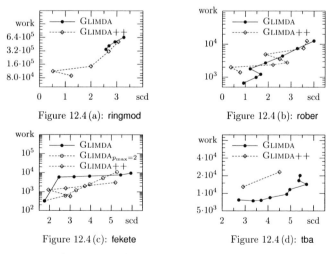

Figure 12.4 (a): **ringmod**

Figure 12.4 (b): **rober**

Figure 12.4 (c): **fekete**

Figure 12.4 (d): **tba**

Figure 12.4: Comparison of GLIMDA and GLIMDA++. Second set of test problems.

The tolerances $\mathsf{rtol} = 10^{-j}$, $j = 2, 3, \ldots, 9$, $\mathsf{atol} = 10^{-3} \cdot \mathsf{rtol}$ and an initial stepsize $h_0 = 10^{-10}$ have been used for the numerical simulations.

Recall that the order 3 method used for GLIMDA is not A-stable. In spite of this, the code performs quite satisfactory. Apart from $\mathsf{rtol} = 10^{-5}$, $\mathsf{atol} = 10^{-8}$ the relation between accuracy and work is almost given by straight line. In particular for loose tolerance GLIMDA is superior to GLIMDA++.

Let us stress that the order 3 method is indeed used for the computations. This was seen in Example 11.6. Hence the lacking A-stability does not seem to be a severe restriction.

- fekete – the Fekete problem (Index 2, $m = 160$, $n = 120$)

This problem taken from [124] computes elliptic Fekete points. The task is to distribute N points $x = \begin{bmatrix} x_1 & \cdots & x_N \end{bmatrix}$ on the unit sphere such that the mutual distance $V(x) = \prod_{i<j} \|x_i - x_j\|_2$ is maximised. The global optimum is difficult to compute as an optimisation problem. Using Lagrange multipliers an equivalent DAE formulation can be found. In order to arrive at the index-2 formulation used here the constraints have to be regularised [10]. The fekete problem was solved for $N = 20$.

The relative tolerances were swept over $\mathsf{rtol} = 10^{-j}$, $j = 0, 1, \ldots, 5$ with $\mathsf{atol} = 10^{-2} \cdot \mathsf{rtol}$. An initial stepsize $h_0 = 10^{-8}$ has been used.

Figure 12.4 (c) shows that both solvers GLIMDA and GLIMDA++ do not perform satisfactory. In particular GLIMDA requires the same amount of work for almost all accuracy requirements. Indeed, the best performance is achieved when restricting the order $1 \leq p \leq 2$ for GLIMDA and thus using A-stable methods only. As a consequence the problems for $p = 3$ might be caused by the lacking A-stability. Unfortunately, [124] contains no information about the exact position of the eigenvalues.

- tba – the Two Bit Adding Unit (index 1, $m = n = 175$)

Given two numbers in base 2 representation the adding unit computes the sum of these numbers. In order to achieve this, voltages are associated with boolean values. By convention, a voltage exceeding $2\,\mathrm{V}$ is interpreted as *true* while values lower than $0.8\,\mathrm{V}$ correspond to *false*. In between the boolean value is undefined.

Let $A_1 A_0$ and $B_1 B_0$ be two 2-bit numbers. These numbers and a carry bit C_{in} are fed into the circuit as input voltages. The circuit performs the addition

$$A_1 A_0 + B_1 B_0 + C_{in} = C\, S_1 S_0.$$

$S_1 S_0$ is the 2-bit representation of the output and C is the updated carry bit. An example computation is indicated in Figure 12.5. The circuit diagram of the Two Bit Adding Unit is available from [124].

It is emphasised in [124] that the equations have the form

$$A\, \dot{q}(x) = f(x, t) \tag{12.3}$$

with $x \in \mathbb{R}^{175}$, but standard solvers require a different formulation. Usually DAEs of the form $My' = f(y, t)$ can be handled. Thus the software part of [124] uses the reformulation

$$\dot{Q} = f(x, t), \qquad 0 = Q - A\, q(x).$$

Introducing the additional variable $Q = A\, q(x)$ doubles the dimension such that in [124] the tba problem is referred to as a DAE of dimension $m = 350$. Since GLIMDA and GLIMDA++ are capable of handling the structure (12.3), this formulation is used directly.

For the tba problem the tolerances rtol $=$ atol $= 10^{-j/2}$, $j = 2, 3, \ldots, 10$ and an initial stepsize $h_0 = 10^{-10}$ have been used.

Again, Figure 12.4 (d) shows a clear superiority of GLIMDA over GLIMDA++. For the latter solver most runs fail as can be seen in Table 12.2.

The summary of failed runs in Table 12.2 shows that GLIMDA is a very robust solver. Only one computation out of 61 runs failed. The computed results are most satisfactory in almost all cases and the lack of A-stability for the order 3 method seems not to play a crucial role. In particular the stiff rober problem does not pose serious difficulties for the solver.

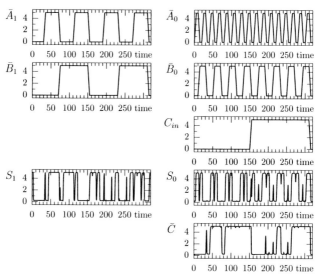

Figure 12.5: The input signals $\bar{A}_1\bar{A}_0$, $\bar{B}_1\bar{B}_0$, C_{in} for the Two Bit Adding Unit tba lead to the output S_1S_0 and the carry bit $\bar{C}$. Notice that the bar denotes logical inversion. For $t = 200$ the input reads $\bar{A}_1\bar{A}_0 = 10$, $\bar{B}_1\bar{B}_0 = 00$, $C_{in} = 1$, such that the addition $01 + 11 + 1 = 101 = C\, S_1S_0$ is performed.

Figure 12.3 shows that GLIMDA is clearly superior to GLIMDA++ for the test problems miller, nand, ringmod_dae and transamp. Higher accuracy is achieved investing a considerably lower amount of work.

The second four test examples ringmod, rober, fekete and tba do not change this assessment, even though the difference between the two solvers is no longer as clear as for the first four problems.

In particular ringmod and fekete seem to indicate a better performance for GLIMDA++ as compared to GLIMDA. However, this is not a severe restriction as the first problem is academic in the sense that the index-2 formulation can be solved much more efficiently, while the situation for fekete can be improved considerably by limiting the maximum order for GLIMDA. It is not clear whether GLIMDA's problems with fekete result from the missing A-stability. In any case, an improved A-stable order 3 method is most desirable.

The numerical experiments of this section indicate that a practical implementation should be based on methods with $s = p$ stages. Although the linear stability properties of methods with $s = p+1$ stages are more desirable and error estimation as well as order control can be realised very elegantly, these methods seem not to benefit from their excellent properties as much as expected. The robust implementation and the reduced work per step for methods with $s = p$ stages is seen to be a clear advantage for the many test problems considered here.

In fact, GLIMDA seems powerful enough to be a tough competitor for classical codes such as DASSL or RADAU. The corresponding comparisons will be performed in the next section.

problem	solver	rtol	reason for failing
ringmod_dae	GLIMDA++	10^{-2}, $10^{-5/2}$	stepsize too small
ringmod	GLIMDA	10^{-2}	stepsize too small
fekete	GLIMDA++	10^{-5}	stepsize too small
tba	GLIMDA++	$10^{-3/2}$, $10^{-5/2}$, 10^{-3}, $10^{-7/2}$, 10^{-4}, $10^{-9/2}$, 10^{-5}	stepsize too small

Table 12.2: Summary of failed runs for GLIMDA and GLIMDA++ (IRKS) applied to miller, nand, ringmod_dae, transamp, ringmod, rober, fekete and tba.

12.3 GLIMDA vs. Standard Solvers

In Chapter 2 classical numerical methods such as linear multistep schemes and Runge-Kutta methods have been considered. It was pointed out that both families of methods have disadvantages for integrated circuit design. Hence, general linear methods have been studied in the context of differential algebraic equations modelling electrical circuit. These studies eventually lead to the test implementation GLIMDA.

The numerical experiments of this section are performed in order to test the code GLIMDA against three of the most successful and highly tuned codes available at [124].

- DASSL solves general implicit DAEs $f(x', x, t) = 0$ having index $\mu \leq 1$. This linear multistep code is based on BDF formulae of order $1 \leq p \leq 5$. DASSL was written by Petzold [10, 128] and can be downloaded from www.netlib.org/ode/ddassl.f.

- RADAU is a standard Runge-Kutta code written by Hairer and Wanner [85, 86]. It is capable of solving differential algebraic equations of the form $M y' = f(y, t)$ with index $\mu \leq 3$. The code uses RadauIIA methods of order $p = 5, 9, 13$ and can be obtained from www.unige.ch/~hairer/prog/stiff/radau.f.

- MEBDFI uses Modified Extended Backward Differentiation Formulae that increase the region of absolute stability as compared to classical BDF methods. The resulting schemes are A-stable up to order 4, but the implementation by Abdulla and Cash uses methods of order $1 \leq p \leq 7$. Similar to DASSL implicit DAEs $f(x', x, t) = 0$ can be solved. The code is applicable if the index satisfies $\mu \leq 3$. MEBDFI is available at www.ma.ic.ac.uk/~jcash/IVP_software/itest/mebdfi.f.

For the numerical experiments performed here, the codes DASSL, RADAU, MEBDFI and the FORTRAN77 problem descriptions have been downloaded from [124]. All simulation results are displayed as work-precision diagrams in Figure 12.6.

It is no surprise that GLIMDA is not competitive with the standard solvers for ringmod and fekete (Figure 12.6 (e) and Figure 12.6 (g)). These issues have already been discussed in the previous section. Observe that RADAU as available from [124] was not able to solve ringmod. DASSL, on the other hand, fails for fekete. A complete summary of failed runs is given in Table 12.3.

For miller and tba GLIMDA shows a performance quite similar to DASSL. Again, RADAU fails completely for tba and Figure 12.6 (h) shows that MEBDFI is not at all competitive with GLIMDA and DASSL for the Two Bit Adding Unit. For the miller problem Figure 12.6 (a) shows a similar situation for GLIMDA, DASSL and MEBDFI, although the three solvers are now much closer together. RADAU, by contrast, is most efficient. For miller this solver seems to benefit from the high order methods being used.

The benefit of high order can be seen clearly in Figure 12.6 (f) for the rober problem. Although GLIMDA is competitive for low accuracy demands, a maximum order of $p = 3$ is not sufficient to keep up with DASSL or RADAU, where orders up to 5 and 13 are used, respectively.

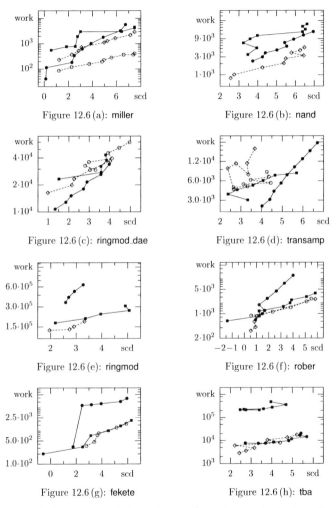

Figure 12.6 (a): miller

Figure 12.6 (b): nand

Figure 12.6 (c): ringmod_dae

Figure 12.6 (d): transamp

Figure 12.6 (e): ringmod

Figure 12.6 (f): rober

Figure 12.6 (g): fekete

Figure 12.6 (h): tba

Figure 12.6: Comparison of GLIMDA (——•——) and the standard solvers DASSL (····◇····), RADAU (····⊙····) and MEBDFI (——■——) problems.

In contrast to these results for ordinary differential equations, computations with a maximum order $p = 3$ seem to be most appropriate for problems from electrical circuit simulation. Indeed, GLIMDA is the most efficient solver for both ringmod_dae and transamp. Figure 12.6 (c) and Figure 12.6 (d) show that GLIMDA is superior for all tolerances. Notice that the gap between GLIMDA and the other solvers increases for low accuracy demands. This is a particularly nice feature as for realistic applications an error of about 2% is sufficient [66].

problem	solver	rtol	reason for failing (as reported by the solver)
miller	MEBDFI	10^{-7}	hmin reduced by a factor of 1.0e10
nand	RADAU	$10^{-1}, \cdots, 10^{-5}$	RADAU can not solve IDEs
ringmod	GLIMDA	10^{-2}	stepsize too small
	DASSL	10^{-2}	failed to converge repeatedly or with abs(h)=hmin
	RADAU	$10^{-2}, \cdots, 10^{-4}$	RADAU can not handle FEVAL IERR
tba	DASSL	10^{-1}	could not converge because ires was equal to minus one
	RADAU	$10^{-1}, \cdots, 10^{-5}$	RADAU can not handle FEVAL IERR
	MEBDFI	10^{-5}	the requested error is smaller than can be handled
transamp	DASSL	$10^{0}, 10^{-1/2}, 10^{-1}$	failed to converge repeatedly or with abs(h)=hmin
	RADAU	$10^{0}, 10^{-1/2}$	step size too small
	MEBDFI	$10^{0}, 10^{-1/2}$	the requested error is smaller than can be handled
ringmod_dae	DASSL	$10^{-2}, 10^{-5/2}$	failed to converge repeatedly or with abs(h)=hmin
	MEBDFI	$10^{-2}, 10^{-5/2}$	corrector convergence could not be achieved
rober	DASSL	$10^{-6}, 10^{-7}, 10^{-9}$	failed to converge repeatedly or with abs(h)=hmin
	RADAU	$10^{-2}, 10^{-3}, 10^{-7}$	step size too small
fekete	DASSL	$10^{0}, \dots, 10^{-5}$	DASSL can not solve higher index problems

Table 12.3: Summary of failed runs for GLIMDAand DASSL, RADAU, MEBDFI

For the nand problem in Figure 12.6 (b) GLIMDA lies in between MEBDFI and DASSL. RADAU was not able to solve this example (see again Table 12.3).

The results of this section show that GLIMDA is competitive with the standard solvers DASSL, MEBDFI and RADAU. Recall that GLIMDA is designed to solve differential algebraic equations

$$A(t)\big[d\big(x(t),t\big)\big]' + b\big(x(t),t\big) = 0$$

with a properly stated leading term and index $\mu \leq 2$. Ordinary differential equations $y' = f(y,t)$ can be cast into this form, such that GLIMDA is capable of handling ODEs as well. Nevertheless, the comparison from Table 12.4 shows that specialised codes such as RADAU should be used. Although GLIMDA shows a reasonable performance for ODEs, other codes are often more efficient.

Table 12.4 attempts a pictorial assessment of the results for ODEs. $\oplus$ indicates superior performance while $\bigcirc$ and $\ominus$ represent reasonable and poor performance, respectively. A hyphen – is used in case of a solver failing for a given problem.

The high potential of GLIMDA is revealed when applying the solver to differential algebraic problems. GLIMDA and MEBDFI are the only codes successful in solving all examples. DASSL and RADAU fail for at least one test problem. Table 12.5 confirms that GLIMDA is indeed competitive if not superior to standard solvers.

	miller	ringmod	rober
GLIMDA	$\bigcirc$	$\ominus$	$\bigcirc$
DASSL	$\bigcirc$	$\oplus$	$\oplus$
RADAU	$\oplus$	–	$\oplus$
MEBDFI	$\bigcirc$	$\oplus$	$\oplus$

Table 12.4: Comparison of different solvers for ordinary differential equations.

	nand	ringmod_dae	fekete	tba	transamp
GLIMDA	$\bigcirc$	$\oplus$	$\ominus$	$\oplus$	$\oplus$
DASSL	$\oplus$	$\bigcirc$	–	$\oplus$	$\ominus$
RADAU	–	$\bigcirc$	$\oplus$	–	$\bigcirc$
MEBDFI	$\ominus$	$\bigcirc$	$\oplus$	$\ominus$	$\bigcirc$

Table 12.5: Comparison of different solvers for differential algebraic equations.

Part V

Summary

Summary

The design and production of today's electronic devices requires extensive numerical testing. Using CMOS technology there is an ongoing trend towards decreasing size and increasing power density for future chip generations. Due to high quality demands and short product cycles, circuit simulation is a key technology in every modern design flow as it allows an efficient layout and parameter optimisation.

The modified nodal analysis (MNA) applied to electrical circuits leads to differential algebraic equations (DAEs) that need to be solved in the time domain. Many classical methods give rise to severe difficulties. Problems due to high computational costs for Runge-Kutta methods and undesired stability behaviour for the trapezoidal rule or BDF schemes have been discussed. General linear methods (GLMs) are suggested as a means to overcome these difficulties.

General linear methods are studied for nonlinear differential algebraic equations having a properly stated leading term. A refined analysis for DAEs of increasing complexity is presented. Restricting attention to DAEs satisfying an additional structural condition, it is shown how to derive a decoupling procedure for nonlinear equations. Using this procedure statements on the existence and uniqueness of solutions are given. Compared to previous results the sufficient conditions derived here are much easier to check for practical applications. The derivative array is not used at all. MNA equations are covered by the results as well as DAEs in Hessenberg form. It is shown how the special structure of DAEs in circuit simulation can be exploited in order to simplify the results considerably.

In order to use the decoupling procedure to study numerical methods for index-2 DAEs, implicit index-1 equations have to be addressed first. A thorough study of general linear methods for implicit index-1 equations is presented. Using rooted tree theory order conditions are derived and it is shown that sufficiently high order p and stage order q guarantee that these order conditions are satisfied. Convergence is proved as well. Special care is taken when assessing the accuracy of stages and stage derivatives. It is shown that methods with a nilpotent stability matrix at infinity and $p = q$ provide the desired accuracy.

Using the decoupling procedure these results are transferred to properly stated index-2 equations. Provided that the DAE is numerically qualified it is shown that a GLM, when applied to the original DAE, behaves as if it was integrating the inherent index-1 system. Hence convergence can be proved for GLMs applied to properly stated index-2 DAEs.

In order to verify the theoretical results, practical GLMs with $s = p + 1$ and $s = p$ stages are constructed. Addressing implementation issues it turns out that error estimation and order control can be realised easily for methods of the former type. Although it is not as straightforward to implement methods with $s = p$ stages, these methods offer reduced costs per step such that they might have higher potential for practical computations.

Using extensive numerical studies it is indeed confirmed that general linear methods can be efficiently used in electrical circuit simulation. In particular the code GLIMDA based on methods with $s = p$ stages is competitive with classical codes such as DASSL or RADAU.

These results motivate a further study of general linear methods for DAEs. GLIMDA is still in an experimental stage and there is room for considerable improvement. In particular, the error estimators and order selection strategies were derived in the context of ODEs. Another highly desired improvement is the derivation of an A-stable order-3 method, if it exists. As indicated by the method of order 2 an adaptive control of the damping behaviour might even be possible.

Bibliography

[1] R. Alexander. Diagonally implicit Runge-Kutta methods for stiff ODEs. *SIAM J. Numer. Anal.*, 14:1006–1021, 1977.

[2] K. Balla and R. März. A unified approach to linear differential algebraic equations and their adjoint equations. Technical Report 00-18, Humboldt Universität zu Berlin, 2000. use `BallaMaerz:2002` instead.

[3] K. Balla and R. März. A unified approach to linear differential algebraic equations and their adjoints. *Z. Anal. Anwendungen*, 21(3):783–802, 2002. ISSN 0232-2064.

[4] A. Bartel. *Partial Differential-Algebraic Models in Chip Design – Thermal and Semiconductor Problems.* PhD thesis, Bergische Universität Wuppertal, 2004. Fortschritt-Berichte VDI, Reihe 20, Nr. 391.

[5] A. Bartel, M. Günther, and A. Kværnø. Multirate methods in electrical circuit simulation. In *Progress in Industrial Mathematics at ECMI 2000*, Mathematics in Industry 1, pages 258–265. Springer, 2002.

[6] F. Bashforth and J.C. Adams. *An attempt to test the theories of capillary action by comparing the theoretical and measured forms of drops of fluid. With an explanation of the method of integration employed in constructing the tables which give the theoretical form of such drops.* Cambridge University Press, 1883.

[7] T.A. Bickart and Z. Picel. High order stiffly stable composite multistep methods for numerical integration of stiff differential equations. *BIT*, 13: 272–286, 1973.

[8] Blas. BLAS – Basic Linear Algebra Subprograms, www.netlib.org/blas.

[9] F. Bornemann. Runge-Kutta methods, trees, and MAPLE. *Selçuk J. Appl. Math.*, 2:3–15, 2001.

[10] K.E. Brenan, S.L. Campbell, and L.R. Petzold. *Numerical solution of initial-value problems in differential-algebraic equations*, volume 14 of *Classics in Applied Mathematics*. Society for Industrial and Applied Mathematics (SIAM), 1996. ISBN 0-89871-353-6.

[11] K.E. Brenan and L.R. Engquist. Backward differentiation approximations of nonlinear differential/algebraic equations. *Math. Comp.*, 51:659–676, 1988.

[12] K. Burrage and J.C. Butcher. Stability criteria for implicit Runge-Kutta methods. *SIAM J. Numer. Anal.*, 16:46–57, 1979.

[13] K. Burrage and J.C. Butcher. Non-linear stability for a general class of differential equation methods. *BIT*, 20:326–340, 1980.

[14] K. Burrage, J.C. Butcher, and F.H. Chipman. An implementation of singly-implicit Runge-Kutta methods. *BIT*, 20:326–340, 1980.

[15] J.C. Butcher. Coefficients for the study of RungeKutta integration processes. *J. Austral. Math. Soc.*, 3:185–201, 1963.

[16] J.C. Butcher. Implicit Runge-Kutta processes. *Math. Comp.*, 18:50–64, 1964.

[17] J.C. Butcher. Integration processes based on Radau quadrature formulas. *Math. Comp.*, 18:233–244, 1964.

[18] J.C. Butcher. On the convergence of numerical solutions of ordinary differential equations. *Math. Comp.*, 20:1–10, 1966.

[19] J.C. Butcher. A stability property of implicit Runge-Kutta methods. *BIT*, 15:358–361, 1975.

[20] J.C. Butcher. Stability properties for a general class of methods for ordinary differential equations. *SIAM J. Numer. Anal.*, 18(1):34–44, 1981.

[21] J.C. Butcher. Linear and non-linear stability for general linear methods. *BIT*, 27:182–189, 1987.

[22] J.C. Butcher. *The numerical analysis of ordinary differential equations, Runge-Kutta and general linear methods*. Wiley, Chichester and New York, 1987.

[23] J.C. Butcher. Diagonally-implicit multi-stage integration methods. *Appl. Numer. Math.*, 11:347–363, 1993.

[24] J.C. Butcher. Miniature 21: R gave me a DAE underneath the Linden tree, 2003. www.math.auckland.ac.nz/~butcher/miniature/miniature21.pdf.

[25] J.C. Butcher. *Numerical methods for ordinary differential equations*. John Wiley & Sons Ltd., Chichester, 2003. ISBN 0-471-96758-0.

[26] J.C. Butcher. General linear methods. to appear in Acta Numerica, 2006.

[27] J.C. Butcher and P. Chartier. The construction of DIMSIMs for stiff ODEs and DAEs. Technical Report 308, The University of Auckland, 1994.

[28] J.C. Butcher and P. Chartier. Parallel general linear methods for stiff ordinary differential and differential algebraic equations. *Appl. Numer. Math.*, 17:213–222, 1995.

[29] J.C. Butcher, P. Chartier, and Z. Jackiewicz. Nordsieck representation of DIMSIMs. *Numer. Algorithms*, 16:209–230, 1997.

[30] J.C. Butcher, P. Chartier, and Z. Jackiewicz. Experiments with a variable-order type 1 DIMSIM code. *Numer. Algorithms*, 22:237–261, 1999.

[31] J.C. Butcher and A.D. Heard. Stability of numerical methods for ordinary differential equations. *Numer. Algorithms*, pages 47–58, 2002.

[32] J.C. Butcher and Z. Jackiewicz. Construction of diagonally implicit general linear methods for ordinary differential equations. *BIT*, 33:452–472, 1996.

[33] J.C. Butcher and Z. Jackiewicz. Construction of diagonally implicit general linear methods of type 1 and 2 for ordinary differential equations. *Appl. Numer. Math.*, 21:385–415, 1996.

[34] J.C. Butcher and Z. Jackiewicz. Construction of high order diagonally implicit multistage integration methods for ordinary differential equations. *BIT*, 33:452–472, 1996.

[35] J.C. Butcher and Z. Jackiewicz. Implementation of diagonally implicit multistage integration methods for ordinary differential equations. *SIAM J. Numer. Anal.*, 34(6):2110–2141, 1997.

[36] J.C. Butcher and Z. Jackiewicz. A reliable error estimation for diagonally implicit multistage integration methods. *BIT*, 41(4):656–665, 2001.

[37] J.C. Butcher and Z. Jackiewicz. A new approach to error estimation for general linear methods. *Numer. Math.*, 95:487–502, 2003.

[38] J.C. Butcher and Z. Jackiewicz. Unconditionally stable general linear methods for ordinary differential equations. *BIT*, 44:557–570, 2004.

[39] J.C. Butcher, Z. Jackiewicz, and H.D. Mittelmann. A nonlinear optimization approach to the construction of general linear methods of high order. *J. Comput. Appl. Math.*, 81:181–196, 1997.

[40] J.C. Butcher and H. Podhaisky. On error estimation in general linear methods for stiff ODEs. Appl. Numer. Math., 2005. accepted for publication.

[41] S.L. Campbell. A procedure for analyzing a class of nonlinear semistate equations that arise in circuit and control problems. *IEEE Trans. Circuits and Systems*, pages 256–261, 1981.

[42] S.L. Campbell. A general form for solvable linear time varying singular systems of differential equations. *SIAM J. Math. Anal.*, 18(4):1101–1115, 1987.

[43] S.L. Campbell and C.W. Gear. The index of general nonlinear DAEs. *Numer. Math.*, 72:173–196, 1995.

[44] S.L. Campbell and E. Griepentrog. Solvability of general differential algebraic equations. *SIAM J. Sci. Comput.*, 16(2):257–270, 1995.

[45] J. Cash. The integration of stiff initial value problems in o.d.e.s using modified extended backward differentiation formulae. *Comput. Math. Appl.*, 9:645–657, 1983.

[46] P. Chartier. General linear methods for differential-algebraic equations of index one and two. Technical Report 1968, Institut national de recherche en informatique et en automatique, 1993.

[47] L.O. Chua, C.A. Desoer, and E.S. Kuh. *Linear and nonlinear circuits.* McGraw Hill Book Company, Singapore, 1987.

[48] C.F. Curtiss and J.O. Hirschfelder. Integration of stiff equations. *Proc. Nat. Acad. Sci.*, 38:235–243, 1952.

[49] G. Dahlquist. A special stability problem for linear multistep methods. *BIT*, pages 27–43, 1963.

[50] G. Dahlquist. Error analysis of a class of methods for stiff nonlinear initial value problems. In *Numerical Analysis, Dundee, Lecture Notes in Mathematics*, 506, pages 60–74, 1976.

[51] G. Dahlquist. G-stability is equivalent to A-stability. *BIT*, 18:384–401, 1978.

[52] K. Dekker and J.G. Verwer. *Stability of Runge-Kutta methods for stiff nonlinear differential equations.* CWI Monographs. North Holland, 1984.

[53] C.A. Desoer and E.S. Kuh. *Basic Circuit Theory.* McGraw Hill Book Company, 1969.

[54] P. Deuflhard, E. Hairer, and J. Zugck. One-step and extrapolation methods for differential-algebraic equations. *Numer. Math.*, 51:501–516, 1987.

[55] H. Döhring. Traktabilitätsindex und Eigenschaften von matrixwertigen Riccati-Typ Algebrodifferentialgleichungen. Master's thesis, Humboldt Universität zu Berlin, 2004.

251

[56] R.C. Dorf. *Introduction to electric circuits*. John Wiley & Sons, 2nd edition, 1989.

[57] J.R. Dormand and P.J. Prince. A family of embedded Runge-Kutta formulae. *J. Comput. Appl. Math.*, 6:19–26, 1980.

[58] B.L. Ehle. On Padé approximations to the exponential function and A-stable methods for the numerical solution of initial value problems. Technical report, Dept. AACS, University of Waterloo, Ontario, Canada, 1969. Research Report CSRR 2010.

[59] E. Eich-Soellner and C. Führer. *Numerical Methods in Multibody Dynamics*. Teubner, Stuttgart, 1998. (since 2002 available as a reprint directly from the authors).

[60] R.F. Enenkel and K.R. Jackson. DIMSEMs – diagonally implicit single-eigenvalue methods for the numerical solution of ODEs on parallel computers. *Adv. Comput. Math.*, 7(1-2):97–133, 1997.

[61] D. Estévez Schwarz. *Consistent initialization for index-2 differential algebraic equations and it's application to circuit simulation*. PhD thesis, Humboldt Universität zu Berlin, 2000. edoc.hu-berlin.de/dissertationen/estevez-schwarz-diana-2000-07-13/PDF/Estevez-Schwarz.pdf.

[62] D. Estévez Schwarz, U. Feldmann, R. März, S. Sturzel, and C. Tischendorf. Finding beneficial DAE structures in circuit simulation. In W. Jäger and H.J. Krebs, editors, *Mathematics – Key technology for the future*, pages 413–428. Springer, Berlin, 2003.

[63] D. Estévez Schwarz and C. Tischendorf. Structural analysis of electric circuits and consequences for MNA. *Internat. J. Circuit Theory Appl.*, 28:131–162, 2000.

[64] U. Feldmann. private communication, 2005.

[65] U. Feldmann and M. Günther. Some remarks on regularization of circuit equations. In W. Mathis, editor, *X. International Symposium on Theoretical Electrical Engineering*, pages 343–348, 1999.

[66] U. Feldmann, M. Günther, and J. ter Maten. *Handbook of Numerical Analysis, Volume XIII: Numerical Methods in Electromagnetics*, chapter *Modelling and Discretization of Circuit Problems*. North-Holland, 2005.

[67] O. Forster. *Analysis 2. Differentialrechnung im $\mathbb{R}^n$, gewöhnliche Differentialgleichungen*. Vieweg, 1984.

[68] F.R. Gantmacher. *The theory of matrices*. Chelsea Pub. Co. New York, 1959.

[69] C.W. Gear. Hybrid methods for initial value problems in ordinary differential equations. *SIAM J. Numer. Anal.*, 2:69–86, 1965.

[70] C.W. Gear. *Numerical initial value problems in ordinary differential equations.* Prentice Hall, 1971.

[71] C.W. Gear. Simultaneous numerical solution of differential-algebraic equations. *IEEE Trans. Circuits and Theory*, CT-18(1):89–95, 1971.

[72] C.W. Gear. Differential-algebraic equation index transformations. *SIAM J. Sci. Statist. Comput.*, 9:39–47, 1988.

[73] C.W. Gear and L.R. Petzold. Differential/algebraic systems and matrix pencils. In B. Kagstrom & A. Ruhe, editor, *Matrix Pencils*, Lecture Notes in Mathematics 973, pages 75–89. Springer Verlag, 1983.

[74] R. Gerstberger and M. Günther. Charge-oriented extrapolation methods in digital circuit simulation. *Appl. Numer. Math.*, 18:115–125, 1995.

[75] E. Griepentrog and R. März. *Differential-algebraic equations and their numerical treatment.* Teubner, Leipzig, 1986.

[76] N. Guglielmi and M. Zennaro. On the asymptotic properties of a family of matrices. *Linear Algebra Appl.*, 322:169–192, 2001.

[77] N. Guglielmi and M. Zennaro. On the zero-stability of variable stepsize multistep methods: the spectral radius approach. *Numer. Math.*, pages 445–458, 2001.

[78] M. Günther. *Ladungsorientierte Rosenbrock-Wanner-Methoden zur numerischen Simulation digitaler Schaltungen.* PhD thesis, TU München, 1995. (ISBN 3-18-316820-0, VDI-Verlag, Düsseldorf).

[79] M. Günther. Simulating digital circuits numerically – A charge oriented row approach. *Numer. Math.*, 79:203–212, 1998.

[80] M. Günther and U. Feldmann. CAD based electrical circuit modelling in industry I: Mathematical structure and index of network equations. *Surveys Math. Indust.*, 8:97–129, 1999.

[81] M. Günther, U. Feldmann, and P. Rentrop. CHORAL – a one-step method as numerical low pass filter in electrical network analysis. In U. van Rienen et. al., editor, *SCEE 2000*, number 18 in Lecture Notes Comp. Sci. Eng., pages 199–215. Springer, Berlin, 2001.

[82] M. Günther, M. Hoschek, and R. Weiner. ROW methods adapted to a cheap Jacobian. *Appl. Numer. Math.*, 37:231–240, 2001.

[83] M. Günther and P. Rentrop. The NAND-gate – a benchmark for the numerical simulation of digital circuits. In W. Mathis and P. Noll, editors, *2. ITG-Diskussionssitzung "Neue Anwendungen Theoretischer Konzepte*

in der Elektrotechnik" - mit Gedenksitzung zum 50. Todestag von Wilhelm Cauer, pages 27–33. VDE-Verlag, Berlin, 1996.

[84] E. Hairer, C. Lubich, and R. Roche. *The numerical solution of differential-algebraic systems by Runge-Kutta methods*, volume 1409 of *Lecture Notes in Mathematics*. Springer-Verlag, Berlin, 1989. ISBN 3-540-51860-6.

[85] E. Hairer and G. Wanner. *Solving ordinary differential equations II: stiff and differential algebraic problems*. Springer, Berlin Heidelberg New York Tokyo, 2 edition, 1996.

[86] E. Hairer and G. Wanner. Stiff differential equations solved by Radau methods. *J. Comput. Appl. Math.*, 111:93–111, 1999.

[87] K. Heun. Neue Methoden zur approximativen Integration der Differentialgleichungen einer unabhängigen Veränderlichen. *Zeit. Math. Phys.*, 45:23–38, 1900.

[88] I. Higueras and R. März. Differential algebraic equations with properly stated leading terms. *Comput. Math. Appl.*, 48(1-2):215–235, 2004.

[89] I. Higueras, R. März, and C. Tischendorf. Stability preserving integration of index-1 DAEs. *Appl. Numer. Math.*, 45(2-3):175–200, 2003. ISSN 0168-9274.

[90] I. Higueras, R. März, and C. Tischendorf. Stability preserving integration of index-2 DAEs. *Appl. Numer. Math.*, 45(2-3):201–229, 2003. ISSN 0168-9274.

[91] R. Horn and C. Johnson. *Topics in Matrix Analysis*. Cambridge University Press, 1991.

[92] E.H. Horneber. *Analyse nichtlinearer RLCÜ-Netzwerke mit Hilfe der gemischten Potentialfunktion mit einer systematischen Darstellung der Analyse nichtlinearer dynamischer Netzwerke*. PhD thesis, Universität Kaiserslautern, 1976.

[93] S.J.Y. Huang. *Implementation of general linear methods for stiff ordinary differential equations*. PhD thesis, The University of Auckland, 2005. www.math.auckland.ac.nz/~butcher/theses/shirleyhuang.pdf.

[94] Z. Jackiewicz. Implementation of DIMSIMs for stiff differential systems. *Appl. Numer. Math.*, 42:251–267, 2002.

[95] Z. Jackiewicz, H. Podhaisky, and R. Weiner. Construction of highly stable two-step W-methods for ordinary differential equations. Technical report, FB Mathematik und Informatik, Martin-Luther Universität Halle-Wittenberg, 2002. www.mathematik.uni-halle.de/reports/sources/2002/02-23report.ps.

[96] P. Kaps and G. Wanner. A study of Rosenbrock-type methods of high order. *Numer. Math.*, 38:279–298, 1981.

[97] L. Kronecker. Algebraische Reduktion der Scharen bilinearer Formen. *Akademie der Wissenschaften Berlin, Werke vol. III*, pages 141–155, 1890.

[98] P. Kunkel and V. Mehrmann. Canonical forms for linear differential algebraic equations with variable coefficients. *J. Comput. Appl. Math.*, 56:225–251, 1994.

[99] P. Kunkel and V. Mehrmann. Regular solutions of nonlinear differential-algebraic equations and their numerical determination. *Numer. Math.*, 79(4):581–600, 1998. ISSN 0029-599X.

[100] P. Kunkel and V. Mehrmann. Index reduction for differential-algebraic equations by minimal extension. *Z. Angew. Math. Mech.*, 84:579–597, 2004.

[101] P. Kunkel and V. Mehrmann. *Differential-Algebraic Equations. Analysis and Numerical Solution.* EMS Publishing House, Zürich, Switzerland, 2006. To appear.

[102] P. Kunkel, V. Mehrmann, and I. Seufer. GENDA: A software package for the numerical solution of general nonlinear differential-algebraic equations. Technical Report 730, Technische Universität Berlin, 2002. www.math.tu-berlin.de/preprints/files/dgenda.ps.gz.

[103] W. Kutta. Beitrag zur näherungsweisen Integration totaler Differentialgleichungen. *Zeit. Math. Phys.*, 46:435–453, 1901.

[104] A. Kværnø. Runge-Kutta methods applied to fully implicit differential-algebraic equations of index 1. *Math. Comp.*, 54(190):583–625, 1990. ISSN 0025-5718.

[105] A. Kværnø. Singly diagonally implicit Runge-Kutta methods with an explicit first stage. *BIT*, 44:489–502, 2004.

[106] A. Kværnø, S.P. Nørsett, and B. Owren. Runge-Kutta research in Trondheim. *Appl. Numer. Math.*, 22(1-3):263–277, 1996.

[107] R. Lamour. Index determination and calculation of consistent initial values for DAEs. *Comput. Math. Appl.*, 50(7):1125–1140, 2005.

[108] R. Lamour. Strangenessindex + 1 = Traktabilitätsindex? 7. Workshop über Deskriptorsysteme, 2005. www.math.tu-berlin.de/numerik/mt/NumMat/Meetings/0503_Descriptor.

[109] Lapack. LAPACK – Linear Algebra PACKage, www.netlib.org/lapack.

[110] W. Liniger and R.A. Willoughby. Efficient integration methods for stiff systems of ordinary differential equations. *SIAM J. Numer. Anal.*, 7: 47–66, 1970.

[111] P. Lötstedt and L.R. Petzold. Numerical solution of nonlinear differential equations with algebraic constraints. I. Convergence results for backward differentiation formulas. *Math. Comp.*, 46(174):491–516, 1986. ISSN 0025-5718.

[112] Maple. MAPLE, Waterloo Maple Inc., www.maplesoft.com.

[113] R. März. Numerical methods for differential-algebraic equations. *Acta Numerica*, pages 141–198, 1992.

[114] R. März. Nonlinear differential-algebraic equations with properly formulated leading terms. Technical Report 01-3, Humboldt Universität zu Berlin, 2001. www.math.hu-berlin.de/publ/pre/2001/P-01-3.ps.

[115] R. März. Differential algebraic systems anew. *Appl. Numer. Math.*, 42 (1-3):315–335, 2002. ISSN 0168-9274.

[116] R. März. The index of linear differential algebraic equations with properly stated leading terms. *Results Math.*, 42:308–338, 2002.

[117] R. März. Differential algebraic systems with properly stated leading term and MNA equations. In *Modeling, simulation, and optimization of integrated circuits (Oberwolfach, 2001)*, volume 146 of *Internat. Ser. Numer. Math.*, pages 135–151. Birkhäuser, Basel, 2003.

[118] R. März. Fine decouplings of regular differential-algebraic equations. *Results Math.*, 46:57–72, 2004.

[119] R. März. Projectors for matrix pencils. Technical report, Humboldt Universität zu Berlin, 2004. www.math.hu-berlin.de/publ/pre/2004/P-04-24. ps.

[120] R. März. Solvability of linear differential algebraic equations with property stated leading terms. *Results Math.*, 45(1–2):88–105, 2004.

[121] R. März. Characterizing differential algebraic equations without the use of derivative arrays. *Comput. Math. Appl.*, 50(7):1141–1156, 2005.

[122] R. März and C. Tischendorf. Solving more general index-2 differential algebraic equations. *Comput. Math. Appl.*, 28(10–12):77–105, 1994.

[123] matlab. MATLAB, The MathWorks, www.mathworks.com.

[124] F. Mazzia and F. Iavernaro. *Test Set for Initial Value Problem Solvers*. Department of Mathematics, University of Bari, 2003. www.dm.uniba.it/ ~testset.

[125] Minpack. MINPACK, www.netlib.org/minpack.

[126] F.R. Moulton. *New methods in exteriour ballistics*. University of Chicago Press, 1926.

[127] A. Nordsieck. On numerical integration of ordinary differential equations. *Math. Comp.*, 16:22–49, 1962.

[128] L.R. Petzold. A description of DASSL: A differential/algebraic system solver. In *Proceedings of IMACS World Congress*, 1982.

[129] H. Podhaisky. *Parallele Zweischritt-W-Methoden*. PhD thesis, Martin-Luther-Universität Halle-Wittenberg, 2002. deposit.ddb.de/cgi-bin/dokserv?idn=964158841.

[130] W.H. Press, S.A. Teukolsky, W.T. Vetterling, and B.P. Flannery. *Numerical recipes in* FORTRAN. Cambridge University Press, 2nd edition, 1992.

[131] A. Prothero and A. Robinson. On the stability and accuracy of one-step methods for solving stiff systems of ordinary differential equations. *Math. Comp.*, 28:145–162, 1974.

[132] R. Pulch. Finite difference methods for multi time scale differential algebraic equations. *Z. Angew. Math. Mech.*, 83:571–583, 2003.

[133] P. Rabier and W. Rheinboldt. *Theoretical and Numerical Analysis of Differential-Algebraic Equations*, volume VIII. Elsevier Science B.V., 2002. edited by P.G. Ciarlet and J.L. Lions.

[134] K. Radhakrishnan and A.C. Hindmarsh. *Description and use of* LSODE, *the Livemore Solver for Ordinary Differential Equations*. Lawrence Livermore National Laboratory Report Series, 1993. URL www.llnl.gov/tid/lof/documents/pdf/240148.pdf.

[135] N. Rattenbury. *Almost Runge-Kutta methods for stiff and non-stiff problems*. PhD thesis, The University of Auckland, 2005.

[136] G. Reissig. *Beiträge zur Theorie und Anwendung impliziter Differentialgleichungen*. PhD thesis, TU Dresden, 1998.

[137] P. Rentrop, M. Roche, and G. Steinebach. The application of Rosenbrock-Wanner type methods with stepsize control in differential-algebraic equations. *Numer. Math.*, 55:545–563, 1989.

[138] R. Riaza and C. Tischendorf. Topological analysis of qualitative features in electrical circuit theory. In *Scientific Computing in Electrical Engineering*, 2004. URL www.matheon.de. to appear.

[139] H.H. Robertson. The solution of a set of reaction rate equations. *Academic Press*, pages 178–182, 1966.

[140] M. Roche. Rosenbrock methods for differential algebraic equations. *Numer. Math.*, 52:45–63, 1988.

[141] S.P. Nørsett. Semi explicit Runge-Kutta methods. Technical report, Dept. Math. Univ. Trondheim, 1974. Report No. 6/74, ISBN 82-7151-009-6.

[142] C. Runge. Über die numerische Auflösung von Differentialgleichungen. *Math. Ann.*, 46:167–178, 1895.

[143] S. Schneider. Convergence of general linear methods on differential-algebraic systems of index 3. *BIT*, 37(2):424–441, 1997.

[144] S. Schulz. General linear methods for linear DAEs. Technical Report 03-10, Humboldt Universität zu Berlin, 2003. www.math.hu-berlin.de/~steffen/publications.html.

[145] S. Schulz. Four lectures on differential algebraic equations. Technical Report 497, The University of Auckland, New Zealand, 2003. www.math.hu-berlin.de/~steffen/publications.html.

[146] I. Schumilina. *Charakterisierung der Algebro-Differentialgleichungen mit Traktabilitätsindex 3*. PhD thesis, Humboldt Universität zu Berlin, 2004.

[147] E.R. Sieber, U. Feldmann, R. Schultz, and H. Wriedt. Timestep control for charge conserving integration in circuit simulation. In R. E. Bank, R. Bulirsch, H. Gajewski, and K. Merten, editors, *Mathematical modelling and simulation of electrical circuits and semiconductor devices*, volume 117 of *Int. Series of Num. Mathematics*. Birkhäuser, Basel, 1994.

[148] G. Söderlind. Automatic control and adaptive time-stepping. *Numer. Algorithms*, 31(1–4):281–310, 2002.

[149] G. Söderlind. Digital filters in adaptive time-stepping. *ACM Trans. Math. Software*, 29(1):1–26, 2003.

[150] Suse. The openSUSE project, www.opensuse.org.

[151] C. Tischendorf. Benchmark: Nand gate. www.math.hu-berlin.de/~caren/nandgatter.ps.

[152] C. Tischendorf. Feasibility and stability behaviour of the BDF applied to index-2 differential algebraic equations. *Z. Angew. Math. Mech.*, 75(12):927–946, 1995. ISSN 0044-2267.

[153] C. Tischendorf. *Solution of index-2 differential algebraic equations and it's application in circuit simulation*. PhD thesis, Humboldt Universität zu Berlin, 1996.

[154] C. Tischendorf. Topological index calculation of differential algebraic equations in circuit simulation. *Surveys Math. Indust.*, 8:187–199, 1999.

[155] C. Tischendorf. Numerische Simulation elektrischer Netzwerke. Lecture Notes, www.math.hu-berlin.de/~caren/schaltungen.pdf, 2000.

[156] S. Voigtmann. General linear methods for nonlinear DAEs in circuit simulation. In *Scientific Computing in Electrical Engineering*, 2004. URL www.matheon.de. to appear.

[157] S. Voigtmann. Accessible criteria for the local existence and uniqueness of DAE solutions. Technical Report 214, MATHEON, 2005. URL www.matheon.de. www.math.hu-berlin.de/~steffen/publications.html.

[158] W. Wright. *General linear methods with inherent Runge-Kutta stability*. PhD thesis, The University of Auckland, New Zealand, 2003. www.math.auckland.ac.nz/~butcher/theses/willwright.pdf.

[159] ZIB. ZIB: Konrad-Zuse-Zentrum für Informationstechnik Berlin, www.zib.de/index.en.html, for LIMEX: www.zib.de/Numerik/numsoft/CodeLib/codes/limex.tar.gz.

[160] G. Zielke. Motivation und Darstellung von verallgemeinerten Matrixinversen. *Beiträge Numer. Math.*, 7:177–218, 1979.

Selbständigkeitserklärung

Hiermit erkläre ich, dass ich die vorliegende Dissertation selbständig verfasst habe und keine anderen als die angegebenen Quellen und Hilfsmittel benutzt worden sind.

Ottobunn, 20.08.2006,   Steffen Voigtmann